Edmund Hlawka
Johannes Schoißengeier
Rudolf Taschner

Geometric and Analytic
Number Theory

With 15 Figures

Springer-Verlag

Berlin Heidelberg New York
London Paris Tokyo
Hong Kong Barcelona
Budapest

Edmund Hlawka
Rudolf Taschner
Technische Universität Wien
Institut für Analysis, Technische Mathematik
und Versicherungsmathematik
Wiedner Hauptstraße 8–10/114
A-1040 Wien, Austria

Johannes Schoißengeier
Universität Wien
Institut für Mathematik
Strudlhofgasse 4
A-1090 Wien, Austria

Translator:

Charles Thomas
Department of Pure Mathematics
and Mathematical Statistics
University of Cambridge
16 Mill Lane
Cambridge CB2 1SB, Great Britain

Title of the original German edition: Geometrische und analytische Zahlentheorie
© Manz-Verlag Wien, 1986

Mathematics Subject Classification (1980): 10-D2, 10E, 10F, 10G, 10H

ISBN-13: 978-3-540-52016-0   e-ISBN-13: 978-3-642-75306-0
DOI: 10.1007/978-3-642-75306-0

Library of Congress Cataloging-in-Publication Data
Hlawka, Edmund.
[Geometrische und analytische Zahlentheorie. English]
Geometric and analytic number theory/Edmund Hlawka, Johannes Schoißengeier, Rudolf
Taschner.
p. cm. – (Universitext)
Translation of: Geometrische und analytische Zahlentheorie.
Includes bibliographical references and index.
ISBN 0-387-52016-3
1. Geometry of numbers. 2. Number theory. I. Schoißengeier, Johannes. II. Taschner, Rudolf J.
(Rudolf Josef) III. Title IV. Title: Analytic number theory. QA241.5.H5313 1991 512'.7–dc20
90-25975 CIP

© Springer-Verlag Berlin Heidelberg 1991
Softcover reprint of the hardcover 1st edition 1991

41/3140-543210   Printed on acid-free paper

Preface to the English Edition

In the English edition, the chapter on the Geometry of Numbers has been enlarged to include the important findings of H. Lenstra; furthermore, tried and tested examples and exercises have been included.

The translator, Prof. Charles Thomas, has solved the difficult problem of transferring the German text into English in an admirable way. He deserves our unreserved praise and special thanks. Finally, we would like to express our gratitude to Springer-Verlag, for their commitment to the publication of this English edition, and for the special care taken in its production.

Vienna, March 1991

E. Hlawka
J. Schoißengeier
R. Taschner

Preface to the German Edition

We have set ourselves two aims with the present book on number theory. On the one hand for a reader who has studied elementary number theory, and who has knowledge of analytic geometry, differential and integral calculus, together with the elements of complex variable theory, we wish to introduce basic results from the areas of the geometry of numbers, diophantine approximation, prime number theory, and the asymptotic calculation of number theoretic functions. However on the other hand for the student who has already studied analytic number theory, we also present results and principles of proof, which until now have barely if at all appeared in text books. For example these include the proof of the irrationality of the Riemann zeta function at the number 3, Newman's application of complex variable theory to the prime number theorem, and Hecke's prime number theorem for the Gaussian integers.

For the choice of material we have been able to rely on lectures held since 1948 in Vienna and since 1967 in Pasadena, but have still had to overcome a number of presentational problems. We thank our fellow workers, colleagues and friends, who stood ready to help us. Our particular thanks go to the publishers Manz and to Dr. Franz Stein for his interest, his suggestions and his patience.

Vienna, Advent 1985

Edmund Hlawka
Johannes Schoißengeier
Rudolf Taschner

Contents

1. The Dirichlet Approximation Theorem

"The integers are the source of all mathematics." Hermann Minkowski prefaced his book on diophantine approximation with this sentence and justifiably – the natural numbers 1,2,3 ... are the only data which the mathematician knows he has to hand. His lack of control over the real numbers, that is points on the continuous real line is already plain from elementary examples – thus the sum $e + \pi$ cannot in the end, that is with complete exactness, be worked out. Even simple questions about the way in which the number $e + \pi$ is put together are unsolved up to now. Therefore the construction of real numbers from natural numbers is no simple problem. The theory of diophantine approximation seeks to understand how well, that is how closely, real numbers can be trapped by relations with the integers.

Theorem 1: Dirichlet's Approximation Theorem. *For each real number α and natural number N one can find a natural number $n \leq N$ and an integer p with*

$$\left| \alpha - \frac{p}{n} \right| < \frac{1}{Nn} \ .$$

In particular

$$|n\alpha - p| < \frac{1}{N} \ .$$

Proof. Dirichlet starts from the $N + 1$ numbers $0 \cdot \alpha, 1 \cdot \alpha, 2 \cdot \alpha, \ldots, N\alpha$ and then reduces them modulo 1, that is, forms the numbers $x_n = n\alpha - [n\alpha]$, $n = 0, 1, 2, \ldots, N$, which lie in the half-open unit interval $[0, 1[$. He subdivides the unit interval into N half-open subintervals of equal length

$$U_m = \left[\frac{m-1}{N}, \frac{m}{N} \right[, \quad m = 1, 2, \ldots, N \ ,$$

and deduces from the fact that the $N+1$ points x_n have to lie in the N intervals U_m, that at least *two* of the numbers x_n must belong to a common interval U_m. Let $x_{n'}, x_{n''}$ be numbers of this kind, which lie in the same interval U_m, and suppose that $n' < n''$. Then from the inequalities

$$\frac{m}{N} \leq n'\alpha - [n'\alpha] < \frac{m+1}{N}$$

$$\frac{m}{N} \leq n''\alpha - [n''\alpha] < \frac{m+1}{N}$$

Dirichlet deduces by substraction that

$$-\frac{1}{N} < n''\alpha - n'\alpha - ([n''\alpha] - [n'\alpha]) < \frac{1}{N} \ .$$

The natural number $n = n'' - n'$ is certainly not larger than N; let the integer $[n''\alpha] - [n'\alpha]$ be p. [1] $\square$

With the help of the Dirichlet approximation theorem one can demonstrate very elegantly one of the most important theorems of elementary number theory, namely the theorem about the linear diophantine equation. Let α denote a rational number, that is, $\alpha = u/v$ with u, v integers and $v \neq 0$. The representation of α as a fraction is not unique, since for each integer $c \neq 0$ we have $\alpha = u/v = cu/cv$. In order to associate with α a uniquely determined representing fraction, among all fractions u/v with $\alpha = u/v$ one chooses $\alpha = a/b$ with b the smallest possible natural denominator, for which there exists an integral numerator with $\alpha = a/b$. a/b describes α in *lowest terms*.

Assume $b \geq 2$. By the Dirichlet approximation theorem for $N = b - 1$ one can find a natural number $n \leq N$ and an integer p with

$$|\alpha n - p| = \left|\frac{a}{b}n - p\right| < \frac{1}{N} = \frac{1}{b-1} \ .$$

Multiplying by b gives

$$|an - bp| < \frac{b}{b-1} = 1 + \frac{1}{b-1} \leq 2 \ .$$

Since $an - bp$ denotes an integer, we must actually have $|an - bp| \leq 1$. The possibility that $an - bp = 0$ is excluded, since it would imply that

$$\alpha = \frac{a}{b} = \frac{p}{n}$$

with $n < b$, contradicting the choice of a/b in lowest terms. Hence the only possibility is $an - bp = \pm 1$. From this one deduces

Corollary 1: Main Theorem on the Linear Diophantine Equation.
For each fraction a/b in lowest terms one can find integers x and y with

$$ax - by = 1 \ .$$

Proof. For $b \geq 2$ we showed above the existence of n and p with $an - bp = \pm 1$. In the case $an - bp = 1$ put $x = n, y = p$; in the case $an - bp = -1$ put $x = -n, y = -p$. If $b = 1$, the equation $ax - y = 1$ is solved immediately by $x = 0, y = -1$. $\square$

Starting from this theorem one can develop all of elementary number theory, similarly to the way we chose in our own book on elementary number theory, starting from the division algorithm with remainder.

Besides the linear diophantine equation the so-called Pell (sometimes Fermat) equation $x^2 - dy^2 = N$ is among the most important equations of number theory. In what follows only the so-called special Pell equation $x^2 - dy^2 = 1$ will be of interest, and here the cases $d = a^2$ with a integral, and $d < 0$ are particularly easy to handle:

(1) $d = a^2 : x^2 - dy^2 = (x - ay)(x + ay) = 1 : x = \pm 1, y = 0$ for $a \neq 0$; $x = \pm 1, y = t$ with t in $\mathbb{Z}$ for $a = 0$.

(2) $d = -1 : x^2 + y^2 = 1 : x = \pm 1, y = 0$ or $x = 0, y = \pm 1$.

(3) $d < -1 : x^2 + |d|y^2 = 1 : x = \pm 1, y = 0$.

Now for the really interesting case, the equation $x^2 - dy^2 = 1$ with a natural number d, which is not the square of a natural number, hence whose root must be irrational. Pell himself could not say anything about this, but the French lawyer, mathematician and founder of modern number theory, Pierre de Fermat, certainly did:

Corollary 2: Fermat's Theorem on the Pell Equation. *If the natural number d is not the square of an integer then the special Pell equation $x^2 - dy^2 = 1$ has infinitely many integral solutions.*

Geometrically this means that there are infinitely many points with integral coordinates on the hyperbola $x^2 - dy^2 = 1$.

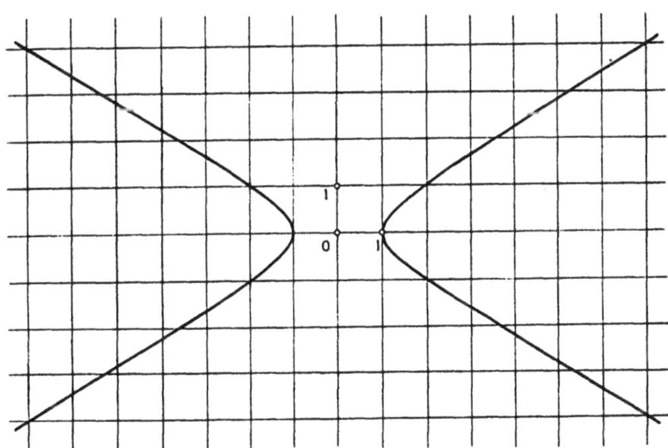

Fig. 1. Hyperbola and lattice of integral coordinates

Proof. In fact it suffices to find a single solution $x = \xi, y = \eta$ with $\eta \neq 0$. If we write $\varepsilon = \xi + \eta\sqrt{d}, \bar{\varepsilon} = \xi - \eta\sqrt{d}$, then $1 = \xi^2 - d\eta^2 = (\xi + \eta\sqrt{d})(\xi - \eta\sqrt{d}) = \varepsilon\bar{\varepsilon}$, and so for the n-th power ε^n the equation $\varepsilon^n \cdot \bar{\varepsilon}^n = 1$ holds. Moreover $\varepsilon^n = (\xi + \eta\sqrt{d})^n$ is of the form $x + y\sqrt{d}$ with integral coefficients, and because of the irrationality of $\sqrt{d}$ the representation is unique. By direct calculation it is easy to check that given $\varepsilon^n = x + y\sqrt{d}$ we have $\bar{\varepsilon}^n = x - y\sqrt{d} = \overline{(\varepsilon^n)}$. In this way the infinitely many powers $\varepsilon^n = x + y\sqrt{d}$ lead to infinitely many solutions

x, y of the special Pell equation (which because $|\varepsilon| \neq 1$ are all distinct from each other).

The proof of the existence of some solution $x = \xi$, $y = \eta \neq 0$ succeeds with the help of the following subsidiary proposition:

Lemma 1. *For each irrational number α there exist infinitely many fractions (in lowest terms) p/n with $|\alpha - p/n| < 1/n^2$.*

Proof. Assume the opposite, so that it must be possible to list the finitely many such fractions with this property as

$$\frac{p_1}{n_1}, \frac{p_2}{n_2}, \ldots, \frac{p_K}{n_K} .$$

Because of the irrationality of α the left hand inequality is satisfied in the formula

$$0 < \left| \alpha - \frac{p_k}{n_k} \right| < \frac{1}{n_k^2} , \quad k = 1, \ldots, K .$$

Let us denote the smallest of the finitely many positive numbers

$$n_k \cdot \left| \alpha - \frac{p_k}{n_k} \right| , \quad k = 1, \ldots, K ,$$

by δ (and if $K = 0$ put $\delta = 1$). On account of the Dirichlet approximation theorem with the natural number $N = [1/\delta] + 1 > 1/\delta$ one can associate a natural number $n \leq N$ and an integer p with $|\alpha - p/n| < 1/nN$ (where without loss of generality p/n may be assumed to be in lowest terms). With $n \leq N$ the relation above leads to

$$\left| \alpha - \frac{p}{n} \right| < \frac{1}{n^2} ,$$

that is p/n belongs to the list above. For some suffix k we have $p = p_k$, $n = n_k$ (in particular $K \geq 1$). From $N > 1/\delta$ it follows that

$$n_k \cdot \left| \alpha - \frac{p_k}{n_k} \right| < \frac{1}{N} < \delta$$

contradicting the definition of δ. This shows that the assumption above is not possible. [2] $\square$

Continuation of the proof of Corollary 2. The irrationality of $\sqrt{d}$ implies the existence of infinitely many fractions (in lowest terms) p/n with

$$0 < \left| \sqrt{d} - \frac{p}{n} \right| < \frac{1}{n^2} .$$

If we put $\alpha = p + n\sqrt{d}$, $\bar{\alpha} = p - n\sqrt{d}$, then we deduce the existence of infinitely many numbers α of the kind above with $|\bar{\alpha}| < 1/n$, and

$$|\alpha| = |p - n\sqrt{d} + 2n\sqrt{d}| \le |\bar{\alpha}| + 2n\sqrt{d} \le \frac{1}{n} + 2n\sqrt{d} \ .$$

Hence

$$|p^2 - dn^2| = |\alpha\bar{\alpha}| \le \left(\frac{1}{n} + 2n\sqrt{d}\right) \cdot \frac{1}{n}$$

$$= \frac{1}{n^2} + 2\sqrt{d} \le 2\sqrt{d} + 1 \ .$$

Since the infinitely many integers $p^2 - dn^2$ lie in the bounded interval $[-2\sqrt{d} - 1, 2\sqrt{d} + 1]$, there must be at least one integer $a \ne 0$ which coincides with infinitely many of the $p^2 - dn^2$. Call two of these infinitely many numbers $\alpha = p + n\sqrt{d}$ and $\beta = q + m\sqrt{d}$ equivalent, if $p \equiv q(\mathrm{mod}\,|a|)$ and $n \equiv m(\mathrm{mod}\,|a|)$. There are at most a^2 equivalence classes arising in this way, hence in at least one of these finitely many equivalence classes there must be infinitely many of the numbers $\alpha, \beta, \ldots$ constructed above. Therefore there certainly exists an equivalent pair $\alpha = p + n\sqrt{d}$, $\beta = q + m\sqrt{d}$ with $\alpha \ne \beta$. Choose the notation in such a way that $|\alpha| > |\beta|$. As a consequence

$$\frac{\alpha - \beta}{a} = \frac{p - q}{a} + \frac{n - m}{a}\sqrt{d}$$

is a number of the form $x + y\sqrt{d}$, and finally

$$\varepsilon = \frac{\alpha}{\beta} = 1 + \frac{a}{\beta} \cdot \frac{\alpha - \beta}{a} = 1 + \frac{\beta\bar{\beta}}{\beta} \cdot \frac{\alpha - \beta}{a}$$

$$= 1 + (q - m\sqrt{d})\left(\frac{p - q}{a} + \frac{n - m}{a}\sqrt{d}\right)$$

is a number of the form $\varepsilon = \xi + \eta\sqrt{d}$ with integral ξ, η for which

$$\xi^2 - d\eta^2 = \varepsilon\bar{\varepsilon} = \frac{\alpha\bar{\alpha}}{\beta\bar{\beta}} = \frac{a}{a} = 1 \ .$$

Since $|\varepsilon| > 1$ we must have $\eta \ne 0$, and so ξ, η form the required solution of the special Pell equation. $\qquad\square$

Lemma 1 is close to the supposition that diophantine approximations can not only be used in the solution theory of diophantine equations, but also for the proof of the irrationality of certain numbers. This is indeed the case:

Proposition 1. *A real number α is irrational if and only if for each $\varepsilon > 0$ one can find integers x, y with $0 < |\alpha x - y| < \varepsilon$.*

Proof. For each irrational number one can certainly find a sequence of infinitely many fractions p_n/q_n with strictly increasing denominators satisfying the condition

$$\left|\alpha - \frac{p_n}{q_n}\right| < \frac{1}{q_n^2} \ .$$

Given an arbitrary $\varepsilon > 0$ choose n so large that $q_n > 1/\varepsilon$, and define x, y by $x = q_n$, $y = p_n$; then one indeed has $0 < |\alpha x - y| < \varepsilon$. If conversely the condition above were satisfied for arbitrary $\varepsilon > 0$, and nevertheless α were rational (i.e. $\alpha = a/b$ with a an integer and b a natural number), then by the assumption above with $\varepsilon = 1/b$, one would have

$$0 < \left| \frac{a}{b} \cdot x - y \right| < \frac{1}{b}, \quad \text{i.e.}$$
$$0 < |ax - by| < 1 \ .$$

This cannot be satisfied for integers x, y. $\qquad\qquad\square$

Example. Set α equal to the Cantor series

$$\alpha = \sum_{n=1}^{\infty} \frac{z_n}{g_1 g_2 \cdots g_n} = \frac{z_1}{g_1} + \frac{z_2}{g_1 g_2} + \frac{z_3}{g_1 g_2 g_3} + \ldots$$

with an increasing sequence of natural numbers $2 \le g_1 \le g_2 \le \ldots \le g_n \le \ldots$, in which the coefficients z_n take the values 0 or 1 independently of each other, subject to the condition that for infinitely many n, $z_n = 1$. Then with the abbreviations $G_N = g_1 g_2 \ldots g_N$ and

$$\sum_{n=1}^{N} \frac{z_n}{g_1 g_2 \cdots g_n} = \frac{P_N}{G_N}$$

the formula for the geometric series gives

$$0 < \left| \alpha - \frac{P_N}{G_N} \right| = \left| \frac{z_{N+1}}{g_1 \cdots g_N g_{N+1}} + \frac{z_{N+2}}{g_1 \cdots g_N g_{N+1} g_{N+2}} + \ldots \right|$$
$$\le \frac{1}{g_1 \cdots g_N g_{N+1}} \left(1 + \frac{1}{g_{N+2}} + \frac{1}{g_{N+2} g_{N+3}} + \ldots \right)$$
$$\le \frac{1}{G_N g_{N+1}} \left(1 + \frac{1}{g_{N+1}} + \frac{1}{g_{N+1}^2} + \ldots \right)$$
$$= \frac{1}{G_N g_{N+1}} \cdot \frac{1}{1 - \dfrac{1}{g_{N+1}}} = \frac{1}{G_N (g_{N+1} - 1)} \ .$$

Assuming that the sequence $\{g_n\}$ is unbounded, since

$$0 < |\alpha G_N - P_N| \le \frac{1}{g_{N+1} - 1}$$

and for large enough N the right hand side can be made arbitrarily small, we obtain an uncountable family of irrational numbers α. In the special case of $g_n = n$ and $z_n = 1$ for all n, we have in addition proved:

Corollary 3. *The base e for the natural logarithm is irrational.*

The actual proof of the irrationality of the second fundamental constant, associated with the circle, namely π, will be put in a context which also covers the irrationality of the numbers

$$\zeta(2) = \sum_{n=1}^{\infty} \frac{1}{n^2} = 1 + \frac{1}{4} + \frac{1}{9} + \frac{1}{16} + \ldots = \frac{\pi^2}{6}$$

and

$$\zeta(3) = \sum_{n=1}^{\infty} \frac{1}{n^3} = 1 + \frac{1}{8} + \frac{1}{27} + \frac{1}{64} + \frac{1}{125} + \ldots \ .$$

The essential core of the proof is obtained from the formulae

$$\zeta(2) = 1 + \frac{1}{2^2} + \ldots + \frac{1}{r^2} + \int_0^1\int_0^1 \frac{(xy)^r}{1-xy}\, dx\, dy$$

and

$$\zeta(3) = 1 + \frac{1}{2^3} + \ldots + \frac{1}{r^3}$$
$$- \frac{1}{2}\int_0^1\int_0^1 \frac{\log(xy)}{1-xy}(xy)^r\, dx\, dy \ .$$

Both formulae are very easily established by working out the more general double integral

$$\int_0^1\int_0^1 \frac{x^{r+\sigma}y^{s+\sigma}}{1-xy}\, dx\, dy$$

for non-negative integers r, s and a non-negative real number σ. Thus, if one expands $(1-xy)^{-1}$ as a geometric series, one obtains

$$\int_0^1\int_0^1 \frac{x^{r+\sigma}y^{s+\sigma}}{1-xy}\, dx\, dy = \sum_{k=0}^{\infty} \frac{1}{(k+r+\sigma+1)(k+s+\sigma+1)} \ ,$$

which for $r = s$, $\sigma = 0$, immediately gives the asserted formula for $\zeta(2)$. If on the other hand one differentiates the formula above according to σ, and then puts $s = r$ and $\sigma = 0$, one obtains

$$\int_0^1\int_0^1 \frac{\log(xy)}{1-xy}(xy)^r\, dx\, dy = \sum_{k=0}^{\infty} \frac{-2}{(k+r+1)^3}$$
$$= -2 \cdot \left(\zeta(3) - \frac{1}{1^3} - \frac{1}{2^3} - \ldots - \frac{1}{r^3}\right) \ ,$$

i.e. the asserted formula for $\zeta(3)$.

If one carries on working out

$$\int_0^1\int_0^1 \frac{x^{r+\sigma}y^{s+\sigma}}{1-xy}\, dx\, dy = \sum_{k=0}^{\infty} \frac{1}{(k+r+\sigma+1)(k+s+\sigma+1)}$$

in the case $r \neq s$ for some $r > s$, the right hand sum simplifies to

$$\sum_{k=0}^{\infty} \frac{1}{r-s}\left(\frac{1}{k+s+\sigma+1} - \frac{1}{k+r+\sigma+1}\right)$$
$$= \frac{1}{r-s}\left(\frac{1}{s+1+\sigma} + \ldots + \frac{1}{r+\sigma}\right) .$$

If one now puts $\sigma = 0$ one recognizes *that*

$$\int_0^1\int_0^1 \frac{x^r y^s}{1-xy}\, dx\, dy , \quad r > s ,$$

represents a rational number, the denominator of which is certainly contained in the square $V(r)^2$ of the lowest common multiple $V(r)$ of all numbers $1,\ldots,r$. If one first differentiates

$$\int_0^1\int_0^1 \frac{x^{r+\sigma} y^{s+\sigma}}{1-xy}\, dx\, dy$$
$$= \frac{1}{r-s}\left(\frac{1}{s+1+\sigma} + \ldots + \frac{1}{r+\sigma}\right)$$

according to σ, and then puts $\sigma = 0$, one sees *that likewise*

$$\int_0^1\int_0^1 \frac{\log(xy)}{1-xy} x^r y^s\, dx\, dy$$
$$= \frac{-1}{r-s}\left(\frac{1}{(s+1)^2} + \ldots + \frac{1}{r^2}\right), \quad r > s ,$$

also describes a rational number, the denominator of which is certainly contained in the cube $V(r)^3$ of the lowest common multiple $V(r)$ of all numbers $1,\ldots,r$.

If now, encouraged by these preparations, one starts from the integral

$$\int_0^1\int_0^1 \frac{(1-y)^n P_n(x)}{1-xy}\, dx\, dy$$

with the polynomial

$$P_n(x) = \frac{1}{n!}\left(\frac{d}{dx}\right)^n \left(x^n(1-x)^n\right)$$

and one observes that $P_n(x)$ has only integral coefficients, then working out the integral yields an expression of the form

$$\int_0^1\int_0^1 \frac{(1-y)^n P_n(x)}{1-xy}\, dx\, dy = \frac{a_n \zeta(2) + b_n}{V(n)^2} ,$$

in which a_n and b_n are integers. Partial integration of the left-hand side n times according to x leads to

$$(-1)^n \int_0^1 \int_0^1 \frac{y^n(1-y)^n x^n(1-x)^n}{(1-xy)^{n+1}} \, dx \, dy$$

$$= \frac{a_n \zeta(2) + b_n}{V(n)^2} \ .$$

If now by means of an extreme value argument one calculates the maximum value of $y(1-y)x(1-x)/(1-xy)$ inside $0 \le x \le 1, 0 \le y \le 1$, one obtains

$$\frac{y(1-y)x(1-x)}{1-xy} \le \left(\frac{\sqrt{5}-1}{2} \right)^5 \ ,$$

which implies that

$$0 < \left| \int_0^1 \int_0^1 \frac{y^n(1-y)^n x^n(1-x)^n}{(1-xy)^{n+1}} \, dx \, dy \right|$$

$$= \frac{|a_n \zeta(2) + b_n|}{V(n)^2} \le \left(\frac{\sqrt{5}-1}{2} \right)^{5n} \int_0^1 \int_0^1 \frac{dx \, dy}{1-xy}$$

$$= \left(\frac{\sqrt{5}-1}{2} \right)^{5n} \zeta(2) \ .$$

From this one obtains the estimate

$$0 < |a_n \zeta(2) + b_n| \le V(n)^2 \left(\frac{\sqrt{5}-1}{2} \right)^{5n} \zeta(2) \ .$$

If one could suppose that

$$\lim_{n \to \infty} V(n)^2 \left(\frac{\sqrt{5}-1}{2} \right)^{5n} = 0$$

one could apply Proposition 1, because one could choose the right-hand side of the inequality obtained above to be arbitrarily small. This requires a more exact knowledge of the number $V(n)$. It is clear that $V(n) \le n^{\pi(n)}$, where $\pi(n)$ denotes the number of primes less than or equal to n, because there can only exist $\pi(n)$ different factors of $V(n)$ and it is impossible for the exponent of each of these factors to exceed n. The *prime number theorem*

$$\lim_{n \to \infty} \frac{\pi(n)}{n/\log n} = 1$$

will be proved in Chapter five of this book. Assuming its validity for the moment, then by taking n to be sufficiently large one has

$$\pi(n) \le \log 3 \cdot \frac{n}{\log n}$$

and therefore can conclude that

$$V(n) \leq n^{n \cdot \log 3/ \log n} = 3^n \ .$$

A consequence of this is

$$0 \leq \lim_{n \to \infty} V(n)^2 \left(\frac{\sqrt{5} - 1}{2} \right)^{5n}$$

$$\leq \lim_{n \to \infty} 9^n \left(\frac{\sqrt{5} - 1}{2} \right)^{5n} \leq \lim_{n \to \infty} \left(\frac{5}{6} \right)^n = 0 \ .$$

This proves

Corollary 4. *The number $\zeta(2) = \pi^2/6$ is irrational.*

In particular we deduce

Corollary 5. *Not only π itself, but also its square π^2, are irrational.*

Similar considerations lead to the irrationality of $\zeta(3)$ – starting from the fact that, because of the initial calculations

$$\int_0^1 \int_0^1 \frac{-\log(xy)}{1 - xy} P_n(x) P_n(y) \, dx \, dy = \frac{a_n \zeta(3) + b_n}{V(n)^3}$$

with a_n and b_n both integral. The identity

$$\frac{-\log(xy)}{1 - xy} = \int_0^1 \frac{1}{1 - (1 - xy)z} \, dz$$

justifies the relation

$$\frac{a_n \zeta(3) + b_n}{V(n)^3} = \int_0^1 \int_0^1 \int_0^1 \frac{P_n(x) P_n(y)}{1 - (1 - xy)z} \, dx \, dy \, dz \ ,$$

which by n-fold partial integration according to x is transformed into

$$\int_0^1 \int_0^1 \int_0^1 \frac{(xyz)^n (1 - x)^n P_n(y)}{(1 - (1 - xy)z)^{n+1}} \, dx \, dy \, dz \ .$$

Next one makes the substitution

$$w = \frac{1 - z}{1 - (1 - xy)z}$$

and carries out an n-fold partial integration according to y in the resulting integral,

$$\int_0^1 \int_0^1 \int_0^1 (1 - x)^n (1 - w)^n \frac{P_n(y)}{1 - (1 - xy)w} \, dx \, dy \, dw \ ,$$

obtaining

$$\frac{a_n\zeta(3)+b_n}{V(n)^3} = \int_0^1\int_0^1\int_0^1 \frac{x^n(1-x)^ny^n(1-y)^nw^n(1-w)^n}{(1-(1-xy)w)^{n+1}}\, dx\, dy\, dw\ .$$

In the same way as before one does the extremal value exercise in order to determine the maximum of $x(1-x)y(1-y)w(1-w)/(1-(1-xy)w)$ in the region $0 \le x \le 1$, $0 \le y \le 1$, $0 \le w \le 1$. Here one has

$$\frac{x(1-x)y(1-y)w(1-w)}{1-(1-xy)w} \le (\sqrt{2}-1)^4\ .$$

The integral above has

$$(\sqrt{2}-1)^{4n} \int_0^1\int_0^1\int_0^1 \frac{1}{1-(1-xy)w}\, dx\, dy\, dw$$
$$= (\sqrt{2}-1)^{4n} \int_0^1\int_0^1 \frac{-\log(xy)}{1-xy}\, dx\, dy$$
$$= 2\zeta(3)\cdot(\sqrt{2}-1)^{4n}$$

as upper bound. If in

$$0 < \frac{|a_n\zeta(3)+b_n|}{V(n)^3} \le 2\zeta(3)(\sqrt{2}-1)^{4n}$$

one chooses n to be sufficiently large, and suppose that $V(n) \le 3^n$, then the right-hand side of the estimate $0 < |a_n\zeta(3)+b_n| \le 2\zeta(3)\cdot 27^n\cdot(\sqrt{2}-1)^{4n} < (\frac{4}{5})^n$ will be arbitrarily small. This proves

Corollary 6: Apéry's Theorem. *The number $\zeta(3) = 1 + 1/2^3 + 1/3^3 + 1/4^3 + \ldots$ is irrational.* [3]

Until now the Dirichlet approximation theorem has only shown how well a *single* real number can be approximately represented by a rational number. Multidimensional diophantine approximations seek to describe several real numbers $\alpha_1, \alpha_2, \ldots, \alpha_L$ as well as possible by rational numbers $p_1/q, p_2/q, \ldots, p_L/q$ having the same denominator. The evident generalisation of the Dirichlet approximation theorem reads:

Proposition 2. *Let $\alpha_1, \ldots, \alpha_L$ be L real numbers and N a natural number, then one can find integers $p_1, \ldots, p_L$ and a natural number $n \le N^L$ with*

$$\left|\alpha_\ell - \frac{p_\ell}{n}\right| < \frac{1}{Nn}, \quad \ell = 1,\ldots,L\ . [4]$$

If one writes the approximation formula of this theorem in the form

$$|\alpha_\ell n - p_\ell| < \frac{1}{N}, \quad \ell = 1,\ldots,L\ ,$$

one can ask oneself, if it might be possible to interchange the roles of denominator n and numerators $p_1,\ldots,p_L$, i.e. if instead of a simultaneous approximation of the L linear forms $\alpha_1 n, \alpha_2 n,\ldots,\alpha_L n$ by L integers $p_1, p_2,\ldots,p_L$ one might achieve the simultaneous approximation of one linear form

$$\sum_{\ell=1}^{L} \alpha_\ell n_\ell = \alpha_1 n_1 + \alpha_2 n_2 + \ldots + \alpha_L n_L$$

by a single integer p. Dirichlet himself showed that this is possible:

Proposition 3. *Let $\alpha_1,\ldots,\alpha_L$ be L real numbers and N a natural number, then one can find a natural number p and L integers $n_1,\ldots,n_L$, which are not all equal to zero, so that for $\ell = 1,\ldots,L$ $|n_\ell| \leq N^{1/L}$ and*

$$\left| \sum_{\ell=1}^{L} \alpha_\ell n_\ell - p \right| < \frac{1}{N}$$

holds. [5]

A generalisation due to Kronecker contains both Propositions 2 and 3:

Theorem 2: Multidimensional Dirichlet Approximation Theorem.
Let $\alpha_{11}, \alpha_{12},\ldots,\alpha_{ML}$ be $M \cdot L$ real numbers and N a natural number, then one can find M integers $p_1,\ldots,p_M$ and L integers $n_1,\ldots,n_L$, where the n_ℓ are not all zero, $|n_\ell| \leq N^{M/L}$, and satisfy the inequalities

$$\left| \sum_{\ell=1}^{L} \alpha_{m\ell} n_\ell - p_m \right| < \frac{1}{N}$$

for all $m = 1,\ldots,M$.

Proof. The M-tuple

$$\left(\sum_{\ell=1}^{L} \alpha_{1\ell} n_\ell - \left[\sum_{\ell=1}^{L} \alpha_{1\ell} n_\ell \right] ,\ldots, \sum_{\ell=1}^{L} \alpha_{M\ell} n_\ell - \left[\sum_{\ell=1}^{L} \alpha_{M\ell} n_\ell \right] \right)$$

lies in the cube $[0,1[^M$, which decomposes into N^M subcubes

$$U_{k_1,\ldots,k_M} = \left[\frac{k_1 - 1}{N}, \frac{k_1}{N} \right[\times \ldots \times \left[\frac{k_M - 1}{N}, \frac{k_M}{N} \right[, \quad 1 \leq k_i \leq N \ .$$

In order to apply the Dirichlet pigeon hole principle one must ensure that the $n_1,\ldots,n_L$ range over more L-tuples than there are subcubes. If we fix $0 \leq n_\ell \leq P$ there are $(P+1)^L$ L-tuples $n_1,\ldots,n_L$; the requirement that $(P+1)^L > N^M$ leads to $P + 1 > N^{M/L}$. It suffices therefore to set $P = \left[N^{M/L} \right]$. Under this assumption once more two of the M-tuples

$$\sum_{\ell=1}^{L} \alpha_{m\ell} n_\ell - \left[\sum_{\ell=1}^{L} \alpha_{m\ell} n_\ell \right]$$

have to lie in one of the subcubes $U_{k_1,\ldots,k_M}$; i.e. for distinct n'_ℓ and n''_ℓ we have

$$\frac{k_m - 1}{N} \le \sum_{\ell=1}^{L} \alpha_{m\ell} n'_\ell - \left[\sum_{\ell=1}^{L} \alpha_{m\ell} n'_\ell \right] < \frac{k_m}{N},$$

$$\frac{k_m - 1}{N} \le \sum_{\ell=1}^{L} \alpha_{m\ell} n''_\ell - \left[\sum_{\ell=1}^{L} \alpha_{m\ell} n''_\ell \right] < \frac{k_m}{N}, \quad m = 1,\ldots,M .$$

From this with

$$n_\ell = n''_\ell - n'_\ell,$$

$$p_m = \left[\sum_{\ell=1}^{L} \alpha_{m\ell} n''_\ell \right] - \left[\sum_{\ell=1}^{L} \alpha_{m\ell} n'_\ell \right]$$

we have

$$\left| \sum_{\ell=1}^{L} \alpha_{m\ell} n_\ell - p_m \right| < \frac{1}{N}$$

for all $m = 1,\ldots,M$, as asserted. From $0 \le n'_\ell \le P \le N^{M/L}$, $0 \le n''_\ell \le P \le N^{M/L}$ we obtain $|n_\ell| \le N^{M/L}$, and since n'_ℓ and n''_ℓ are distinct, not all the n_ℓ are equal to zero. In the case $L = 1$ one can even assume that $n''_1 > n'_1$ with n_1 a natural number. This proves the theorem above and also finally Proposition 2 ^{3}. $\qquad\qquad\qquad\Box$

Finally one can ask how small linear forms

$$\left(\sum_{\ell=1}^{L} \alpha_{1\ell} x_\ell, \ldots, \sum_{\ell=1}^{L} \alpha_{M\ell} x_\ell \right)$$

can occur.

Corollary 7. *For each M-tuple*

$$\left(\sum_{\ell=1}^{L} \alpha_{1\ell} x_\ell, \ldots, \sum_{\ell=1}^{L} \alpha_{M\ell} x_\ell \right)$$

of linear forms formed from the elements $\alpha_{m\ell}$ of the real $M \times L$ matrix, one can find integers $x_1,\ldots,x_L$, not all equal to zero, which possess the properties

$$|x_\ell| \le N^{M/L} \quad for \quad \ell = 1,\ldots,L$$

and

$$\left| \sum_{\ell=1}^{L} \alpha_{m\ell} x_\ell \right| \le 2L \cdot A \cdot N^{M/L - 1} \quad for \quad m = 1,\ldots,M .$$

As before N denotes an arbitrarily chosen natural number and A stands for the maximum absolute value among the elements $\alpha_{m\ell}$ in the matrix above.

In this estimate what is important above all else is the appearance of more variables than linear forms, i.e. in all cases $M < L$.

Proof. It is simplest to begin with the M linear forms

$$\sum_{\ell=1}^{L} \kappa \alpha_{m\ell} x_\ell = \kappa \cdot \sum_{\ell=1}^{L} \alpha_{m\ell} x_\ell$$

where the size of the positive constant κ will be fixed later. Theorem 2 ensures the existence of an M-tuple of integers $p_1, \ldots, p_M$ with

$$\left| \sum_{\ell=1}^{L} \kappa \alpha_{m\ell} x_\ell - p_m \right| < \frac{1}{N} \; ,$$

where not all the integers $x_1, \ldots, x_L$ are equal to zero, and the $x_1, \ldots, x_L$ satisfy the inequalities $|x_\ell| \leq N^{M/L}$. The idea of the proof rests on the consideration, that one can choose κ in such a way that $|p_m| < 1$, so that all p_m must equal 0. This succeeds with the help of the estimate

$$|p_m| = \left| p_m - \sum_{\ell=1}^{L} \kappa \alpha_{m\ell} x_\ell + \sum_{\ell=1}^{L} \kappa \alpha_{m\ell} x_\ell \right|$$

$$\leq \left| p_m - \sum_{\ell=1}^{L} \kappa \alpha_{m\ell} x_\ell \right| + \kappa \cdot \left| \sum_{\ell=1}^{L} \alpha_{m\ell} x_\ell \right|$$

$$< \frac{1}{N} + \kappa \cdot \sum_{\ell=1}^{L} |\alpha_{m\ell}| \, |x_\ell| \leq \frac{1}{N} + \kappa L \cdot A \cdot N^{M/L} \; .$$

One achieves $p_m = 0$ by setting

$$\frac{1}{N} + \kappa L \, A \, N^{M/L} = 1 \; ,$$

i.e.

$$\kappa = \left(1 - \frac{1}{N} \right) \frac{1}{L \, A \, N^{M/L}} \; ,$$

assuming that $A > 0$. However the case $A = 0$ is trivial. As a consequence one now has

$$\left| \kappa \sum_{\ell=1}^{L} \alpha_{m\ell} x_\ell \right| < \frac{1}{N}$$

$$\left| \sum_{\ell=1}^{L} \alpha_{m\ell} x_\ell \right| < \frac{1}{\kappa N} = \frac{L \, A \, N^{-1+M/L}}{1 - \dfrac{1}{N}}$$

with which for $N \geq 2$ the assertion is already proved, given that it is certainly true that

$$\frac{1}{1 - \frac{1}{N}} \leq 2 \ .$$

For $N = 1$ it suffices to put $x_\ell = 1$, since the assertion then follows from

$$\left| \sum_{\ell=1}^{L} \alpha_{m\ell} x_\ell \right| \leq \sum_{\ell=1}^{L} |\alpha_{m\ell}| \leq L A \ . \qquad \square$$

In practice the application of Corollary 7 makes it possible to solve systems of homogeneous linear equations

$$\sum_{\ell=1}^{L} a_{m\ell} x_\ell = 0, \quad m = 1, 2, \ldots, M \ ,$$

with more unknowns than equations in integers x_ℓ, when not all the x_ℓ vanish. The coefficient matrix of the $a_{m\ell}$ must here consist of integers. It is therefore enough to arrive at

$$\left| \sum_{\ell=1}^{L} a_{m\ell} x_\ell \right| < 1 \quad \text{for all } m = 1, \ldots, N \ .$$

Then (with $\alpha_{m\ell} = a_{m\ell}$), because of

$$\left| \sum_{\ell=1}^{L} a_{m\ell} x_\ell \right| \leq 2L \, A \, N^{M/L - 1}$$

it is enough to require $2L \, A \, N^{M/L - 1} < 1$, i.e. $N > (2L \, A)^{L/(L-M)}$. This result ties up the following theorem:

Corollary 8: Carl Ludwig Siegel's Lemma. *If the coefficients $a_{m\ell}$ of the M linear forms*

$$\sum_{\ell=1}^{L} a_{m\ell} x_\ell, \quad m = 1, \ldots, M \ ,$$

are integral, and if the number L of the variables $x_1, \ldots, x_L$ exceeds M, then one can find integers x_ℓ, which are not all zero and which solve the equations

$$\sum_{\ell=1}^{L} a_{1\ell} x_\ell = 0, \ldots, \sum_{\ell=1}^{L} a_{M\ell} x_\ell = 0 \ .$$

If one denotes by A the greatest absolute value occurring among the elements $a_{m\ell}$ in the coefficient matrix, then one can even arrive at

$$|x_\ell| \leq \left(\left[(2L \, A)^{\frac{L}{L-M}} \right] + 1 \right)^{\frac{M}{L}} \ .$$

Siegel applied this result in the proof of the transcendence of certain real numbers.

Exercises on Chapter 1

An asterix indicates a harder exercise.

1. Find integers p, q with $1 \leq q \leq 50$ so that $|\pi q - p| < \frac{1}{50}$.

2. For each natural number $N > 1$ find some real number α, so that, for all integers p, q with $1 \leq q < N$, we have $|\alpha q - p| \geq \frac{1}{N}$. (This means that the inequality $1 \leq q \leq N$ in Theorem 1 cannot be sharpened.)

3. For each real number x let $\langle x \rangle$ denote the distance from x to the nearest integer. Then

$$\eta(\alpha) := \sup \left\{ \gamma \in \mathbb{R} : \underline{\lim}_{q \to \infty} q^{\gamma} \langle q\alpha \rangle = 0 \right\}$$

is called the *type of* α. Show that:
(i)   If α is rational, then $\eta(\alpha) = \infty$.
(ii)   For all real α, $\eta(\alpha) \geq 1$.

4. Let $(g_n)_{n \geq 1}$ be a monotonically increasing sequence of natural numbers, $2 < g_2$ and $\lim_{n \to \infty} g_n = \infty$. For $n \geq 1$ let $z_n \in \{0, 1\}$, and for $n \geq 0$ let

$$G_n = g_1 \ldots g_n , \ P_n = G_n \sum_{k=1}^{n} \frac{z_k}{g_1 \ldots g_k} .$$

Write $\alpha = \sum_{k=1}^{\infty} \frac{z_k}{g_1 \ldots g_k}$. Show that for $n \geq 0$, $z_{n+1} = [G_{n+1}\alpha] - P_n g_{n+1}$.

5. Let $(g_n)_{n \geq 1}$ be a monotonically increasing sequence of natural numbers, $2 < g_2$ and $\lim_{n \to \infty} g_n = \infty$. For $n \geq 1$ let $z_n, w_n \in \{0, 1\}$ and

$$\sum_{k=1}^{\infty} \frac{z_k}{g_1 \ldots g_k} = \sum_{k=1}^{\infty} \frac{w_k}{g_1 \ldots g_k} .$$

Show that, for $n \geq 1$, $w_n = z_n$.

6. Let $\varphi : \mathbb{N} \to (0, 1]$ and α be irrational. φ is called an *approximation function* for α, if there exist infinitely many rational numbers $\frac{p}{q}$ with $|\alpha q - p| < \varphi(q)$. By $A(\varphi)$ we denote the set of all $\alpha \in \mathbb{R} \setminus \mathbb{Q}$ having φ as an approximation function. Show that if $\alpha \in A(\varphi)$ and $(p_n, q_n)_{n \geq 1}$ is a sequence of distinct pairs in $\mathbb{Z} \times \mathbb{N}$ with $|\alpha q_n - p_n| < \varphi(q_n)$, then $\lim_{n \to \infty} q_n = \infty$.

* 7. Let $\varphi : \mathbb{N} \to (0, 1]$. Show that $A(\varphi)$ has the same cardinality as $\mathbb{R}$.

8. Let $\varphi : \mathbb{N} \to (0, 1]$. Show that $A(\varphi)$ is dense in $\mathbb{R}$.

* 9. Let $\varphi : \mathbb{N} \to (0, 1]$ and $\sum_{q=1}^{\infty} \varphi(q) < \infty$. Show that $A(\varphi)$ has measure zero, (Chinchin's theorem).

10. Let α be irrational. For all natural numbers k suppose there exists a rational number $\frac{p}{q}$ with $q > 1$ and $|\alpha - \frac{p}{q}| < q^{-k}$. Then α is called a *Liouville number*.

Denote the set of all Liouville numbers by $\mathcal{L}$. For $k \geq 1$ let $\varphi_k : \mathbb{N} \to (0,1]$ be given by $\varphi_k(q) = q^{-k}$. Show that the following statements are equivalent:

(i) $\alpha \in \mathcal{L}$

(ii) $\alpha \in \bigcap\limits_{k=1}^{\infty} A(\varphi_k)$

(iii) $\alpha \notin \mathbb{Q}$ and $\eta(\alpha) = \infty$.

* 11. Let α be a real algebraic irrational number, which satisfies a polynomial of degree n with integral coefficients. Show that there exists some $c > 0$, so that, for all $\frac{p}{q} \in \mathbb{Q}$, $|\alpha - \frac{p}{q}| \geq cq^{-n}$ (Liouville's theorem).

12. Show that each Liouville number is transcendental, and that $\mathcal{L}$ is a dense, uncountable subset of $\mathbb{R}$ with measure zero.

13. Show that $\mathbb{Q}^* \cdot \mathcal{L} = \mathcal{L}$ and that $\mathbb{Q} + \mathcal{L} = \mathcal{L}$.

* 14. Show that $\mathcal{L} + \mathcal{L} = \mathbb{R}$ (Erdös' theorem).

15. For $n \in \mathbb{N}$ let $\mathcal{F}_n = \{\frac{a}{b} : (a,b) \in \mathbb{Z} \times \mathbb{N}, b \leq n\}$. $\mathcal{F}_n$ is called the *nth Farey series*. Show that $\mathcal{F}_n$ is neither bounded above nor below, and has no point of accumulation. Hence for $\frac{a}{b} \in \mathcal{F}_n, (-\infty, \frac{a}{b}) \cap \mathcal{F}_n$ has a greatest (resp. $(\frac{a}{b}, \infty) \cap \mathcal{F}_n$ a smallest) element. This is called the *lower* (resp. *upper*) *neighbour of* $\frac{a}{b}$ *in* $\mathcal{F}_n$.

16. Let $\frac{a}{b}, \frac{c}{d}, \frac{e}{f}$ be fractions with $b, d, f \in \mathbb{N}$ and let $|ad - bc| = 1$. Show that there exist integers x, y with $ax + cy = e$, $bx + dy = f$, and that $\frac{e}{f}$ then lies (properly) between $\frac{a}{b}$ and $\frac{c}{d}$, when these are positive.

17. Let $\frac{a}{b}, \frac{c}{d}$ be fractions with $b, d \in \mathbb{N}$ and $|ad - bc| = 1$. Show:
   (i)   If $\max(b,d) \leq n < b + d$, then $\frac{a}{b}$ is a neighbour of $\frac{c}{d}$ in $\mathcal{F}_n$.
   (ii)   If $b + d \leq n$, then $\frac{a}{b}$ is not a neighbour of $\frac{c}{d}$ in $\mathcal{F}_n$.

18. Let $\frac{a}{b}, \frac{c}{d}$ be fractions in $\mathcal{F}_n$, $b, d \in \mathbb{N}$ and $g.c.d.(a,b) = g.c.d.(c,d) = 1$. Show that the following statements are equivalent:
   (i) $\frac{a}{b}$ and $\frac{c}{d}$ are neighbours in $\mathcal{F}_n$.
   (ii) $n < b + d$ and $|ad - bc| = 1$.

* 19. Let α be a real and n a natural number. In lowest terms let the fractions $\frac{a}{b}$ and $\frac{c}{d}$ be neighbours in $\mathcal{F}_n$, and so chosen that $\frac{a}{b} \leq \alpha < \frac{c}{d}$. Then one of the three fractions $\frac{p}{q} \in \left\{\frac{a}{b}, \frac{c}{d}, \frac{a+c}{b+d}\right\}$ satisfies the inequality

$$\left|\alpha - \frac{p}{q}\right| < \frac{1}{\sqrt{5}q^2} \ .$$

20. Let α be irrational. Then there exist infinitely many fractions $\frac{p}{q}$ with

$$\left|\alpha - \frac{p}{q}\right| < \frac{1}{\sqrt{5}q^2} \qquad \text{(Hurwitz' theorem)} \ .$$

Hints for the Exercises on Chapter 1

7. Let $Z = \{(z_n)_{n \geq 1} : \text{for all } n \geq 1, z_n \in \{0,1\} \text{ and for infinitely many } n, z_n = 1\}$. This set has the same cardinality as $\mathbb{R}$. One can choose a monotone

increasing sequence of natural numbers $(q_n)_{n\geq 1}$ with $\lim_{n\to\infty} q_n = \infty$ and $q_{n+1} > 1 + \frac{1}{\varphi(q_1 \cdots q_n)}$ for $n \geq 0$. By (5) the map $\psi : Z \to A(\varphi)$, $\psi((z_n)_{n\geq 1}) = \sum_{n=1}^{\infty} \frac{z_n}{q_1 \cdots q_n}$ is injective.

9. It is enough to show that for each $g \in \mathbb{N}$, $A(\varphi) \cap [-g, g]$ has measure zero. For this show

$$A(\varphi) \cap [-g, g] \subseteq \bigcap_{N=1}^{\infty} \bigcup_{q=N}^{\infty} \bigcup_{|p| \leq gq} \left(\frac{p}{q} - \frac{\varphi(q)}{q}, \frac{p}{q} + \frac{\varphi(q)}{q} \right) .$$

11. Let $m(x) \in \mathbb{Z}[x]$ be irreducible, $m(\alpha) = 0$. Put $c_1 = \sup_{|x-\alpha|\leq 1} |m'(x)|$ and $c = \min(1, c_1^{-1})$. Then without loss of generality one can suppose that $\left| \alpha - \frac{p}{q} \right| \leq 1$. By the mean-value theorem there exists some x_0 between α and $\frac{p}{q}$ with

$$q^{-n} \leq \left| m\left(\frac{p}{q} \right) \right| = \left| \alpha - \frac{p}{q} \right| |m'(x_0)| .$$

14. By (13) it suffices to show that each $\alpha \in (0,1)$, $\alpha \notin \mathbb{Q} \cup \mathcal{L}$, is a sum of two Liouville numbers. Let $\alpha = \sum_{k=1}^{\infty} a_k 2^{-k}$ be the dyadic expansion of α. It is possible to define a sequence $(n_k)_{k\geq 1}$ of natural numbers by $n_k! \leq k < (n_k+1)!$ and for $k \geq 1$ to put

$$b_k = \frac{1}{2}(1 - (-1)^{n_k}) a_k , \quad c_k = \frac{1}{2}(1 + (-1)^{n_k}) a_k ,$$

$$\beta = \sum_{k=1}^{\infty} b_k 2^{-k} , \qquad \gamma = \sum_{k=1}^{\infty} c_k 2^{-k} .$$

Show that with $q_k = 2^{(2k)!-1}$ and $p_k = q_k \sum_{1 \leq j < (2k)!} b_j 2^{-j}$, we have

$$\left| \beta - \frac{p_k}{q_k} \right| \leq q_k^{-k} .$$

18. In order to prove (i) $\Rightarrow$ (ii) choose w.l.o.g. $b \leq d$ and $x_0, y_0 \in \mathbb{Z}$ so that $dx_0 - cy_0 = \text{sgn}(ad - bc)$. For $t \in \mathbb{Z}$ let $x_t = x_0 + ct$, $y_t = y_0 + dt$ and choose t maximal w.r.t. the condition $y_t \leq d$. By (17) $\frac{x_t}{y_t}$ is a neighbour of $\frac{c}{d}$ in $\mathcal{F}_d$ and indeed an upper neighbour, if it is $\frac{a}{b}$. Hence $\frac{x_t}{y_t} = \frac{a}{b}$.

19. W.l.o.g. let $\frac{a}{b} \leq \alpha < \frac{a+b}{c+d} < \frac{c}{d}$. One argues by contradiction. From (18) by addition of suitable inequalities one obtains

$$\frac{1}{b(b+d)} \geq \frac{1}{\sqrt{5}} \left(\frac{1}{b^2} + \frac{1}{(b+d)^2} \right) \quad \text{and} \quad \frac{1}{bd} \geq \frac{1}{\sqrt{5}} \left(\frac{1}{b^2} + \frac{1}{d^2} \right) .$$

However there exists no rational number $x = \frac{d}{b}$, which satisfies

$$\sqrt{5} \geq x + 1 + \frac{1}{x+1} \quad \text{and} \quad \sqrt{5} \geq x + \frac{1}{x} .$$

2. The Kronecker Approximation Theorem

By the Dirichlet approximation theorem, for each real number α and each natural number N there exists an integer p and a natural number $n \leq N$ with $|\alpha n - p| < 1/N$. If instead of the real number α and the real line $\mathbb{R}$ one considers only the *fractional part* $\alpha - [\alpha] = \{\alpha\}$ and the half-open interval $[0,1[$, i.e. one *identifies* two real numbers if they differ only by an integer, then from the *algebraic* point of view this implies a reduction to numbers modulo 1 and passage from $\mathbb{R}$ to $\mathbb{R}/\mathbb{Z}$. From the *geometric* point of view it involves rolling up the real line onto a circle of circumference 1, where the circle replaces the unit interval $[0,1[$. *Topologically* open sets on the circle are identified as open sets on $[0,1[$, i.e. neighbourhoods in $[0,1[$ are defined by the quotient topology in $\mathbb{R}/\mathbb{Z}$. Considered in this way Dirichlet's theorem states

Proposition 1. *In each neighbourhood of zero there are infinitely many numbers of the form $\alpha n(\mathrm{modulo}\ 1)$ where n runs through the natural numbers and α is real and arbitrary.*

Here one must distinguish more carefully between rational and irrational α. The rationals $\alpha = a/b$ form the singular case: for all integral multiples $n = kb$ of b one has $\alpha n \equiv 0(\mathrm{mod}\ 1)$. If the highest common factor $(a,b) = 1$, the remaining multiples αn in any case cannot lie in each neighbourhood of the zero point. In the regular case of irrational α on the other hand it can never happen that $\alpha n \equiv 0(\mathrm{mod}\ 1)$ for $n \neq 0$. That one limits oneself in the Dirichlet theorem to neighbourhoods of the zero point characterises the theorem as a statement about a homogeneous approximation problem – similar to the theory of systems of linear equations

$$\sum_{\ell=1}^{L} \alpha_{m\ell} x_\ell = \beta_m, \quad m = 1,\ldots,M \ ,$$

which is said to be homogeneous if $\beta_1 = \ldots = \beta_M = 0$. For a system of singular linear equations with arbitrary values β_m on the right-hand side there in general exists no solution. One obtains a similar result in approximation theory by passing from the singular case of a rational number to the inhomogeneous approximation problem.

Proposition 2. *For rational numbers $\alpha = a/b$ it is impossible to find for each real β and each positive number ε integers n, p for which the inequality*

$$|\alpha n - \beta - p| < \varepsilon$$

holds. [7]

Proof. If this were the case (assuming $b > 0$) then $|an - bp - b\beta| < b\varepsilon$, which for $\beta = 1/b\sqrt{8}$ leads to the realisation that there must be some integer $g = an - bp$ with $|g - 1/\sqrt{8}| < b\varepsilon$. However since

$$\frac{1}{\sqrt{8}} = \frac{1}{2\sqrt{2}} < \frac{1}{2}$$

$1/\sqrt{8}$ is at least a distance $1/\sqrt{8}$ from the nearest integer, and moreover

$$\frac{1}{\sqrt{8}} \leq \left| g - \frac{1}{\sqrt{8}} \right| < b\varepsilon$$

it follows that $\varepsilon > 1/b\sqrt{8}$. But this contradicts the assumption that one can take $\varepsilon > 0$ to be arbitrarily small. $\qquad\square$

On the other hand analogously to systems of linear equations, in approximation theory for the regular case of an irrational α the inhomogeneous approximation problem is always solvable. This is the core of the statement of the Kronecker theorem.

Theorem 1: Kronecker's Approximation Theorem. *For each irrational number α, each real number β, each preassigned arbitrarily small number $\varepsilon > 0$, and arbitrarily large number Ω, there exist integers p and n with $|n| \geq \Omega$ and $|\alpha n - \beta - p| < \varepsilon$.*

If we reduce modulo 1 the integral part p falls away from $\alpha n - \beta - p$, and we can formulate the Kronecker approximation theorem in the following simpler manner.

Corollary 1. *For each irrational number α infinitely many of the numbers αn modulo 1 lie in each arbitrarily small ε-neighbourhood of an arbitrary element β from $[0, 1[$.*

Even simpler:

Corollary 2. *For each irrational number α the set of numbers αn reduced modulo 1 is dense in the whole interval $[0, 1[$.*

Proof. For the proof of the Kronecker theorem one starts from the fact that there exists an integer g and a natural number q with

$$0 < |\alpha q - g| < \varepsilon \ .$$

The left-hand inequality holds because of the irrationality of α, and one obtains the right-hand inequality by means of the Dirichlet theorem by choosing $N \geq 1/\varepsilon$. Now form the equations $n = kq$, $p = kg + c$, where the integers k and c are to be determined in the course of the proof. Since one must have

$$|\alpha n - \beta - p| = |k(\alpha q - g) - \beta - c|$$
$$= |\alpha q - g| \cdot \left| k - \frac{\beta + c}{\alpha q - g} \right|$$

remaining smaller than ε, fix k by means of the equation

$$k = \left[\frac{\beta + c}{\alpha q - g} \right] + 1 \ .$$

Finally choose c so that $|n| \geq \Omega$. Because $|n| = q|k|$ it suffices to restrict oneself to $|k| \geq \Omega$, that is,

$$\left| \frac{\beta + c}{\alpha q - g} \right| \geq \Omega + 1 \ .$$

The inequality

$$\left| \frac{\beta + c}{\alpha q - g} \right| \geq \frac{|c|}{|\alpha q - g|} - \frac{|\beta|}{|\alpha q - g|}$$

shows that the condition above is guaranteed, if one ensures that $|c| \geq (\Omega + 1)|\alpha q - g| + |\beta|$. In this way not only is

$$|\alpha n - \beta - p| < \varepsilon$$

guaranteed, but also $|n| \geq \Omega$, which was to be proved. $\qquad\square$

Remarks. (1) It is possible to require that either $n \geq \Omega$, or $n \leq -\Omega$. In order to achieve $n \geq \Omega$ one has to take care of $k > 0$. The proof allows an arbitrary preliminary choice of c. For $n \geq \Omega$ it suffices to let c have the same sign as $\alpha q - g$. Then from

$$k = \left[\frac{\beta + c}{\alpha q - g} \right] + 1 \geq \frac{c}{\alpha q - g} - \left| \frac{\beta}{\alpha q - g} \right|$$
$$\geq \left(\Omega + \left| \frac{\beta}{\alpha q - g} \right| \right) - \left| \frac{\beta}{\alpha q - g} \right| = \Omega$$

one does indeed have $k > 0$. If on the other hand one wants to achieve $n \leq -\Omega$, one requires that c have the same sign as $g - \alpha q$. Then

$$k = \left[\frac{\beta + c}{\alpha q - g} \right] + 1 \leq - \left| \frac{c}{\alpha q - g} \right| + \left| \frac{\beta}{\alpha q - g} \right| + 1 \ ,$$
$$-k \geq \left| \frac{c}{\alpha q - g} \right| - \left| \frac{\beta}{\alpha q - g} \right| - 1 \geq \Omega \ ,$$
$$k \leq -\Omega \ ,$$

which implies that $k < 0$.

One might think that more simply one can justify the free choice of sign by replacing

$$|\alpha n - \beta - p| < \varepsilon$$

by

$$|\alpha(-n) - (-\beta) - (-p)| < \varepsilon \ .$$

However this justification fails because n depends on β – the earlier argument still stands.

(2) One could have shown immediately that $n \geq \Omega$, had one known that, for each irrational α, each real β and each $\varepsilon > 0$, there exists an integer p and a natural number n with $|\alpha n - \beta - p| < \varepsilon$, i.e. the introduction of c into the proof would not have been necessary. Then one would have been able to replace β by $\beta' = \beta - \alpha\Omega$ and for integral p' and natural number n' to obtain

$$|\alpha n' - \beta' - p'| = |\alpha(n' + \Omega) - \beta - p'| < \varepsilon \ .$$

When $n = n' + \Omega, p = p'$ this implies that $n \geq \Omega$ (supposing that Ω is a natural number). This argument fails in the generalisation to several dimensions.

(3) The Kronecker theorem gives arbitrarily large *lower* bounds for n, while one is accustomed to upper bounds for n from the Dirichlet theorem. Indeed it is not possible to give a constant $K(\varepsilon)$ dependent only on ε, so that for arbitrary irrational α and real β one can always find an integer p and a natural number $n \leq K(\varepsilon)$ with $|\alpha n - \beta - p| < \varepsilon$. For example, if ϑ is some irrational number lying between 0 and 1 and $\beta = \frac{3}{4}$, then from the existence of $n \leq K(1/8) = K$ and p with

$$\left| \frac{\vartheta n}{2K} - \frac{3}{4} - p \right| < \frac{1}{8} : - \frac{1}{8} < \frac{\vartheta n}{2K} - \frac{3}{4} - p < \frac{1}{8},$$

$$- \frac{7}{8} < \frac{\vartheta}{2K} - \frac{7}{8} \leq \frac{\vartheta n}{2K} - \frac{1}{8} - \frac{3}{4} < p$$

$$< \frac{1}{8} - \frac{3}{4} + \frac{\vartheta n}{2K} < -\frac{5}{8} + \frac{1K}{2K} = -\frac{1}{8}$$

it would follow that the integer p would have to satisfy the impossible inequality

$$-7/8 < p < -1/8 \ .$$

Instead of a single irrational number α the multidimensional version of the Kronecker theorem makes use of L real numbers $\alpha_1, \ldots, \alpha_L$ which are *linearly independent over* $\mathbb{Z}$. By this we mean that for all integers $x_1, \ldots, x_L$ the integral nature of the expression

$$\sum_{\ell=1}^{L} x_\ell \alpha_\ell$$

must imply that $x_1 = x_2 = \ldots = x_L = 0$. For example, if a single number α is dependent over $\mathbb{Z}\,(L = 1)$ there exists an integer n different from zero so that $n\alpha = p$ is likewise integral, i.e. $\alpha = p/n$ is rational. Conversely a single irrational number is linearly independent over $\mathbb{Z}$.

Theorem 2: Multidimensional Kronecker Approximation Theorem. *Let $\alpha_1, \ldots, \alpha_L$ be real numbers linearly independent over $\mathbb{Z}$, $\beta_1, \ldots, \beta_L$ arbitrary real numbers and ε, Ω arbitrary positive numbers, then one can find integers n and $p_1, \ldots, p_L$ with*

$$|\alpha_1 n - \beta_1 - p_1| < \varepsilon, \ldots, |\alpha_L n - \beta_L - p_L| < \varepsilon \; ,$$

where $|n| \geq \Omega$ and the sign of n can be arbitrarily chosen. [8]

Before giving the proof let us give a short sketch of the importance of this theorem. Again consider the numbers $\alpha_1, \ldots, \alpha_L$ modulo 1, that is we are interested only in their fractional parts. Instead of the space $\mathbb{R}^L$ of all real L-tuples we now work in the space $\mathbb{R}^L/\mathbb{Z}^L$ of L-tuples reduced modulo 1. The relation

$$\alpha_\ell n - \beta_\ell - p_\ell \equiv \alpha_\ell n - \beta_\ell \;(\text{mod } 1) \qquad \ell = 1, \ldots, L \; ,$$

shows that the integers $p_1, \ldots, p_L$ only play a marginal role in the theorem. The following statement is important:

Corollary 3. *Given L numbers $\alpha_1, \ldots, \alpha_L$ linearly independent over $\mathbb{Z}$, there are infinitely many L-tuples $(\alpha_1 n, \ldots, \alpha_L n)$, $n = 1, 2, \ldots$ (modulo 1) in each arbitrarily small ε-neighbourhood of an L-tuple $(\beta_1, \ldots, \beta_L)$ from $\mathbb{R}^L/\mathbb{Z}^L$.*

Here one considers that as a set $\mathbb{R}^L/\mathbb{Z}^L$ agrees with the L-dimensional unit cube $[0, 1[^L$, but is equipped with the quotient topology of $\mathbb{R}^L/\mathbb{Z}^L$. From the geometric point of view this implies the identification of pairs of opposite $(L-1)$-dimensional bounding surfaces. For this reason $\mathbb{R}^L/\mathbb{Z}^L$ is called the L-dimensional torus.
Seen in this way the Kronecker theorem states

Corollary 4. *For L numbers $\alpha_1, \ldots, \alpha_L$ linearly independent over $\mathbb{Z}$ the set of L-tuples $(\alpha_1 n, \ldots, \alpha_L n)$ reduced modulo 1 is everywhere dense in the L-dimensional torus $[0, 1[^L$.*

Proof. In 1934 Estermann carried out the full inductive proof of the Kronecker theorem; for $L = 1$ we have the conclusion of Theorem 1. The inductive assumption is that Theorem 2 holds as stated for $L' = L - 1$, and in order to carry out the inductive step we take L numbers $\alpha_1, \ldots, \alpha_L$ linearly independent over $\mathbb{Z}$, real numbers $\beta_1, \ldots, \beta_L$ and arbitrarily chosen positive ε and Ω. With the help of a positive number δ, whose magnitude will be given in the

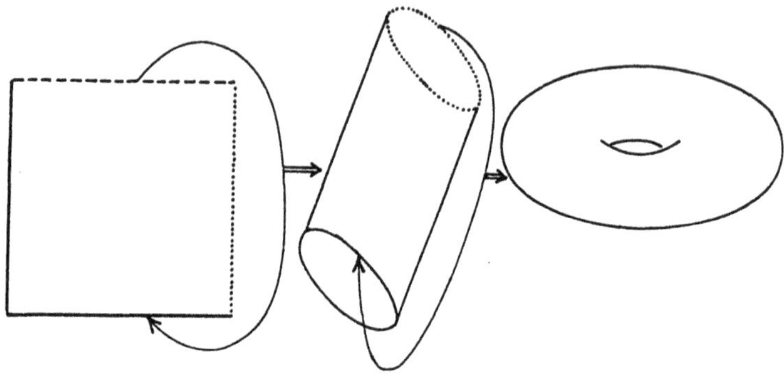

Fig. 2. Construction of the torus from the unit square

course of the proof, one can find integers $g_1, \ldots, g_L$ and a natural number q with

$$0 < |\alpha_\ell q - g_\ell| < \delta, \quad \ell = 1, \ldots, L \ .$$

The left-hand inequalities follow from the irrationality of the α_ℓ, the right-hand from the multidimensional version of the Dirichlet approximation theorem. One has only to take $N \geq 1/\delta$.

As for a single dimension put $n = kq$, $p_1 = kg_1 + c_1, \ldots, p_L = kg_L + c_L$ with as yet undetermined k and $c_1, \ldots, c_L$. We then have

$$|\alpha_\ell n - \beta_\ell - p_\ell| = |k(\alpha_\ell q - g_\ell) - \beta_\ell - c_\ell|$$

$$= |\alpha_\ell q - g_\ell| \left| k - \frac{\beta_\ell + c_\ell}{\alpha_\ell q - g_\ell} \right| \ .$$

For the last component $\ell = L$ Estermann argues analogously to the one-dimensional case. He puts

$$k = \frac{\beta_L + c_L}{\alpha_L q - g_L} + \vartheta, \quad |c_L| \geq \Omega |\alpha_L q - g_L| + |\beta_L| \ ,$$

where ϑ with $0 < \vartheta \leq 1$ increases the fraction as written to the next larger integer. In this way he obtains

$$|\alpha_L n - \beta_L - p_L| < \delta, \quad |n| \geq \Omega$$

and also that n is positive or negative depending on whether c_L has the same sign as $\alpha_L q - g_L$ or not. For $\ell' = 1, 2, \ldots, L - 1 = L'$ Estermann now considers

$$\alpha_{\ell'} n - \beta_{\ell'} - p_{\ell'} = \alpha_{\ell'} kq - \beta_{\ell'} - kg_{\ell'} - c_{\ell'}$$

$$= \alpha_{\ell'} q \left(\frac{\beta_L + c_L}{\alpha_L q - g_L} + \vartheta \right) - \left(\frac{\beta_L + c_L}{\alpha_L q - g_L} + \vartheta \right) g_{\ell'}$$

$$- \beta_{\ell'} - c_{\ell'}$$

$$= c_L \left(\frac{\alpha_{\ell'} q - g_{\ell'}}{\alpha_L q - g_L} \right) - \left(\beta_{\ell'} - \frac{\beta_L(\alpha_{\ell'} q - g_{\ell'})}{\alpha_L q - g_L} \right)$$

$$- c_{\ell'} + \vartheta \left(\alpha_{\ell'} q - g_{\ell'} \right)$$

and defines the numbers $\tilde{\alpha}_1, \ldots, \tilde{\alpha}_{L-1}$ and $\tilde{\beta}_1, \ldots, \tilde{\beta}_{L-1}$ by

$$\tilde{\alpha}_{\ell'} = \frac{\alpha_{\ell'} q - g_{\ell'}}{\alpha_L q - g_L}, \quad \tilde{\beta}_{\ell'} = \beta_{\ell'} - \frac{\beta_L(\alpha_{\ell'} q - g_{\ell'})}{\alpha_L q - g_L} \ .$$

The $\tilde{\alpha}_1, \ldots, \tilde{\alpha}_{L-1}$ are linearly independent over $\mathbb{Z}$. For with integral parameters $x_1, \ldots, x_L$ the equation

$$\sum_{\ell'=1}^{L-1} x_{\ell'} \tilde{\alpha}_{\ell'} + x_L$$

$$= x_L + \left(\sum_{\ell'=1}^{L-1} x_{\ell'} \alpha_{\ell'} q - \sum_{\ell'=1}^{L-1} x_{\ell'} g_{\ell'} \right) \frac{1}{a_L q - g_L} = 0$$

leads to

$$\sum_{\ell=1}^{L} q x_\ell \alpha_\ell = \sum_{\ell=1}^{L} x_\ell g_\ell \ ,$$

which because of the linear independence of $\alpha_1, \ldots, \alpha_L$ over $\mathbb{Z}$, implies that $q x_1 = \ldots = q x_L = 0$ and then also $x_1 = \ldots = x_L = 0$. Besides this one has

$$|\alpha_{\ell'} n - \beta_{\ell'} - p_{\ell'}|$$

$$\leq \left| c_L \tilde{\alpha}_{\ell'} - \tilde{\beta}_{\ell'} - c_{\ell'} \right| + |\alpha_{\ell'} q - g_{\ell'}|$$

$$< \left| c_L \tilde{\alpha}_{\ell'} - \tilde{\beta}_{\ell'} - c_{\ell'} \right| + \delta \ .$$

Given the inductive assumption there exists an integer $\tilde{n}$ with

$$|\tilde{n}| \geq \Omega |\alpha_L q - g_L| + |\beta_L|$$

and integers $\tilde{p}_{\ell'}$, for which

$$\left| \tilde{\alpha}_{\ell'} \tilde{n} - \tilde{\beta}_{\ell'} - \tilde{p}_{\ell'} \right| < \delta \ .$$

Moreover the sign of $\tilde{n}$ can be freely chosen. Since c_L only has to satisfy the condition

$$|c_L| \geq \Omega |\alpha_L q - g_L| + |\beta_L|$$

and one may suppose the sign of n as already given, *it is possible to set $c_L = \tilde{n}$,* and $c_{\ell'} = \tilde{p}_{\ell'}$. In this way one obtains not only $|n| \geq \Omega$,

$$|\alpha_L n - \beta_L - p_L| < \delta \ ,$$

but also

$$|\alpha_\ell n - \beta_\ell - p_\ell| < 2\delta$$

for all ℓ. Since there is no problem in taking δ to be smaller than $\varepsilon/2$, Estermann has completed his proof of the theorem. $\qquad\qquad\qquad\qquad\qquad\qquad\Box$

Bohl, Sierpinski and Weyl found another proof of the Kronecker approximation theorem by passing from the sequence

$$(\{\alpha_1 n\}, \ldots, \{\alpha_L n\}) = (\alpha_1 n - [\alpha_1 n], \ldots, \alpha_L n - [\alpha_L n]),$$
$$n = 1, 2, \ldots$$

in $\mathbb{R}^L/\mathbb{Z}^L$ to arbitrary sequences ω_ℓ with elements $(\omega_1(n), \ldots, \omega_L(n))$, $n = 1, 2, \ldots$.

If $J = [\alpha_1, \beta_1[\times \ldots \times [\alpha_L, \beta_L[$ is some parallelepiped in $[0, 1[^L$ aligned along the axes and by $A(\omega_\ell, N, J)$ one counts the number of sequence elements lying in J from among

$$(\omega_1(1), \ldots, \omega_L(1)), (\omega_1(2), \ldots, \omega_L(2)), \ldots, (\omega_1(N), \ldots, \omega_L(N)) \ ,$$

then ω_ℓ is said to be *uniformly distributed* in $[0, 1[^L$ if the quantity $\frac{A(\omega_\ell, N, J)}{N}$ tends as $N \to \infty$ to the volume $\mathrm{vol}(J) = (\beta_1 - \alpha_1) \ldots (\beta_L - \alpha_L)$ of the arbitrarily chosen solid. A sequence ω_ℓ in $\mathbb{R}^L$ is said to be *uniformly distributed* modulo 1, if the mod 1 reduced sequence $\{\omega_\ell\}$ with typical element

$$(\omega_1(n) - [\omega_1(n)], \ldots, \omega_L(n) - [\omega_L(n)]),$$
$$n = 1, 2, 3, \ldots,$$

is uniformly distributed. Since each non-empty open subset has positive volume, the following statement is immediate:

Proposition 3. *Each uniformly distributed sequence is dense in* $[0, 1[^L$.

The assertion

Proposition 4. *If* $\alpha_1, \ldots, \alpha_L$ *are linearly independent over* $\mathbb{Z}$, *the sequence of L-tuples* $(\alpha_1 n, \ldots, \alpha_L n)$, $n = 1, 2, \ldots$ *is uniformly distributed* modulo 1,

contains the conclusion of Kronecker's theorem. This asserts only the density of this sequence. Weyl's method of proof rest on a reformulation of the definition of "uniformly distributed".

Let c_J denote the characteristic function of the parallelepiped J in $[0, 1[^L$, i.e. $c_J(x_\ell) = 1$ if $(x_1, \ldots, x_L)$ belongs to J and $c_J(x_\ell)$ equals zero otherwise. Then

$$\frac{A(\omega_\ell, N, J)}{N} = \frac{1}{N} \sum_{n=1}^{N} c_J(\omega_\ell(n)) \ .$$

If for an arbitrary function f on $[0, 1[^L$ one defines the positive linear functional $\#_{\omega_\ell}{}^N$ by

$$\#_{\omega_\ell}{}^N (f) = \frac{1}{N} \sum_{n=1}^{N} f(\omega_\ell(n)) \ ,$$

then the definition of uniform distribution asserts that for all characteristic functions $f = c_J$ of parallelepipeds one has

$$\lim_{n \to \infty} \#_{\omega_\ell}{}^N (f) = \mathrm{vol}(J) = \int_0^1 \cdots \int_0^1 f(x_\ell)dx_1 \ldots dx_L \ .$$

The linearity of the functional implies that

$$\lim_{N \to \infty} \#_{\omega_\ell}{}^N (f) = \int_0^1 \cdots \int_0^1 f(x_1, \ldots, x_L)dx_1 \ldots dx_L$$

for all step functions. *This relation then holds for all Riemann integrable functions.*

Thus if f is Riemann integrable (and real-valued), then we can find two step functions f_1, f_2 with $f_1 \le f \le f_2$, the integrals of which differ from each other by an arbitrarily small amount:

$$\int_0^1 \cdots \int_0^1 f_2(x_1, \ldots, x_L)dx_1 \ldots dx_L$$
$$- \int_0^1 \cdots \int_0^1 f_1(x_1, \ldots, x_L)dx_1 \ldots dx_L < \varepsilon \ .$$

From

$$\lim_{N \to \infty} \frac{1}{N} \sum_{n=1}^{N} f_1(\omega_\ell(n)) = \lim_{N \to \infty} \#_{\omega_\ell}{}^N (f_1)$$
$$= \int_0^1 \cdots \int_0^1 f_1(x_1, \ldots, x_L)dx_1 \ldots dx_L$$
$$\ge \int_0^1 \cdots \int_0^1 f(x_1, \ldots, x_L)dx_1 \ldots dx_L - \varepsilon$$

it follows that for sufficiently large N

$$\frac{1}{N} \sum_{n=1}^{N} f_1(\omega_\ell(n)) = \#_{\omega_\ell}{}^N (f_1)$$
$$> \int_0^1 \cdots \int_0^1 f(x_1, \ldots, x_L)dx_1 \ldots dx_L - \varepsilon \ ,$$

which forces

$$\frac{1}{N}\sum_{n=1}^{N} f\left(\omega_1(n),\ldots,\omega_L(n)\right) = {}^{\#}\omega_\ell{}^N\,(f)$$

$$> \int_0^1 \cdots \int_0^1 f(x_1,\ldots,x_L)dx_1\ldots dx_L - 2\varepsilon \ .$$

Analogously one obtains

$$\frac{1}{N}\sum_{n=1}^{N} f\left(\omega_1(n),\ldots,\omega_L(n)\right) = {}^{\#}\omega_\ell{}^N\,(f)$$

$$< \int_0^1 \cdots \int_0^1 f(x_1,\ldots,x_L)dx_1\ldots dx_L + 2\varepsilon \ .$$

With this calculation one has already shown part of the basic theorem from Weyl's theory of uniform distribution.

Theorem 3: Weyl's Criterion. *A sequence ω_ℓ in $\mathrm{I\!R}^L$ is uniformly distributed* modulo 1, *if for all integral L-tuples $(h_1,\ldots,h_L)$ which are different from $(0,\ldots,0)$ one has*

$$\lim_{N\to\infty}\frac{1}{N}\sum_{n=1}^{N} e^{2\pi i \sum_{\ell=1}^{L} h_\ell \omega_\ell(n)} = 0 \ .$$

It is clear that in the presence of the uniform distribution of ω_ℓ modulo 1 the criterion holds, for the exponential function, because of its periodicity, does not distinguish between the sequence and its reduction modulo 1. If not all $h_1,\ldots,h_L$ are equal to zero, then

$$\int_0^1 \cdots \int_0^1 e^{2\pi i \sum_{\ell=1}^{L} h_\ell x_\ell} dx_1\ldots dx_L = 0$$

holds.

It remains only to prove that, if the sequence satisfies the criterion, then it is indeed uniformly distributed modulo 1. Of course the equation set out in the criterion holds when $h_1 = \ldots = h_L = 0$,

$$\lim_{N\to\infty}\frac{1}{N}\sum_{n=1}^{N} e^{2\pi i \sum_{\ell=1}^{L} h_\ell \omega_\ell(n)}$$

$$= \int_0^1 \cdots \int_0^1 e^{2\pi i \sum_{\ell=1}^{L} h_\ell x_\ell} dx_1\ldots dx_L \ .$$

If it also holds for all other integral h_ℓ, then because of the linearity of the functional it holds for all trignometric polynomials. By the Weierstrass approximation theorem, for each real-valued continuous f (with period 1 in each component) one can construct a real-valued trignometric polynomial f_ε with $|f - f_\varepsilon| < \varepsilon$, ε assumed to be arbitrarily small. The functions $f_1 = f_\varepsilon - \varepsilon$,

$f_2 = f_\epsilon + \epsilon$ are two trignometric polynomials which catch f between them, and for which the integrals

$$\int_0^1 \cdots \int_0^1 f_1(x_1,\ldots,x_L)dx_1 \ldots dx_L \ ,$$

$$\int_0^1 \cdots \int_0^1 f_2(x_1,\ldots,x_L)dx_1 \ldots dx_L$$

differ from each other by 2ϵ. Analogously to the argument carried out above one obtains

$$\lim_{N\to\infty} {}^{\#}\omega_\ell{}^N(f) = \int_0^1 \cdots \int_0^1 f(x_1,\ldots,x_L)dx_1 \ldots dx_L \ .$$

If $J = [\alpha_1,\beta_1[\times \ldots \times [\alpha_L,\beta_L[$ denotes a subparallelepiped of $[0,1[^L$ and c_J is the characteristic function of J, extended with period 1 to all of $\mathbb{R}^L$, then one can give continuous functions f_1, f_2 with $f_1 \leq c_J \leq f_2$ and make the difference between the integrals

$$\int_0^1 \cdots \int_0^1 f_2(x_1,\ldots,x_L)dx_1 \ldots dx_L$$

$$- \int_0^1 \cdots \int_0^1 f_1(x_1,\ldots,x_L)dx_1 \ldots dx_L$$

arbitrarily small. Thus, if in the case $\beta_\ell - \alpha_\ell < 1$ for a sufficiently small ϵ one puts

$$g_\ell(x) = \begin{cases} \dfrac{x - \alpha_\ell}{\epsilon}, & \alpha_\ell \leq x < \alpha_\ell + \epsilon \\[2mm] 1, & \alpha_\ell + \epsilon \leq x < \beta_\ell - \epsilon \\[2mm] \dfrac{\beta_\ell - x}{\epsilon}, & \beta_\ell - \epsilon \leq x < \beta_\ell \\[2mm] 0, & \beta_\ell \leq x < 1 + \alpha_\ell \ , \end{cases}$$

$$G_\ell(x) = \begin{cases} \dfrac{x - (\alpha_\ell - \epsilon)}{\epsilon}, & \alpha_\ell - \epsilon \leq x < \alpha_\ell \\[2mm] 1, & \alpha_\ell \leq x < \beta_\ell \\[2mm] \dfrac{(\beta_\ell + \epsilon) - x}{\epsilon}, & \beta_\ell \leq x < \beta_\ell + \epsilon \\[2mm] 0, & \beta_\ell + \epsilon \leq x < 1 + \alpha_\ell - \epsilon \ , \end{cases}$$

extends this function to $\mathbb{R}$ with period 1, and for $\beta_\ell - \alpha_\ell = 1$ writes $g_\ell(x) = G_\ell(x) = 1$, then for

$$f_1(x_1, \ldots, x_L) = g_1(x_1) \cdot \ldots \cdot g_L(x_L) \ ,$$
$$f_2(x_1, \ldots, x_L) = G_1(x_1) \cdot \ldots \cdot G_L(x_L)$$

one does indeed obtain continuous functions with the desired property. Because of this property one obtains

$$\lim_{N \to \infty} \frac{1}{N} \sum_{n=1}^{N} c_J(\omega_\ell(n)) = \lim_{N \to \infty} \#\omega_\ell^N(c_J)$$

$$= \int_0^1 \cdots \int_0^1 c_J(x_1, \ldots, x_L) dx_1 \ldots dx_L$$

$$= \mathrm{vol}(J) \ ,$$

which proves the uniform distribution of ω_ℓ modulo 1.

In the example of the sequence

$$(\omega_1(n), \ldots, \omega_L(n)) = (\alpha_1 n, \ldots, \alpha_L n)$$
$$n = 1, 2, \ldots,$$

with $\alpha_1, \ldots, \alpha_L$ linearly independent over $\mathbb{Z}$, for integral $h_1, \ldots, h_L$ the integrality of $h_1 \alpha_1 + \ldots + h_L \alpha_L$ is only possible if $h_1 = \ldots = h_L = 0$. In all other cases, because of the sum formula for the geometric series, we have

$$\sum_{n=1}^{N} e^{2\pi i \sum_{\ell=1}^{L} h_\ell \omega_\ell(n)} = \sum_{n=1}^{N} e^{2\pi i \sum_{\ell=1}^{L} h_\ell \alpha_\ell \cdot n}$$

$$= e^{2\pi i \sum_{\ell=1}^{L} h_\ell \alpha_\ell} \cdot \frac{1 - e^{2\pi i \sum_{\ell=1}^{L} h_\ell \alpha_\ell N}}{1 - e^{2\pi i \sum_{\ell=1}^{L} h_\ell \alpha_\ell}} \ .$$

The denominator is different from zero and independent of N; the numerator is bounded in absolute value by 2. This proves Proposition 4 and hence once again Theorem 2. $\qquad \square$

Weyl tried to obtain further examples of uniformly distributed sequences modulo 1. For these we first prove the following subsidiary result:

Lemma 1: Fundamental Inequality of van der Corput. *Let f denote a complex-valued function defined on the natural numbers with $|f(n)| = 1$ for all $n = 1, 2, \ldots, N$, then for all $Q \leq N$ one has the inequality*

$$\left| \sum_{n=1}^{N} f(n) \right|^2$$

$$\leq \left(\frac{N}{Q} + 1 \right) \left(N + 2 \cdot \sum_{q=1}^{Q} \left(1 - \frac{q}{Q} \right) \left| \sum_{n=1}^{N-q} \overline{f(n)} f(n+q) \right| \right) \ .$$

Proof. For the sake of simplicity suppose that for all integers $k \leq 0$ and $k > N$, $f(k)$ is put equal to zero, so that we have

$$Q \sum_{n=1}^{N} f(n) = \sum_{r=1}^{Q} \sum_{k=-\infty}^{\infty} f(k+r)$$

$$= \sum_{k=-\infty}^{\infty} \sum_{r=1}^{Q} f(k+r) = \sum_{k=-Q}^{N-1} \sum_{r=1}^{Q} f(k+r) \ .$$

Applying the Schwarz inequality gives

$$\left| Q \sum_{n=1}^{N} f(n) \right|^2 \leq \sum_{k=-Q}^{N-1} 1 \cdot \sum_{k=-Q}^{N-1} \left| \sum_{r=1}^{Q} f(k+r) \right|^2$$

$$= (N+Q) \cdot \sum_{k=-\infty}^{\infty} \left(\sum_{r=1}^{Q} f(k+r) \sum_{s=1}^{Q} \overline{f(k+s)} \right)$$

$$= (N+Q) \cdot \sum_{r,s=1}^{Q} \sum_{k=-\infty}^{\infty} f(k+r)\overline{f(k+s)}$$

$$= (N+Q) \left[\sum_{\substack{r,s=1 \\ r=s}}^{Q} \sum_{k=-\infty}^{\infty} f(k+r)\overline{f(k+s)} \right.$$

$$+ \sum_{\substack{r,s=1 \\ r<s}}^{Q} \sum_{k=-\infty}^{\infty} f(k+r)\overline{f(k+s)}$$

$$\left. + \sum_{\substack{r,s=1 \\ r>s}}^{Q} \sum_{k=-\infty}^{\infty} f(k+r)\overline{f(k+s)} \right] \ .$$

The first summand in the square brackets is

$$\sum_{\substack{r,s=1 \\ r=s}}^{Q} \sum_{k=-\infty}^{\infty} f(k+r)\overline{f(k+s)}$$

$$= \sum_{r=1}^{Q} \sum_{k=-r+1}^{N-r} |f(k+r)|^2 = NQ \ .$$

For the second summand we make the substitution $s = r + q$, $q \leq Q - r$, so that

$$\sum_{\substack{r,s=1\\r<s}}^{Q} \sum_{k=-\infty}^{\infty} f(k+r)\overline{f(k+s)}$$

$$= \sum_{r=1}^{Q} \sum_{q=1}^{Q-r} \sum_{k=-\infty}^{\infty} f(k+r)\overline{f(k+r+q)}$$

$$= \sum_{q=1}^{Q} \sum_{r=1}^{Q-q} \sum_{k=-\infty}^{\infty} f(k+r)\overline{f(k+r+q)}$$

$$= \sum_{q=1}^{Q} \sum_{r=1}^{Q-q} \sum_{n=-\infty}^{\infty} f(n)\overline{f(n+q)}$$

$$= \sum_{q=1}^{Q} (Q-q) \sum_{n=-\infty}^{\infty} f(n)\overline{f(n+q)}$$

$$= \sum_{q=1}^{Q} (Q-q) \sum_{n=1}^{N-q} f(n)\overline{f(n+q)} \ .$$

The third summand is complex conjugate to the second, hence

$$\left| Q \sum_{n=1}^{N} f(n) \right|^2$$

$$\leq (N+Q)\left(NQ + 2 \cdot \sum_{q=1}^{Q}(Q-q) \cdot \mathrm{Re}\left(\sum_{n=1}^{N-q} \overline{f(n)}f(n+q) \right) \right)$$

$$\leq (N+Q)\left(NQ + 2 \cdot \sum_{q=1}^{Q}(Q-q) \left| \sum_{n=1}^{N-q} \overline{f(n)}f(n+q) \right| \right)$$

holds. Division by Q^2 proves the fundamental inequality. [10] From this follows very quickly

Theorem 4: Main Theorem of the Theory of Uniform Distribution.
If the difference sequences $\Delta_q\omega$ with $\Delta_q\omega = \omega(n+q) - \omega(n)$ of a real-valued sequence ω are uniformly distributed modulo 1 for all natural numbers q, then ω itself is also uniformly distributed modulo 1.

Proof. The fundamental inequality

$$\left| \frac{1}{N} \sum_{n=1}^{N} e^{2\pi i h \omega(n)} \right|^2$$

$$\leq \left(\frac{1}{Q} + \frac{1}{N} \right)\left(1 + 2 \cdot \sum_{q=1}^{Q}\left(1 - \frac{q}{Q} \right) \left| \frac{1}{N} \sum_{n=1}^{N-q} e^{2\pi i h \Delta_q \omega(n)} \right| \right)$$

and the assumption of the theorem, namely that for integral $h \neq 0$ one has

$$\lim_{N \to \infty} \frac{1}{N} \sum_{n=1}^{N-q} e^{2\pi i h \Delta_q \omega(n)}$$

$$= \lim_{N \to \infty} \frac{N-q}{N} \cdot \frac{1}{N-q} \sum_{n=1}^{N-q} e^{2\pi i h \Delta_q \omega(n)} = 0$$

imply that

$$\overline{\lim_{N \to \infty}} \frac{1}{N^2} \left| \sum_{n=1}^{N} e^{2\pi i h \omega(n)} \right|^2 \leq \frac{1}{Q} \ .$$

Q can be chosen to be arbitrarily large. Because

$$\lim_{N \to \infty} \frac{1}{N} \sum_{n=1}^{N} e^{2\pi i h \omega(n)} = 0$$

the Weyl criterion demonstrates the validity of the theorem. $\square$

As an illustrative example let p be a polynomial of degree $K \geq 1$, and consider the resulting one dimensional sequence ω with $\omega(n) = p(n)$. In the case $K = 1$ the Kronecker approximation theorem shows that an irrational coefficient of the linear member is necessary and sufficient for the uniform distribution of ω modulo 1. Let us inductively assume that these sequences are uniformly distributed modulo 1 for all degrees between 1 and K, so long as the leading coefficient of the polynomial is irrational. In

$$p(x) = A_{K+1} x^{K+1} + A_K x^K + \ldots$$

let A_{K+1} denote an irrational number. For each integer $q \neq 0$

$$p(x + q) - p(x)$$
$$= A_{K+1} \left((x + q)^{K+1} - x^{K+1} \right) + A_K \left((x + q)^K - x^K \right) + \ldots$$
$$= A_{K+1} \left(\sum_{k=1}^{K+1} \binom{K+1}{k} q^k x^{K+1-k} \right)$$
$$+ A_K \left(\sum_{k=1}^{K} \binom{K}{k} q^k x^{K-k} \right) + \ldots$$

forms a polynomial of degree K with the irrational leading coefficient

$$a_K = A_{K+1}(K+1)q \ .$$

Since by the inductive assumption $\Delta_q \omega$ with $\Delta_q \omega(n) = p(n+q) - p(n)$ is uniformly distributed modulo 1 for all natural numbers q, by the Main Theorem this also holds for ω with $\omega(n) = p(n)$. Thus

Corollary 5: Weyl's Theorem. *A non-constant polynomial p having an irrational coefficient for the term of highest degree yields a uniformly distributed sequence ω modulo 1 with $\omega(n) = p(n)$.* [9]

In particular it follows from this that for an irrational α and integers $h_1, \ldots, h_L$, which are not all zero,

$$\lim_{N \to \infty} \frac{1}{N} \sum_{n=1}^{N} e^{2\pi i \sum_{\ell=1}^{L} h_\ell \alpha n^\ell} = 0 \ .$$

This holds since the polynomial

$$p(x) = h_1 \alpha x + h_2 \alpha x^2 + \ldots + h_L \alpha x^L$$

yields a uniformly distributed sequence modulo 1. If one interprets the formula above as

$$\lim_{N \to \infty} \frac{1}{N} \sum_{n=1}^{N} e^{2\pi i \sum_{\ell=1}^{L} h_\ell \omega_\ell(n)} = 0$$

with the L-dimensional sequence ω_ℓ given by $\omega_\ell(n) = \alpha n^\ell$, then as an immediate consequence we have:

Corollary 6. *For each irrational number α the L-dimensional sequence $(\alpha n, \alpha n^2, \ldots, \alpha n^L)$, $n = 1, 2, \ldots$ describes a uniformly distributed sequence modulo 1.*

Corollary 7. *For each irrational number α, arbitrary real numbers $\beta_1, \ldots, \beta_L$ and arbitrary positive numbers ε, Ω, one can always find integers n and $p_1, \ldots, p_L$ with*

$$
\begin{aligned}
|\alpha n &- \beta_1 &- p_1| &< \varepsilon, \\
|\alpha n^2 &- \beta_2 &- p_2| &< \varepsilon, \\
&\cdots\cdots\cdots\cdots\cdots\cdots\cdots\cdots\cdots\cdots\cdots, \\
|\alpha n^L &- \beta_L &- p_L| &< \varepsilon
\end{aligned}
$$

where $n \geq \Omega$.

Exercises on Chapter 2

1. Find two discrete subgroups $\mathfrak{G}$ and $\mathfrak{H}$ in $\mathbb{R}^n$ so that $\mathfrak{G} + \mathfrak{H}$ is dense in $\mathbb{R}^n$.

2. Let $\alpha \notin \mathbb{Q}, \epsilon > 0$ and β some real number. Show that there exist infinitely many pairs $(p, q) \in \mathbb{Z} \times \mathbb{N}$ with $0 < \alpha q - p - \beta < \epsilon$.

3. Suppose given the digits $z_1, \ldots, z_N \in \{0, 1, \ldots, 9\}, z_1 \neq 0$. Show that there exists some natural number n so that the numeral representation of 2^n in base 10 begins with $z_1 \ldots z_N$.

* 4. Let α, β be real numbers, $(p, q) \in \mathbb{Z} \times \mathbb{N}$ with g.c.d.$(p, q) = 1$, and suppose that $\left| \alpha - \frac{p}{q} \right| < \frac{1}{\sqrt{5}q^2}$. Show that there exist integers x, y with $|x| \leq \frac{q}{2}$ and $|\alpha x - y - \beta| < \frac{\sqrt{5}+5}{10q}$ (Chebyshev's theorem).

5. Let α be an irrational and β a real number. Show that there exist infinitely many pairs $(p, q) \in \mathbb{Z} \times \mathbb{N}$ with $|\alpha q - p - \beta| < \frac{2\sqrt{5}+5}{10q}$.

6. Let $M \subseteq \mathbb{R}$. M is called *linearly independent over* $\mathbb{Z}$, if each finite subset of M is linearly independent over $\mathbb{Z}$. Apply the transcendence of e in order to show that $\{\log p : p \text{ prime}\}$ is linearly independent over $\mathbb{Z}$.

* 7. Let $M = \{g^{1/2} : g \in \mathbb{N}, g > 1 \text{ and square free}\}$. Show that M is linearly independent over $\mathbb{Z}$ (Besicovitch's theorem).

8. Let x be a real number. Then $\{x\} = x - [x]$ is called the *fractional part of* x. Show that $(\{\log n\})_{n \geq 1}$ is dense in $[0, 1]$.

9. Let $(x_n)_{n \geq 1}$ be a sequence in $(0, 1)$ and $\epsilon > 0$. Find an open set M of measure $\leq \epsilon$, so that $M \subseteq (0, 1)$ and $\lim_{N \to \infty} \frac{1}{N} \sum_{n=1}^{N} c_M(x_n) = 1$. What can be said about M in the case that $(x_n)_{n \geq 1}$ is uniformly distributed?

10. Let $\left(\frac{p_n}{q_n} \right)$ be a mod 1 uniformly distributed sequence of rational numbers with g.c.d. $(p_n, q_n) = 1$ for $n \geq 1$. Show that

$$\lim_{N \to \infty} \frac{1}{N} \sum_{n=1}^{N} \frac{1}{q_n} = 0 \ .$$

11. Let $\alpha \notin \mathbb{Q}$ and $0 \leq x \leq 1$. Show that

$$\lim_{N \to \infty} \frac{1}{N} \sum_{n=1}^{N} c_{[0,x)}(\{\sin 2\pi n\alpha\})$$

exists, and calculate this limiting value. Deduce from this that $(\sin 2\pi n\alpha)_{n \geq 1}$ is not uniformly distributed mod 1.

12. Let α be a real number. Show that $\alpha \notin \mathbb{Q}$ if and only if $(\{\sin 2\pi n\alpha\})_{n \geq 1}$ is dense in $[0, 1]$.

* 13. If $(x_n)_{n \geq 1}$ is uniformly distributed mod 1, then so is $(x_n + \log n)_{n \geq 1}$ (Rindler, 1972).

14. Show that $(\log n)_{n \geq 1}$ is not uniformly distributed mod 1.

15. Let $(x_n)_{n \geq 1}$ be uniformly distributed mod 1 and $(y_n)_{n \geq 1}$ a sequence of real numbers with $\lim_{N \to \infty} \frac{1}{N} \sum_{n=1}^{N} |y_n| = 0$. Show that $(x_n + y_n)$ is uniformly distributed mod 1.

16. Let $F_0 = F_1 = 1$ and for $n \geq 1$ $\quad F_{n+1} = F_n + F_{n-1}$. The numbers $(F_n)_{n \geq 1}$ are called *Fibonacci* numbers. By applying the transcendence of e show that $(\log F_n)_{n \geq 1}$ is uniformly distributed mod 1.

* 17. Let $f : [1, \infty) \to \mathbb{R}$ be differentiable, $f' > 0$ and f' monotone decreasing. Suppose further that $\lim_{x \to \infty} f'(x) = 0$ and $\lim_{x \to \infty} x f'(x) = \infty$. Show that $\left(f(n) \right)_{n \geq 1}$ is uniformly distributed mod 1.

18. Let k be a natural number and $f : [1,\infty) \to \mathbb{R}$ be k-fold differentiable. Let $f^{(k)}$ be monotone decreasing, $\lim_{x\to\infty} f^{(k)}(x) = 0$ and $\lim_{x\to\infty} x f^{(k)}(x) = \infty$. Show that $\left(f(n) \right)_{n\geq 1}$ is uniformly distributed mod 1 (Fejer's theorem).

19. Let σ be a positive, non-integral number and $\alpha \neq 0$ real. Show that $(\alpha n^\sigma)_{n\geq 1}$ is uniformly distributed mod 1.

* 20. Let $x \in [0,1]$. Calculate

$$\underline{\lim}_{N\to\infty} \frac{1}{N} \sum_{n=1}^{N} c_{[0,x)}(\{\log n\}) \quad \text{and}$$

$$\overline{\lim}_{N\to\infty} \frac{1}{N} \sum_{n=1}^{N} c_{[0,x)}(\{\log n\}) \ .$$

Hints for the Exercises on Chapter 2

4. Choose $b' \in \mathbb{Z}$ so that $|b' - \beta q| \leq 1/2$ and $x_0, y_0 \in \mathbb{Z}$ so that $b' = p x_0 - q y_0$. Let $x_t = x_0 + qt, y_t = y_0 + pt$, and choose t so that $|x_t| \leq q/2$.

5. As in (4), but choose $t \in \mathbb{Z}$ so that $1 \leq x_t \leq q$.

7. Let $\left\{ g_1^{1/2}, \ldots, g_k^{1/2} \right\}$ be linearly dependent over $\mathbb{Z}$, and k chosen as small as possible. Put $K = \mathbb{Q}\left(\sqrt{g_1}, \ldots, \sqrt{g_{k-1}} \right)$, $\sqrt{g_k} = \sum_{i=1}^{k-1} a_i \sqrt{g_i} + a_k$, with $a_i \in \mathbb{Q}(1 \leq i \leq k)$. For each homomorphism $\sigma : K \to \mathbb{C}, 1 \leq i \leq k$, put $\epsilon_i(\sigma) \in \{-1, 1\}$ when $\sigma(\sqrt{g_i}) = \epsilon_i(\sigma)\sqrt{g_i}$. Show that K is normal over $\mathbb{Q}$, $\epsilon_i(\sigma) = \epsilon_k(\sigma)$ for $1 \leq i \leq k$ and $\sigma \in G$ (= Galois group of K over $\mathbb{Q}$). ϵ_k is a homomorphism with trivial kernel.

13. Note that for each convergent sequence of complex numbers $(a_n)_{n\geq 1}$ the sequence $\left(\frac{1}{N} \sum_{n=1}^{N} a_n \right)_{n\geq 1}$ converges to the same limit. Now apply Abel transformation from Chapter 4, the Weyl criterion and the inequality

$$\left| e^{ix} - e^{iy} \right| \leq |x - y| \quad \text{for} \quad x, y \in \mathbb{R} \ .$$

17. Apply the Weyl criterion, the Euler sum formula from Chapter 4 and the second mean-value theorem from integral calculus.

18. Use induction on k. Apply the main theorem from the theory of uniform distribution.

20. Let $g_x : [0,1] \to \mathbb{R}$,

$$g_x(t) = 1 - e^{-t} \frac{e - e^x}{e - 1}, \quad \text{for} \quad 0 \leq t < x$$

$$g_x(t) = e^{1-t} \frac{e^x - 1}{e - 1}, \quad \text{for} \quad x \leq t \leq 1 \ .$$

Show that for some constant $c > 0$,

$$\left| \sum_{n=1}^{N} c_{[0,x)}(\{\log n\}) - N g_x(\{\log N\}) \right| \leq c \log N \ ,$$

so that the sequences $(g_x(\{\log N\}))_{N \geq 1}$ and $\left(\frac{1}{N} \sum_{n=1}^{N} c_{[0,x)}(\{\log n\}) \right)_{N \geq 1}$ have the same accumulation points. For this note that

$$\sum_{n=1}^{N} c_{[0,x)}(\{\log n\}) = \sum_{0 \leq p \leq \log N} \sum_{n=1}^{N} c_{[p,p+x)}(\log n) \ .$$

Finally apply (8).

3. Geometry of Numbers

So far multidimensional diophantine approximation has concerned only points of the number spaces $\mathbb{R}^N$ or $\mathbb{Z}^N$. If, making use of the language of geometry, one passes from number spaces to vector spaces, the concept of an integer in the one-dimensional case and that of an integral N-tuple in $\mathbb{R}^N$ generalises to the natural concept of a lattice. If N linearly independent vectors $\mathfrak{a}_1, \ldots, \mathfrak{a}_N$ are given in an N-dimensional vector space $\mathfrak{V}$, by the *lattice* $\mathfrak{G} = \mathfrak{G}(\mathfrak{a}_1, \ldots, \mathfrak{a}_N)$ one understands the collection of all vectors $\mathfrak{g}$, which may be represented as linear combinations

$$\mathfrak{g} = g_1 \mathfrak{a}_1 + \ldots + g_N \mathfrak{a}_N = \sum_{n=1}^{N} g_n \mathfrak{a}_n$$

of the $\mathfrak{a}_n$ with integral coefficients g_n.

The linear independence of the $\mathfrak{a}_1, \ldots, \mathfrak{a}_N$ ensures the uniqueness of the coefficients g_n. If one restricts oneself to the vector space $\mathbb{R}^N$ and specifies $\mathfrak{a}_n$ to be the column vector with 1 in the nth position and zero elsewhere

$$\mathfrak{a}_1 = \begin{pmatrix} 1 \\ 0 \\ \vdots \\ 0 \end{pmatrix}, \ \mathfrak{a}_2 = \begin{pmatrix} 0 \\ 1 \\ \vdots \\ 0 \end{pmatrix}, \ldots, \ \mathfrak{a}_N = \begin{pmatrix} 0 \\ 0 \\ \vdots \\ 1 \end{pmatrix},$$

one obtains the lattice $\mathbb{Z}^N$ described at the beginning. For $N = 2, 3$ one can visualize lattices (Fig.3, Fig.4).

The vectors $\mathfrak{a}_1, \ldots, \mathfrak{a}_N$ form the *lattice basis* of $\mathfrak{G}$. If the vector space is spanned by the basis vectors $\mathfrak{e}_1, \ldots, \mathfrak{e}_N$, one obtains the lattice basis from the change of the basis formula:

$$\mathfrak{a}_n = \sum_{m=1}^{N} a_{mn} \mathfrak{e}_m .$$

If once and for all one fixes a "standard basis" $\mathfrak{e}_1, \ldots, \mathfrak{e}_N$ of the vector space $\mathfrak{V}$, the lattice is specified by giving the transformation matrix with the a_{mn} as elements.

Each *lattice vector* $\mathfrak{g}$, that is each element of the lattice described by

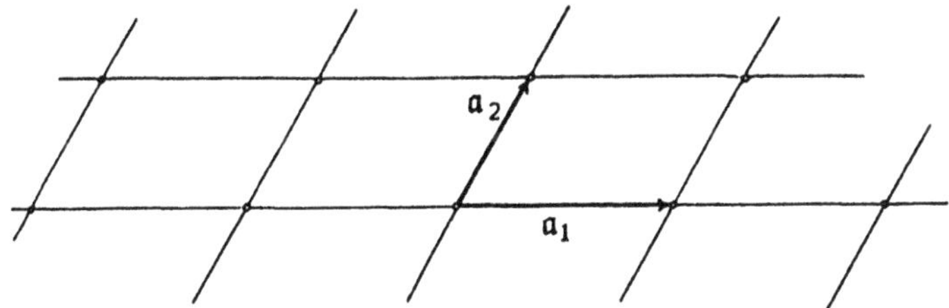

Fig. 3. Two-dimensional lattice

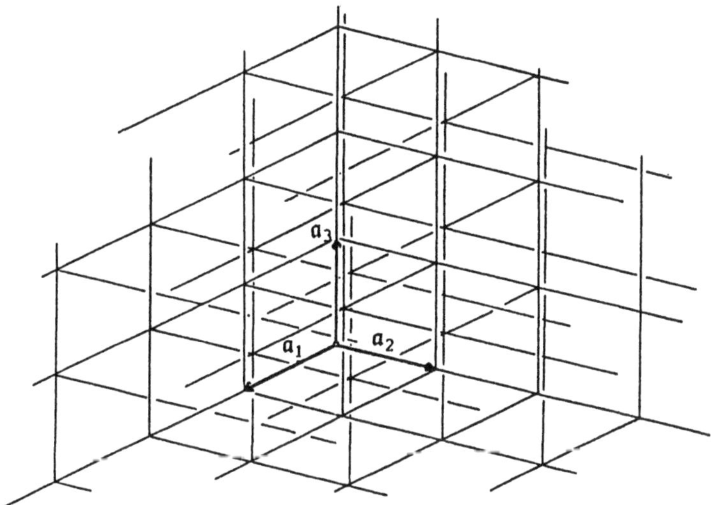

Fig. 4. Three-dimensional lattice

$$\mathfrak{g} = \sum_{n=1}^{N} g_n \mathfrak{a}_n$$

is determined with respect to the basis $\mathfrak{e}_1, \ldots, \mathfrak{e}_N$ by the equation

$$\mathfrak{g} = \sum_{n=1}^{N} g_n \sum_{m=1}^{N} a_{mn} \mathfrak{e}_m = \sum_{m=1}^{N} \left(\sum_{n=1}^{N} a_{mn} g_n \right) \mathfrak{e}_m \ .$$

Proposition 1. *The components of the lattice vectors* $\mathfrak{g}$ *in* $\mathfrak{G}$ *are obtained by matrix multiplication*

$$
\begin{pmatrix}
a_{11} & a_{12} & \dots & a_{1N} \\
a_{21} & a_{22} & \dots & a_{2N} \\
\multicolumn{4}{c}{\dotfill} \\
a_{N1} & a_{N2} & \dots & a_{NN}
\end{pmatrix}
\begin{pmatrix}
g_1 \\
g_2 \\
\dots \\
g_N
\end{pmatrix}
= (a_{mn})(g_n)
$$

where the matrix (a_{mn}) specifies $\mathfrak{G}$, and (g_n) is an arbitrary column vector from $\mathbb{Z}^N$.

If one defines the *integral unimodular matrix*

$$
\begin{pmatrix}
u_{11} & u_{12} & \dots & u_{1N} \\
u_{21} & u_{22} & \dots & u_{2N} \\
\multicolumn{4}{c}{\dotfill} \\
u_{N1} & u_{N2} & \dots & u_{NN}
\end{pmatrix}
= (u_{mn})
$$

by the properties that all elements are integral and the determinant

$$
\det(u_{mn}) =
\begin{vmatrix}
u_{11} & u_{12} & \dots & u_{1N} \\
u_{21} & u_{22} & \dots & u_{2N} \\
\multicolumn{4}{c}{\dotfill} \\
u_{N1} & u_{N2} & \dots & u_{NN}
\end{vmatrix}
= \pm 1
$$

then the passage from the matrix (a_{mn}) to the matrix

$$
(a'_{mp}) = (a_{mn})(u_{np}) = \left(\sum_{n=1}^{N} a_{mn} u_{np} \right)
$$

implies no change in the lattice specified by (a_{mn}). For on the one hand, if (g_n) belongs to $\mathbb{Z}^N$ so does

$$
(u_{np})(g_p) = \left(\sum_{p=1}^{N} u_{np} g_p \right)
$$

and as a consequence

$$
\sum_{m=1}^{N} (a'_{mp})(g_p) = \sum_{m=1}^{N} (a_{mn})(u_{np})(g_p)\, \mathfrak{e}_m
$$
$$
= \sum_{m=1}^{N} (a_{mn}) \left(\sum_{p=1}^{N} u_{np} g_p \right) \mathfrak{e}_m
$$
$$
= \sum_{m=1}^{N} \sum_{n=1}^{N} \sum_{p=1}^{N} a_{mn} u_{np} g_p \mathfrak{e}_m
$$

also belongs to the lattice. On the other hand the inverse matrix of (u_{mn}) (equal to $(\tilde{u}_{mn})$) is also integral and unimodular, since all minors of the matrix (u_{mn}) are integers, and remain so on dividing by $\det(u_{mn}) = \pm 1$. The multiplication rule for determinants shows that

$$\det\left(\tilde{u}_{mn}\right) = \frac{1}{\det(u_{mn})} = \pm 1 \ .$$

The passage from (a'_{mn}) to $(a_{mn}) = \left(a'_{mp}\right)\left(\tilde{u}_{pn}\right)$ reverses the argument above, which indeed implies the equality of the two lattices. Even more is actually true:

Proposition 2. *The lattice bases*

$$\mathfrak{a}_n = \sum_{m=1}^{N} a_{mn}\mathfrak{e}_m, \quad \mathfrak{b}_n = \sum_{m=1}^{N} b_{mn}\mathfrak{e}_m$$

generate the same lattice $\mathfrak{G}$ if and only if one basis goes over to the other by means of an integral unimodular transformation.

Proof. If $\mathfrak{a}_n$ and $\mathfrak{b}_n$ form bases of the same lattice and there exists a change of basis, $\mathfrak{b}_n = \sum_{m=1}^{N} u_{mn}\mathfrak{a}_m$, then the integrality of all the coefficients u_{mn} follows because all elements $\mathfrak{b}_n$ belong to the lattice $\mathfrak{G}(\mathfrak{a}_1,\ldots,\mathfrak{a}_N)$ as lattice vectors. There must also be a reverse transformation

$$\mathfrak{a}_p = \sum_{n=1}^{N} \tilde{u}_{np}\mathfrak{b}_n$$

and here also $(\tilde{u}_{np})$ can only denote an integral matrix. From the equations

$$\mathfrak{a}_p = \sum_{n=1}^{N} \tilde{u}_{np}\mathfrak{b}_n = \sum_{n=1}^{N} \tilde{u}_{np} \sum_{m=1}^{N} u_{mn}\mathfrak{a}_m$$
$$= \sum_{m=1}^{N} \left(\sum_{n=1}^{N} u_{mn}\tilde{u}_{np}\right) \mathfrak{a}_m$$

it follows that the matrix product $(u_{mn})(\tilde{u}_{np})$ agrees with the unit matrix, from which by the multiplication theorem for determinants we have

$$\det\left(u_{mn}\right)\det\left(\tilde{u}_{np}\right) = 1 \ .$$

Determinants of integral matrices can only be integral, so the only possibility is $\det\left(u_{mn}\right) = \pm 1$, showing the validity of the assertion. $\qquad\square$

In particular it follows from these considerations that the determinants of the coefficient matrices (a_{mn}), (b_{mn}) of the two lattice bases $\mathfrak{a}_1,\ldots,\mathfrak{a}_N$ and $\mathfrak{b}_1,\ldots,\mathfrak{b}_N$ can differ *only in sign*, provided that the bases generate the same lattice. Hence the absolute value of these determinants depends only on the lattice $\mathfrak{G}$ itself; it is called the *lattice constant*

$$d(\mathfrak{G}) = |\det\left(a_{mn}\right)| = |\det\left(b_{mn}\right)|$$

of $\mathfrak{G}$.

Geometrically one comes to grips with the notion of lattice constant, when one associates with the vector space $\mathfrak{V}$ an underlying Euclidean point space $\mathbb{P}$, for which the standard basis $\mathfrak{e}_1,\ldots,\mathfrak{e}_N$ serves as an orthonormal system. By this we mean that to each point P from $\mathbb{P}$ and each vector $\mathfrak{a}$ from $\mathfrak{V}$ we can associate a new point $Q = P + \mathfrak{a}$, in such a way that for any two points P and Q from $\mathbb{P}$ the difference $\mathfrak{a} = P - Q = \overrightarrow{PQ}$ lies in $\mathfrak{V}$, and for each three points P, Q, R from $\mathbb{P}$ the parallelogram rule $\overrightarrow{PQ} + \overrightarrow{QR} = \overrightarrow{PR}$ holds. If one chooses some point O from $\mathbb{P}$ as origin, then O together with the basis $\mathfrak{e}_1,\ldots,\mathfrak{e}_N$ form a Cartesian coordinate system and the map of an arbitrary point P from $\mathbb{P}$ to its coordinate N-tuple $(x_1,\ldots,x_N)$ by means of the formula

$$P = O + x_1\mathfrak{e}_1 + \ldots + x_N\mathfrak{e}_N$$

is uniquely invertible. Hence one can write $P = (x_1,\ldots,x_N)$. With the help of the coordinate system $(O,\mathfrak{e}_1,\ldots,\mathfrak{e}_N)$ $\mathbb{P}$ is identified with the space $\mathbb{R}^N$. In particular if $\mathfrak{a}_1,\ldots,\mathfrak{a}_N$ denotes a lattice basis of $\mathfrak{G}$, then the *fundamental parallelepiped* of this lattice basis is parametrised by the set F of all points

$$P = O + \xi_1\mathfrak{a}_1 + \ldots + \xi_N\mathfrak{a}_N = O + \sum_{n=1}^{N} \xi_n\mathfrak{a}_n; \quad 0 \leq \xi_n < 1 \ .$$

The parameters $\xi_1,\ldots,\xi_N$ run independently of each other through the points of the interval $[0,1[$.

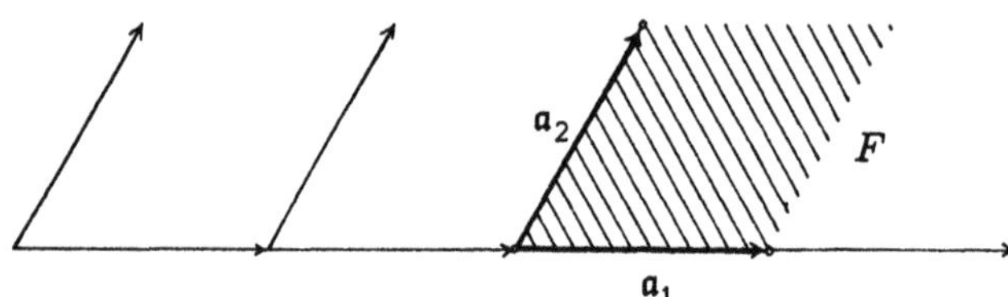

Fig. 5. Fundamental parallelepiped of a plane lattice

Because of

$$\sum_{m=1}^{N} x_m\mathfrak{e}_m = \sum_{n=1}^{N} \xi_n\mathfrak{a}_n = \sum_{n=1}^{N} \xi_n \sum_{m=1}^{N} a_{mn}\mathfrak{e}_m$$
$$= \sum_{m=1}^{N} \left(\sum_{n=1}^{N} a_{mn}\xi_n \right) \mathfrak{e}_m$$

one obtains the coordinates $P = (x_1,\ldots,x_N)$ of the points P from F as

$$x_m = \sum_{n=1}^{N} a_{mn}\xi_n \ .$$

If one starts from the fact that the standard basis $\mathfrak{e}_1,\ldots,\mathfrak{e}_N$ spans the unit cube of volume one, then the fundamental parallelepiped has volume

$$\begin{aligned}
\mathrm{vol}(F) &= \int_F \cdot\mathrel{\cdot}\cdot \int dx_1\ldots dx_N \\
&= \int_0^1 \cdots \int_0^1 |\det(a_{mn})|\, d\xi_1\ldots d\xi_N = |\det(a_{mn})| \\
&= d(\mathfrak{G})\ .
\end{aligned}$$

Proposition 3. *The volumes of the fundamental parallelepipeds of a lattice are independent of the chosen lattice basis and agree with the lattice constants.*

Besides fundamental parallelepipeds we will now consider general sets of points M from $\mathbb{P}$. Given a vector $\mathfrak{a}$ from $\mathfrak{V}$ let $M + \mathfrak{a}$ denote the set of all points $Q = P + \mathfrak{a}$, as P runs through the points of M; this is the translate of the set M by the vector $\mathfrak{a}$.

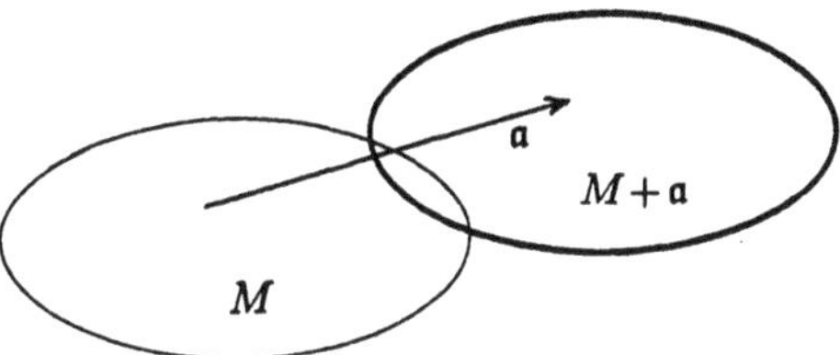

Fig. 6. Displacement of a set by a vector

More generally, if $\mathfrak{S}$ denotes some set of vectors from $\mathfrak{V}$, then by $M + \mathfrak{S}$ one understands the set of all points $Q = P + \mathfrak{a}$ for an arbitrary P from M and an arbitrary $\mathfrak{a}$ from $\mathfrak{S}$. For example, if $\mathfrak{S} = \mathfrak{G}$ is a lattice, then the sum $M + \mathfrak{G}$ is called the *figure lattice* associated to M by $\mathfrak{G}$ (Fig. 7).

If one defines the characteristic function c_M of the set M by $c_M(P) = 1$ if P belongs to M, and $c_M(P) = 0$ if P does not belong to M, then from the considerations above it follows that a point P belongs to the figure lattice $M + \mathfrak{G}$ (respectively does not belong) if and only if

$$\sum_{\mathfrak{g}\in\mathfrak{G}} c_M(P+\mathfrak{g}) \geq 1 \qquad \left(\text{resp. } \sum_{\mathfrak{g}\in\mathfrak{G}} c_M(P+\mathfrak{g}) = 0\right)\ .$$

If for all points from $\mathbb{P}$ the formula

$$\sum_{\mathfrak{g}\in\mathfrak{G}} c_M(P+\mathfrak{g}) \geq 1\ ,$$

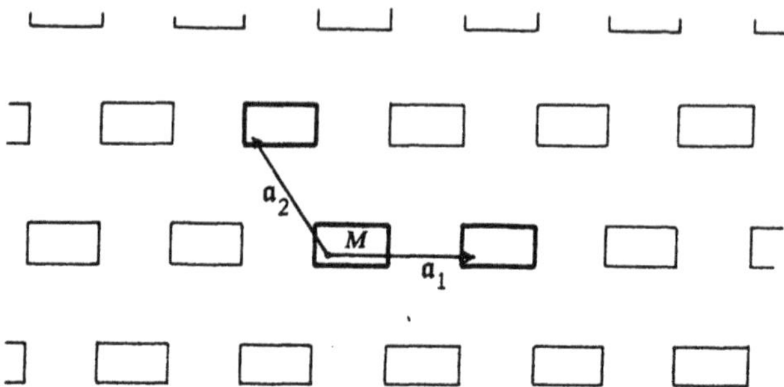

Fig. 7. Figure lattice associated to a set by a lattice

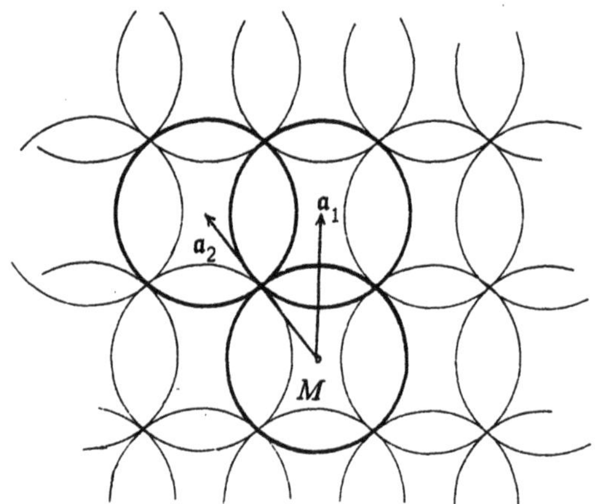

Fig. 8. Example of a covering lattice

holds, then one calls $\mathfrak{G}$ a *covering lattice* for M, because in this case $M + \mathfrak{G} = \mathbb{P}$.

If on the other hand, for all points P from $\mathbb{P}$ the formula

$$\sum_{\mathfrak{g} \in \mathfrak{G}} c_M(P + \mathfrak{g}) \leq 1 \ ,$$

holds, the lattice $\mathfrak{G}$ is said to *fill up* M; in this case an arbitrary point P can belong to at most one of the sets $M + \mathfrak{g}$ with $\mathfrak{g} \in \mathfrak{G}$.

A set M is called a *fundamental domain* of a lattice $\mathfrak{G}$, if $\mathfrak{G}$ simultaneously covers and fills out M. Not only fundamental parallelepipeds but differently structured sets can serve as fundamental domains; (none the less they may all be assumed to be measurable).

Formally one defines fundamental domains M by the property

$$\sum_{\mathfrak{g} \in \mathfrak{G}} c_M(P + \mathfrak{g}) = 1$$

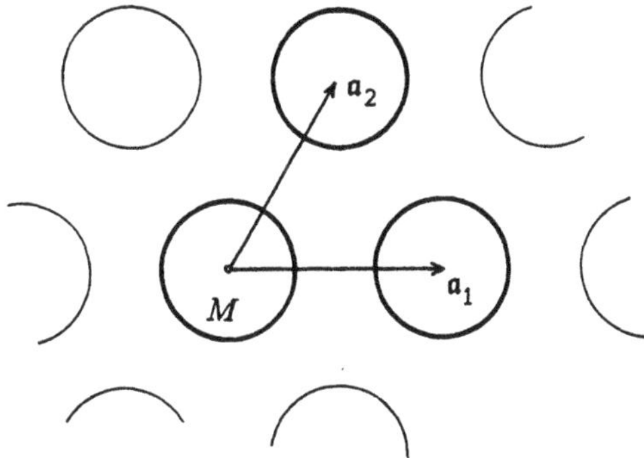

Fig. 9. Example of a filling lattice

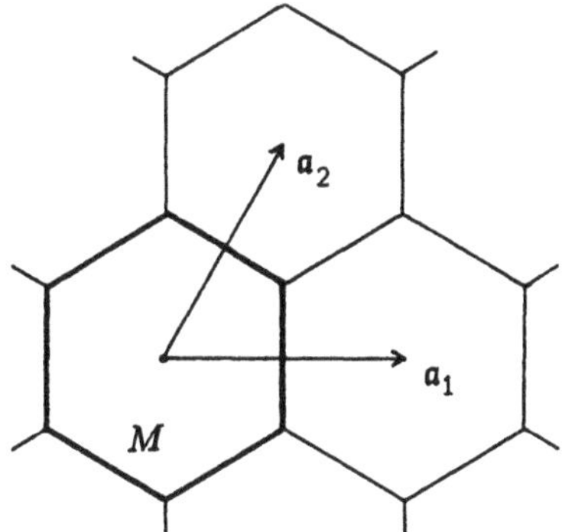

Fig. 10. Example of a fundamental domain

for all points P from $\mathbb{P}$. With the help of the Cartesian coordinate system $(O, e_1, \ldots, e_N)$ an arbitrary integrable function f on $\mathbb{R}^N$ can be considered as a function on $\mathbb{P}$ by setting $f(P) = f(x_1, \ldots, x_N)$, where $(x_1, \ldots, x_N)$ are the coordinates of P. The integral is defined by

$$\int_{\mathbb{P}} f(P)dP = \int_{-\infty}^{\infty} \cdots \int_{-\infty}^{\infty} f(x_1, \ldots, x_N)dx_1 \ldots dx_N \ ,$$

and more generally by

$$\int_{M} f(P)dP = \int_{-\infty}^{\infty} \cdots \int_{-\infty}^{\infty} c_M(x_1, \ldots, x_N)f(x_1, \ldots, x_N)dx_1 \ldots dx_N \ .$$

The transformations

$$\int_{\mathbb{P}} f(P)dP = \sum_{\mathfrak{g}\in\mathfrak{G}} \int_{M+\mathfrak{g}} f(P)dP = \sum_{\mathfrak{g}\in\mathfrak{G}} \int_M f(P-\mathfrak{g})dP$$

$$= \sum_{\mathfrak{g}\in\mathfrak{G}} \int_M f(P+\mathfrak{g})dP = \int_M \left(\sum_{\mathfrak{g}\in\mathfrak{G}} f(P+\mathfrak{g}) \right) dP \ ,$$

hold for a fundamental domain M (in the third equation one uses the fact that $-\mathfrak{g}$ as well as $\mathfrak{g}$ runs through $\mathfrak{G}$). For example, if for some positive number c one has

$$\int_{\mathbb{P}} f(P)dP \geq c \cdot d(\mathfrak{G})$$

then for at least one point P of the fundamental parallelepiped F one must have

$$\sum_{\mathfrak{g}\in\mathfrak{G}} f(P+\mathfrak{g}) \geq c \ .$$

If conversely

$$\sum_{\mathfrak{g}\in\mathfrak{G}} f(P+\mathfrak{g}) < c$$

holds for all points P from F, the calculation

$$\int_{\mathbb{P}} f(P)dP = \int_F \left(\sum_{\mathfrak{g}\in\mathfrak{G}} f(P+\mathfrak{g}) \right) < c \cdot \int_F dP$$

$$= c \cdot d(\mathfrak{G})$$

would give a contradiction to the assumption.

If in this example one restricts attention to $f = c_M$, where M is a bounded measurable set, the volume of M is given by

$$\int_{\mathbb{P}} c_M(P)dP = \mathrm{vol}(M) \ .$$

The relation

$$\sum_{\mathfrak{g}\in\mathfrak{G}} c_M(P+\mathfrak{g}) \geq \frac{\mathrm{vol}(M)}{d(\mathfrak{G})}$$

says that at least $\mathrm{vol}(M)/d(\mathfrak{G})$ points of the form $P+\mathfrak{g}$ from $\mathfrak{G}$ lie in the set M. (If $\mathrm{vol}(M)/d(\mathfrak{G})$ is not an integer, replace it by the next larger integer $z \geq \mathrm{vol}(M)/d(\mathfrak{G})$). $P+\mathfrak{G}$ is called the *point lattice* with origin P and the elements of $P+\mathfrak{G}$ are the *lattice points* of this point lattice. With this terminology one formulates the following theorem:

Proposition 4: Blichfeldt's Theorem. *For each lattice $\mathfrak{G}$ and each nonempty, bounded, measurable set M one can find a point P, so that the number z of lattice points of the point lattice $P+\mathfrak{G}$ based at P lying inside the set M satisfies $z \geq \mathrm{vol}(M)/d(\mathfrak{G})$.*

Since in Blichfeldt's theorem little is assumed about the set M, it follows that the point P serving as origin for the point lattice $P + \mathfrak{G}$ remains completely unspecified. If one wishes to avoid this lack of precision, then far-reaching assumptions on the set M are necessary. A set M is said to be *symmetric* with respect to a point Q if $P = Q + \mathfrak{a}$ belonging to the set implies that $P' = Q - \mathfrak{a}$ also does. Q is called the *centre of symmetry* of this set.

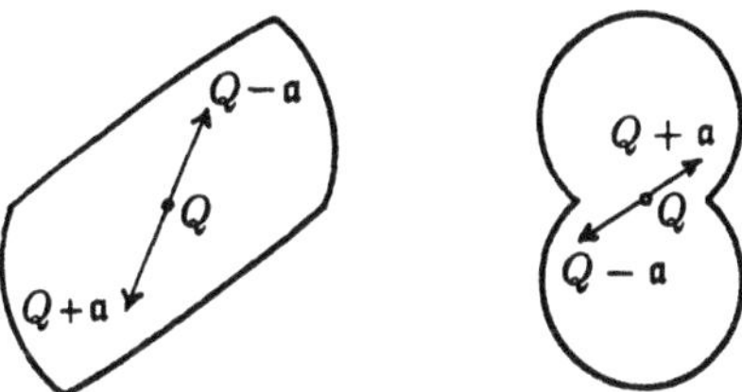

Fig. 11. Symmetric sets

A set M is said to be *star-shaped* with respect to a point Q if $P = Q + \mathfrak{a}$ belonging to the set implies that $P_t = Q + t\mathfrak{a}$ does also for all $0 \le t \le 1$.

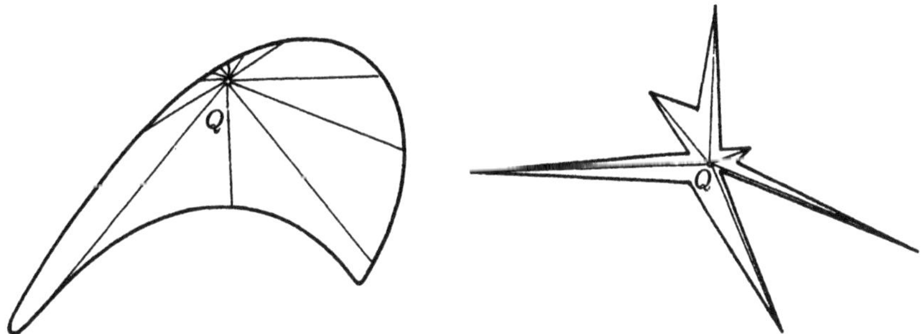

Fig. 12. Star-shaped sets

A set M is called *convex* if it is star-shaped with respect to *each of its points*. Another criterion for convex sets reads:

Proposition 5. *A set M is convex if and only if for each two points P and Q belonging to it so does the line segment consisting of all points $R_t = P + t\overrightarrow{PQ}$, $0 \le t \le 1$.*

Theorem 1: Minkowski's Lattice Point Theorem. *If P denotes the centre of symmetry of the symmetric, bounded and convex set K, and $\mathfrak{G}$ is a lattice with*

$$\mathrm{vol}(K) > 2^N \cdot d(\mathfrak{G})$$

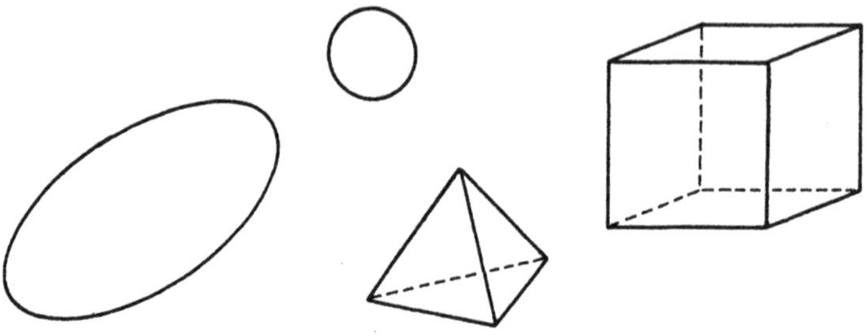

Fig. 13. Convex sets

then K contains at least one lattice point $Q = P + \mathfrak{g}$, distinct from P, $\mathfrak{g} \in \mathfrak{G}$, from the point lattice $P + \mathfrak{G}$ based at P.

Proof. The idea of the proof is to apply Blichfeldt's theorem for the set M of all $P + \frac{1}{2}\mathfrak{a}$, as $P + \mathfrak{a}$ runs through all points from K. Since

$$\mathrm{vol}(M) = \frac{1}{2^N}\mathrm{vol}(K) > \frac{1}{2^N} \cdot 2^N \cdot d(\mathfrak{G}) = d(\mathfrak{G})$$

by Proposition 4 at least two points of the form $R + \mathfrak{g}', R + \mathfrak{g}''$ lie in M for distinct $\mathfrak{g}', \mathfrak{g}''$ from $\mathfrak{G}$. Since $P + (\overrightarrow{PR} + \mathfrak{g}'), P + (\overrightarrow{PR} + \mathfrak{g}'')$ lie in M, by definition the points $P + 2(\overrightarrow{PR} + \mathfrak{g}')$ and $P + 2(\overrightarrow{PR} + \mathfrak{g}'')$ are in K. Because K is symmetric the points

$$Q' = P + 2(\overrightarrow{PR} + \mathfrak{g}'), \ Q'' = P + 2(-\overrightarrow{PR} - \mathfrak{g}'')$$

lie in K, and because K is convex their mid-point

$$\begin{aligned}
Q &= Q' + \frac{1}{2}\overrightarrow{Q'Q''} \\
&= P + 2(\overrightarrow{PR} + \mathfrak{g}') + \frac{1}{2}(2(-\overrightarrow{PR} - \mathfrak{g}'') - 2(\overrightarrow{PR} + \mathfrak{g}')) \\
&= P + (\mathfrak{g}' - \mathfrak{g}'') = P + \mathfrak{g}
\end{aligned}$$

is also in K. Since $\mathfrak{g} = \mathfrak{g}' - \mathfrak{g}'' \neq \mathfrak{o}$ this point is certainly a distinct lattice point from P. $\qquad\qquad\square$

From the example of the cube K, i.e. the set of all

$$P + t_1\mathfrak{e}_1 + \ldots + t_N\mathfrak{e}_N \ ; \ -1 < t_1 < 1, \ldots, -1 < t_N < 1 \ ,$$

one sees that the requirement that $\mathrm{vol}(K) > 2^N \cdot d(\mathfrak{G})$ cannot be weakened – for $\mathfrak{G}$ one chooses the lattice determined by the identity matrix. However it is possible to formulate a variant of this theorem with the not so restrictive inequality $\mathrm{vol}(K) \geq 2^N \cdot d(\mathfrak{G})$.

Corollary 1. *If P denotes the centre of symmetry of the symmetric, compact, convex set K, and if $\mathfrak{G}$ is a lattice with*

$$\mathrm{vol}(K) \geq 2^N \cdot d(\mathfrak{G}) \ ,$$

then K contains at least one lattice point $Q = P + \mathfrak{g}$ distinct from P, $\mathfrak{g} \in \mathfrak{G}$, from the point lattice $P + \mathfrak{G}$ based at P.

Proof. Let $\lambda > 1$ be an arbitrary real number, then at least in K_λ, the set of all $P + \lambda \mathfrak{a}$ with $P + \mathfrak{a}$ from K, by Minkowski's theorem there exists at least one lattice point $Q_\lambda = P + \mathfrak{g}_\lambda$ distinct from P. For each $\lambda > 1$ there are only finitely many such lattice points available. Hence it is possible to construct a sequence λ_k converging to 1, for which $Q_{\lambda_k} = Q \neq P$ always denotes the same lattice point. It follows from the compactness of K, that the intersection of all the K_{λ_k} must agree with K, and hence Q lies in K. $\qquad\square$

The applications of Minkowski's theorem show its importance. A few examples of such applications will make this very clear.

Examples: (1) Rectangular solid: K consists of all points

$$P = O + t_1 \mathfrak{e}_1 + \ldots + t_N \mathfrak{e}_N \ , -\lambda_n < t_n < \lambda_n \ ,$$

and its volume equals

$$\mathrm{vol}(K) = 2^N \cdot \lambda_1 \cdot \ldots \cdot \lambda_N \ .$$

The lattice $\mathfrak{G}$ is generated by the vectors

$$\mathfrak{a}_n = \sum_{m=1}^{N} a_{mn} \mathfrak{e}_m \ , n = 1, \ldots, N \ ,$$

and hence is specified by the matrix (a_{mn}). If the lattice point

$$Q = O + \mathfrak{g} = O + \sum_{n=1}^{N} g_n \mathfrak{a}_n = O + \sum_{m=1}^{N} \sum_{n=1}^{N} a_{mn} g_n \mathfrak{e}_m$$

lies in K, then this implies

$$\left| \sum_{n=1}^{N} a_{mn} g_n \right| < \lambda_m \ , m = 1, \ldots, N \ .$$

Since $d(\mathfrak{G}) = |\det(a_{mn})|$ holds, Minkowski's theorem implies

Corollary 2: Minkowski's Theorem on Linear Forms. *If (a_{mn}) denotes an $N \times N$ matrix with non-vanishing determinant and if for positive numbers $\lambda_1, \ldots, \lambda_N$ one has*

$$\lambda_1 \cdot \ldots \cdot \lambda_N > |\det(a_{mn})| \ ,$$

then there exist integers $g_1, \ldots, g_N$ which are not all zero, and which satisfy the inequalities

$$\left| \sum_{n=1}^{N} a_{1n} g_n \right| < \lambda_1, \ldots, \left| \sum_{n=1}^{N} a_{Nn} g_n \right| < \lambda_N \ .$$

A simple sharpening of the linear form theorem reads

Corollary 3. *If (a_{mn}) is an $N \times N$ matrix with non-vanishing determinant, and if for positive numbers $\lambda_1, \ldots, \lambda_N$ one has*

$$\lambda_1 \cdot \ldots \cdot \lambda_N \geq |\det(a_{mn})| \ ,$$

then there exist integers $g_1, \ldots, g_N$, which are not at all zero and which satisfy the inequalities

$$\left| \sum_{n=1}^{N} a_{1n} g_n \right| \leq \lambda_1,$$

$$\left| \sum_{n=1}^{N} a_{2n} g_n \right| < \lambda_2, \ldots, \left| \sum_{n=1}^{N} a_{Nn} g_n \right| < \lambda_N \ .$$

Proof. If one replaces λ_1 by $\lambda_1(1 + \varepsilon)$ application of the linear form theorem establishes the existence of integers $g_1(\varepsilon), \ldots, g_N(\varepsilon)$, which are not all zero, and which solve the inequalities

$$\left| \sum_{n=1}^{N} a_{1n} g_n(\varepsilon) \right| < \lambda_1 \cdot (1 + \varepsilon),$$

$$\left| \sum_{n=1}^{N} a_{2n} g_n(\varepsilon) \right| < \lambda_2, \ldots, \left| \sum_{n=1}^{N} a_{Nn} g_n(\varepsilon) \right| < \lambda_N \ .$$

For each $\varepsilon > 0$ there can exist only finitely many such $g_n(\varepsilon)$, and hence it is possible to construct a null sequence ε_k with the property that $g_n = g_n(\varepsilon_k)$ denotes an integer independent of ε_k. Passage to the limit $\varepsilon_k \to 0$ leaves the relations

$$\left| \sum_{n=1}^{N} a_{mn} g_n \right| < \lambda_m, \qquad m = 2, \ldots, N$$

unchanged, while

$$\left| \sum_{n=1}^{N} a_{1n} g_n \right| < \lambda_1 \cdot (1 + \varepsilon_k)$$

leads to

$$\left| \sum_{n=1}^{N} a_{1n}g_n \right| \leq \lambda_1$$

as asserted. $\square$

This sharpening of the linear form theorem proves the following assertion:

Corollary 4: Product Theorem for Homogeneous Linear Forms. *If (a_{mn}) denotes an $N \times N$ matrix with non-vanishing determinant and $N > 1$, then there exist integers $g_1, \ldots, g_N$, which are not all zero and for which*

$$\left| \left(\sum_{n=1}^{N} a_{1n}g_n \right) \left(\sum_{n=1}^{N} a_{2n}g_n \right) \cdots \left(\sum_{n=1}^{N} a_{Nn}g_n \right) \right| < |\det(a_{mn})|$$

holds.

Proof. For

$$\lambda_1 = \ldots = \lambda_N = \sqrt[N]{|\det(a_{mn})|}$$

we do indeed have

$$\left| \sum_{n=1}^{N} a_{1n}g_n \right| \leq \sqrt[N]{|\det(a_{mn})|},$$

$$\left| \sum_{n=1}^{N} a_{pn}g_n \right| < \sqrt[N]{|\det(a_{mn})|}, \qquad p = 2, \ldots, N \ ,$$

assuming that the integers g_n are chosen as in Corollary 3. Taking the product of the left- and right-hand sides of this chain of inequalities concludes the proof. $\square$

For example, let $\alpha_1, \ldots, \alpha_S$ be arbitrary real numbers and for $N = S + 1$ let the matrix (a_{mn}) be given as

$$(a_{mn}) = \begin{pmatrix} 1 & 0 & \ldots & 0 & -\alpha_1 \\ 0 & 1 & \ldots & 0 & -\alpha_2 \\ \cdots\cdots\cdots\cdots\cdots\cdots\cdots \\ 0 & 0 & \ldots & 1 & -\alpha_S \\ 0 & 0 & \ldots & 0 & 1 \end{pmatrix} .$$

Clearly $\det(a_{mn}) = 1$; if one now chooses $\lambda_1 = \ldots = \lambda_S = 1/M$ and $\lambda_N = M^S$ for some natural number M, then Corollary 3 implies the existence of integers $g_1, \ldots, g_S$ and $g_N = q$ with

$$|g_1 - \alpha_1 q| < \frac{1}{M}, \ldots, |g_S - \alpha_S q| < \frac{1}{M}, \qquad |q| \leq M^S .$$

Clearly $q \neq 0$, since otherwise $|g_s| < 1/M$ will also imply that $g_s = 0$. *Minkowski's linear form theorem thus implies the Dirichlet approximation theorem.*

(2) The double pyramid: K_r consists of all points

$$P = O + t_1 e_1 + \ldots + t_N e_N;$$
$$|t_1| \leq r - |t_N|, \ldots, |t_{N-1}| \leq r - |t_N|,$$
$$|t_N| \leq r \ .$$

As above for the sake of simplicity of notation write $S = N - 1$, so that K_r is given by

$$P = O + \sum_{s=1}^{S} x_s e_s + x_N e_N; \qquad |x_s| + |x_N| \leq r,$$

$$|x_N| \leq r \ .$$

The volume of K_r equals

$$\mathrm{vol}(K_r) = \int\limits_{|x_N| \leq r} dx_N \int\limits_{|x_s| \leq r - |x_N|} \ldots \int dx_1 \ldots dx_s$$

$$= \int\limits_{|x_N| \leq r} (r - |x_N|)^S \cdot 2^S dx_N$$

$$= 2^S \cdot \left(\int\limits_0^r (r - x_N)^S dx_N + \int\limits_{-r}^0 (r + x_N)^S dx_N \right)$$

$$= 2^S \cdot \left(\frac{r^N}{N} + \frac{r^N}{N} \right) = \frac{2^N}{N} \cdot r^N \ .$$

The condition of Minkowski's theorem takes the form $\frac{2^N}{N} \cdot r^N \geq 2^N \cdot d(\mathfrak{G})$, and hence $r \geq \sqrt[N]{N \cdot d(\mathfrak{G})}$. The equality sign is admissable because K is compact. If the lattice $\mathfrak{G}$ is specified by the matrix (a_{mn}), this implies the existence of integers $g_1, \ldots, g_S, g_N = q$, which are not all zero and which satisfy the inequalities

$$\left| \sum_{n=1}^{N} a_{sn} g_n \right| + \left| \sum_{n=1}^{N} a_{Nn} g_n \right| \leq \sqrt[N]{N \cdot d(\mathfrak{G})}$$

$$= \sqrt[N]{N \cdot |\det(a_{mn})|}, \qquad s = 1, \ldots, S = N - 1 \ .$$

In order to achieve at a more powerful approximation theorem, further modifications are necessary. If in the arithmetic-geometric inequality

$$\sqrt[N]{|\xi_1 \cdot \ldots \cdot \xi_N|} \leq \frac{|\xi_1| + \ldots + |\xi_N|}{N}$$

one puts $\xi_1 = \ldots = \xi_S = \frac{\varphi}{S}, \xi_N = \psi \, (\varphi \geq 0, \psi \geq 0)$, then

$$\left(\frac{\varphi}{S} \right)^S \psi \leq \frac{1}{N^N} \left(S \cdot \frac{\varphi}{S} + \psi \right)^N = \left(\frac{\varphi + \psi}{N} \right)^N \ .$$

For each $s = 1, \ldots, S$ this implies

$$\left| \sum_{n=1}^{N} a_{sn} g_n \right|^S \left| \sum_{n=1}^{N} a_{Nn} g_n \right|$$

$$\leq S^S \cdot \left(\frac{\left| \sum_{n=1}^{N} a_{sn} g_n \right| + \left| \sum_{n=1}^{N} a_{Nn} g_n \right|}{N} \right)^N$$

$$\leq S^S \cdot \frac{N d(\mathfrak{G})}{(S+1)^{S+1}} = \left(\frac{S}{S+1} \right)^S \cdot d(\mathfrak{G}) .$$

If the lattice $\mathfrak{G}$ is specified by the matrix

$$(a_{mn}) = \begin{pmatrix} t & 0 & \ldots & 0 & -\alpha_1 t \\ 0 & t & \ldots & 0 & -\alpha_2 t \\ \ldots\ldots\ldots\ldots\ldots\ldots\ldots \\ 0 & 0 & \ldots & t & -\alpha_S t \\ 0 & 0 & \ldots & 0 & t^{-S} \end{pmatrix} ,$$

we have $d(\mathfrak{G}) = |\det(a_{mn})| = t^S/t^S = 1$. Hence there exist integers $g_1, \ldots, g_S, g_N = q$, which are not all zero, and which have the property

$$|t(g_s - \alpha_s q)|^S \left| \frac{1}{t^S} \cdot q \right| \leq \left(\frac{S}{S+1} \right)^S ,$$

$$|g_s - \alpha_s q|^S |q| \leq \left(\frac{S}{S+1} \right)^S , \qquad s = 1, \ldots, S .$$

Besides

$$|t(g_s - \alpha_s q)| + \left| \frac{q}{t^S} \right| \leq \sqrt[N]{N} .$$

This shows that for t sufficiently large $q \neq 0$, because if $q = 0$ one would have a contradiction for $t > \sqrt[N]{N}$. Moreover $|q| \leq \sqrt[N]{N} \cdot t^S$ provides an upper bound for q. If without loss of generality one supposes that q is a natural number, then the result of these calculations reads

Corollary 5. *For arbitrary real numbers* $\alpha_1, \ldots, \alpha_S$ *and an additional real number* $t > {}^{s+1}\!\!\sqrt{S+1}$ *one can find integers* $g_1, \ldots, g_S$ *and a natural number* $q \leq t^S \cdot {}^{s+1}\!\!\sqrt{S+1}$ *with the property*

$$|\alpha_1 q - g_1| \leq \frac{S}{S+1} \cdot \frac{1}{q^{1/S}}, \ldots, |\alpha_S q - g_S| \leq \frac{S}{S+1} \cdot \frac{1}{q^{1/S}} ,$$

i.e. with

$$\left| \alpha_s - \frac{g_s}{q} \right| \leq \frac{S}{S+1} \cdot \frac{1}{q^{1+1/S}} .$$

(3) The octahedron: K_r consists of all points

$$P = O + x_1 e_1 + \ldots + x_N e_N;$$
$$|x_1| + \ldots + |x_N| \leq r \ .$$

For $N = 1$ one of course has that $\mathrm{vol}(K_r) = \mathrm{vol}(K_r; N = 1) = 2r$: recursively one obtains the volume of the N-dimensional octahedron for $N > 1$ from

$$
\begin{aligned}
\mathrm{vol}(K_r; N) &= \underset{|x_1|+\ldots+|x_N|\leq r}{\int \ldots \int} dx_1 \ldots dx_N \\[2mm]
&= \underset{|x_N|\leq r}{\int} dx_N \cdot \underset{|x_1|+\ldots+|x_{N-1}|\leq r-|x_N|}{\int \ldots \int} dx_1 \ldots dx_{N-1} \\[2mm]
&= \underset{|x_N|\leq r}{\int} \mathrm{vol}\big(K_{r-|x_N|}; N-1\big) \cdot dx_N \\[2mm]
&= \mathrm{vol}(K_1; N-1) \cdot \underset{|x_N|\leq r}{\int} (r - |x_N|)^{N-1} \, dx_N \\[2mm]
&= \mathrm{vol}(K_1; N-1) \cdot \frac{2r^N}{N} \ .
\end{aligned}
$$

In particular this shows that

$$\mathrm{vol}(K_1; N) = \frac{2}{N} \cdot \mathrm{vol}(K_1; N-1) \ ,$$

i.e. because $\mathrm{vol}(K_1; N = 1) = 2$,

$$\mathrm{vol}(K_1; N) = \frac{2^N}{N!} \ .$$

If, noting that

$$\mathrm{vol}(K_r) = \frac{(2r)^N}{N!}$$

one applies Minkowski's theorem, the essential condition reads

$$\frac{(2r)^N}{N!} \geq 2^N \cdot d(\mathfrak{G}) \ ,$$

i.e.

$$r \geq \sqrt[N]{N! \cdot d(\mathfrak{G})} \ .$$

If (a_{mn}) specifies the lattice of $\mathfrak{G}$, then Minkowski's theorem guarantees the existence of integers $g_1, \ldots, g_N$, which are not all zero and which solve

$$\left| \sum_{n=1}^{N} a_{1n} g_n \right| + \left| \sum_{n=1}^{N} a_{2n} g_n \right| + \ldots + \left| \sum_{n=1}^{N} a_{Nn} g_n \right|$$

$$\leq \sqrt[N]{N! \cdot d(\mathfrak{G})} \ .$$

Here applying the arithmetic-geometric inequality gives

$$\left| \left(\sum_{n=1}^{N} a_{1n}g_n \right) \left(\sum_{n=1}^{N} a_{2n}g_n \right) \cdots \left(\sum_{n=1}^{N} a_{Nn}g_n \right) \right|$$

$$\leq \left(\frac{\left| \sum_{n=1}^{N} a_{1n}g_n \right| + \left| \sum_{n=1}^{N} a_{2n}g_n \right| + \ldots + \left| \sum_{n=1}^{N} a_{Nn}g_n \right|}{N} \right)^{N}$$

$$\leq \frac{N! \cdot d(\mathfrak{G})}{N^N} \ .$$

Hence the product theorem for homogeneous linear forms admits the following sharpening:

Corollary 6. *If* (a_{mn}) *denotes an* $N \times N$ *matrix with non-vanishing determinant and if* $N > 1$, *then there exist integers* $g_1, \ldots, g_N$ *which are not all zero and for which*

$$\left| \left(\sum_{n=1}^{N} a_{1n}g_n \right) \left(\sum_{n=1}^{N} a_{2n}g_n \right) \cdots \left(\sum_{n=1}^{N} a_{Nn}g_n \right) \right| \leq \frac{N!}{N^N} |\det(a_{mn})|$$

holds.

(4) The ball for $N = 4$: application of Minkowski's theorem to the ball K_r in four-dimensional space leads to a proof of a famous number theoretic result of Lagrange. This states

Corollary 7: Lagrange's Theorem. *Each natural number can be expressed as the sum of at most four integral squares.*

Proof. The proof proceeds through several steps. In the first one solves the geometric problem of finding the volume of the ball K_r. This is the set of points

$$P = O + x_1 \mathfrak{e}_1 + \ldots + x_N \mathfrak{e}_N;$$
$$x_1^2 + \ldots + x_N^2 \leq r^2 \ .$$

For dimensions $N = 1$ and $N = 2$ one has $\mathrm{vol}(K_1; N = 1) = 2$, $\mathrm{vol}(K_1; N = 2) = \pi$. By integration

$$\mathrm{vol}(K_r; N)$$

$$= \int \cdots \int_{x_1^2 + \ldots + x_N^2 \leq r^2} dx_1 \ldots dx_N$$

$$= \iint_{x_{N-1}^2 + x_N^2 \leq r^2} dx_{N-1} dx_N \cdot \int \cdots \int_{x_1^2 + \ldots x_{N-2}^2 \leq r^2 - (x_{N-1}^2 + x_N^2)} dx_1 \ldots dx_{N-2}$$

$$= \iint_{x_{N-1}^2 + x_N^2 \leq r^2} \mathrm{vol}\left(K_{\sqrt{r^2 - (x_{N-1}^2 + x_N^2)}}; N - 2 \right) dx_{N-1} dx_N$$

$$= \mathrm{vol}(K_1; N - 2) \cdot \iint_{x_{N-1}^2 + x_N^2 \leq r^2} \sqrt{r^2 - (x_{N-1}^2 + x_N^2)}^{N-2} dx_{N-1} dx_N \ .$$

If one introduces the polar coordinates $x_N = \rho\cos\varphi$, $x_{N-1} = \rho\sin\varphi$; $dx_{N-1}dx_N = \rho d\rho d\varphi$, $0 < \rho \leq r$, $0 \leq \varphi \leq 2\pi$, then

$$\mathrm{vol}(K_r; N) = \mathrm{vol}(K_1; N - 2) \cdot \int_0^r \left(r^2 - \rho^2\right)^{\frac{N-2}{2}} \rho d\rho \int_0^{2\pi} d\varphi$$

$$= 2\pi \cdot \mathrm{vol}(K_1; N - 2) \cdot \int_{\rho=0}^{\rho=r} \left(r^2 - \rho^2\right)^{\frac{N}{2}-1} d(\rho^2) \cdot \frac{1}{2}$$

$$= \pi \cdot \mathrm{vol}(K_1; N - 2) \cdot \frac{r^N}{N/2} = \frac{2\pi}{N} \cdot r^N \cdot \mathrm{vol}(K_1; N - 2) \ .$$

Recursively one has proved that

$$\mathrm{vol}(K_1; N) = \frac{\pi^{N/2}}{\Gamma\left(\frac{N}{2} + 1\right)} \ ,$$

where in this connection one has assumed only the validity of the formulae

$$\Gamma(x + 1) = x \cdot \Gamma(x), \ \Gamma(1) = 1, \ \Gamma\left(\frac{1}{2}\right) = \sqrt{\pi}$$

for the Γ-function. In particular one has

$$\mathrm{vol}(K_1; N = 4) = \frac{\pi^2}{2}, \ \mathrm{vol}(K_r; N = 4) = \frac{\pi^2}{2} \cdot r^4 \ .$$

In the second step of our proof we meet preliminaries from elementary number theory.

Lemma 1. *For each prime number p there exist integers a, b with the property*

$$a^2 + b^2 + 1 \equiv 0 (\mathrm{mod} \ p) \ .$$

In the case $p = 2$ one can for example put $a = 1$ and $b = 0$. If $p \neq 2, (p-1)/2$ is a natural number and the $(p+1)/2$ natural numbers

$$0^2, 1^2, \ldots, \left(\frac{p-1}{2}\right)^2$$

are pairwise non-congruent modulo p. From $h^2 \equiv k^2 (\mathrm{mod}\ p)$, $h^2 - k^2 \equiv 0 (\mathrm{mod}\ p)$, $(h-k)(h+k) \equiv 0 (\mathrm{mod}\ p)$ and $0 \leq h \leq \frac{p-1}{2}$, $0 \leq k \leq \frac{p-1}{2}$ it follows that $h = k$. The $(p+1)/2$ numbers

$$-1 - 0^2, -1 - 1^2, \ldots, -1 - \left(\frac{p-1}{2}\right)^2$$

are also pairwise non-congruent modulo p. Since the union of both sets contains $p+1 > p$ numbers, one number from the first set must be congruent to a number from the second, that is

$$a^2 \equiv -1 - b^2 (\mathrm{mod}\ p) \ ,$$

as asserted.

By the Chinese remainder theorem one can carry over the congruence $a^2 + b^2 + 1 \equiv 0 (\mathrm{mod}\ p)$ from prime numbers p to square-free numbers $p > 1$. The solvability of the simultaneous congruences

$$a^2 + b^2 + 1 \equiv 0 (\mathrm{mod}\ p_1), \ \ldots, \ a^2 + b^2 + 1 \equiv 0 (\mathrm{mod}\ p_L)$$

for $p = p_1 \cdot \ldots \cdot p_L$ has the solvability of the congruence

$$a^2 + b^2 + 1 \equiv 0 (\mathrm{mod}\ p)$$

as a consequence. Hence Lemma 1 also holds for square free $p > 1$.

Clearly Lagrange's theorem has been proved if in addition to

$$1 = 1^2 + 0^2 + 0^2 + 0^2$$

one has expressed each square free natural number $p > 1$ as a sum of four squares.

For 1 there is nothing more to show; for p one starts from the lattice $\mathfrak{G}$ specified by

$$(a_{mn}) = \begin{pmatrix} 1 & 0 & 0 & 0 \\ 0 & 1 & 0 & 0 \\ a & b & p & 0 \\ -b & a & 0 & p \end{pmatrix} \qquad (N = 4)$$

with $d(\mathfrak{G}) = p^2$. As a convex set one chooses some ball K_r, which satisfies Minkowski's condition. Thus

$$\mathrm{vol}(K_r) = \frac{\pi^2}{2} \cdot r^4 \geq 2^4 \cdot d(\mathfrak{G}) = 2^4 p^2$$

and one must have

$$r^4 = \frac{2^5 p^2}{\pi^2}, \; r = \frac{\sqrt[4]{2^5}\sqrt{p}}{\sqrt{\pi}} \; .$$

The existence of a lattice point in this ball distinct from O

$$P = O + \sum_{n=1}^{4} a_{1n} g_n \mathfrak{e}_1 + \ldots + \sum_{n=1}^{4} a_{4n} g_n \mathfrak{e}_4$$

$$= O + g_1 \mathfrak{e}_1 + g_2 \mathfrak{e}_2 + (a g_1 + b g_2 + p g_3) \mathfrak{e}_3$$
$$+ (-b g_1 + a g_2 + p g_4) \mathfrak{e}_4$$

says that

$$g_1^2 + g_2^2 + (a g_1 + b g_2 + p g_3)^2 + (-b g_1 + a g_2 + p g_4)^2$$
$$\leq \frac{4\sqrt{2}}{\pi} \cdot p < \frac{4 \cdot \frac{3}{2}}{3} \cdot p = 2p \; .$$

For the integer

$$k = g_1^2 + g_2^2 + (a g_1 + b g_2 + p g_3)^2$$
$$+ (-b g_1 + a g_2 + p g_4)^2 < 2p \; ,$$

it follows from

$$(a g_1 + b g_2)^2 + (-b g_1 + a g_2)^2 = (a^2 + b^2)(g_1^2 + g_2^2)$$

that

$$k \equiv g_1^2 + g_2^2 + (a g_1 + b g_2)^2 + (-b g_1 + a g_2)^2$$
$$= (g_1^2 + g_2^2)(1 + a^2 + b^2) \pmod{p} \; .$$

Exactly as was shown in Lemma 1, by a suitable choice of a and b one can ensure that $1 + a^2 + b^2 \equiv 0 \pmod{p}$, i.e. $k \equiv 0 \pmod{p}$. As a sum of squares k is non-negative; since not all g_1, g_2, g_3, g_4 equal zero, k is a natural number. Because of divisibility by p and $k < 2p$ one even has $k = p$. Thus p is indeed the sum of four integral squares and Lagrange's theorem is proved. $\qquad\square$

7 cannot be expressed as the sum of less than 4 integral squares, and so 4 is the smallest number for which Lagrange's theorem holds.

(5) The ball for $N = 2$: a second number theoretic application of the Minkowski theorem, using the 2-dimensional ball, concerns the representation of integers as the sum of two squares. In this example one can take the vector space $\mathfrak{V}$ to be the set $\mathbb{C}$ of complex numbers with the standard basis $\mathfrak{e}_1 = 1$ and $\mathfrak{e}_2 = i$. The space of points $\mathbb{P}$ is taken to be the Gaussian number plane $\mathbb{C}$ with 0 as the origin of the coordinate system. In this example the ball K_r consists of all points

$$P = O + x_1 \mathfrak{e}_1 + x_2 \mathfrak{e}_2 = x_1 + i x_2;$$
$$x_1^2 + x_2^2 \leq r^2 \; .$$

If p denotes a prime number of the form $p \equiv 1 \pmod{4}$ the lattice $\mathfrak{G}$ is specified by

$$(a_{mn}) = \begin{pmatrix} 1 & 0 \\ a & p \end{pmatrix}$$

with an undetermined integer a. Here one has $d(\mathfrak{G}) = p$ and one can choose the ball K_r to satisfy the condition of Minkowski's theorem as a convex set, provided

$$\text{vol}(K_r) = \pi r^2 \geq 2^2 d(\mathfrak{G}) = 4p \; ,$$

i.e.

$$r^2 = \frac{4p}{\pi} \; .$$

The existence of a lattice point distinct from O

$$\begin{aligned} P &= O + (a_{11}g_1 + a_{12}g_2)\mathfrak{e}_1 + (a_{21}g_1 + a_{22}g_2)\mathfrak{e}_2 \\ &= g_1 + i(ag_1 + pg_2) \end{aligned}$$

inside this ball implies

$$g_1^2 + (ag_1 + pg_2)^2 \leq \frac{4p}{\pi} < 2p \; .$$

The left hand side

$$g_1^2 + (ag_1 + pg_2)^2 = g_1^2(1 + a^2) + 2apg_1g_2 + p^2g_2^2$$

is a natural number divisible by p, so long as one requires $1 + a^2 \equiv 0$, $a^2 \equiv -1$ (mod p). This is always possible, since -1 is a quadratic residue modulo p (by a basic result from the theory of quadratic residues). Hence for $x = g_1$, $y = ag_1 + pg_2$, the sum of squares $x^2 + y^2$ agrees with p, that is $p = x^2 + y^2$.

Corollary 8: Fermat-Euler Theorem. *A prime number p is expressible as the sum of two squared natural numbers if and only if $p = 2$ or $p \equiv 1(\text{mod } 4)$. In this case the sum is unique up to the order of the summands. If $p \equiv 3(\text{mod } 4)$, p cannot be written as the sum of two squared natural numbers.*

Proof. The essential part of the proposition – the existence of the expression for $p \equiv 1(\text{mod } 4)$ – has already been proved. Moreover $2^2 = 1^2 + 1^2$. From $0^2 \equiv 0, 1^2 \equiv 1, 2^2 \equiv 0, 3^2 \equiv 1(\text{mod } 4)$ it follows that $x^2 \equiv 0$ or 1, $y^2 \equiv 0$ or $1(\text{mod } 4)$. Hence for $x^2 + y^2$ the only possible residues (mod 4) are $0, 1, 2$ (mod 4), which excludes such an expression for $p \equiv 3(\text{mod } 4)$.

The *uniqueness* of the expression is obtained using a completely new idea – besides the vector space operations $\mathbb{C}$ also admits a *multiplication*. $\mathbb{C}$ is a field and the lattice $\mathbb{Z}(i)$ spanned by the standard basis $\mathfrak{e}_1 = 1, \mathfrak{e}_2 = i$, is a subring of this field called the *Gaussian integers*. To each element $\xi = a + ib$ from $\mathbb{C}$ one can associate a *norm*

$$N(\xi) = \xi \cdot \bar{\xi} = (a + ib)(a - ib) = a^2 + b^2$$

which is clearly positive for all $\xi \neq 0$, integral for all ξ from $\mathbb{Z}(i)$, and homomorphic with respect to multiplication: $N(\xi\eta) = N(\xi)N(\eta)$.

An *integral element* $\xi = a + ib$, i.e. an integral Gaussian number from $\mathbb{Z}(i)$ is said to be divisible by an integral Gaussian number $\eta = c + id$, if there exists an integral element $\zeta = x + iy$ with $\xi = \eta\zeta$. The number ξ is always divisible by the units $1, i, -1, -i$, and by its *associated* elements $\xi' = i\xi$, $\xi'' = -\xi$, $\xi''' = -i\xi$. If ξ is not a unit and is divisible by no further elements ξ is said to be *prime*.

By decomposing all its factors into prime elements each $\xi \neq 0$ from $\mathbb{Z}(i)$ can be written as a product

$$\xi = \varepsilon\varphi_1\varphi_2\ldots\varphi_R = \varepsilon \cdot \prod_{r=1}^{R} \varphi_r$$

of a unit ε and prime elements $\varphi_1, \varphi_2, \ldots, \varphi_R$. (If ξ is itself a unit one takes $R = 0$.) This expression is essentially unique – up to associates and order. For the proof we use the following technical proposition:

Lemma 2. *If for integral elements ξ, η one has $0 < N(\eta) \leq N(\xi)$, then for η or for one of its associates η' one can achieve $N(\xi - \eta') < N(\xi)$.*

Proof. We may choose the element η' associated to η (where possibly $\eta = \eta'$) so that the triangle of points ξ, η' and 0 in the complex plane has an angle at the origin, whose magnitude equals at most half a right-angle, $\pi/4$. This is possible because the associates of η form a square with 0 at its centre. Since the root of the norm agrees geometrically with the distance from 0, the inequality $N(\xi - \eta') < N(\xi)$ is obvious.   □

Proposition 6. *The representation of integers other than zero from $\mathbb{Z}(i)$ as products of prime elements is unique up to the order of the factors and associates of the individual prime elements.*

Proof. If ξ is such that the norm $N(\xi)$ is minimal and there are two distinct decompositions

$$\xi = \varphi_1\varphi_2\ldots\varphi_R = \psi_1\psi_2\ldots\psi_S \ ,$$

then none of the prime factors $\varphi_1, \varphi_2, \ldots, \varphi_R$ can agree or be associated with any of the prime factors $\psi_1, \psi_2, \ldots, \psi_S$. Otherwise, dividing by this element, one would obtain two prime decompositions of an element ξ/φ_r with smaller norm. In particular without loss of generality we can suppose that $N(\varphi_1) \leq N(\psi_1)$ and by Lemma 2 (if necessary replacing φ_1 by an associated element) obtain $N(\psi_1 - \varphi_1) < N(\psi_1)$. The element

$$\eta = (\psi_1 - \varphi_1)\psi_2\ldots\psi_S = \varphi_1(\varphi_2\ldots\varphi_R - \psi_2\ldots\psi_S)$$

has norm smaller than that of ξ : $N(\eta) < N(\xi)$. However since φ_1 cannot divide the difference $\psi_1 - \varphi_1$, η possesses two decompositions as above, with φ_1 appearing as a prime factor in the second but in contrast not in the first.

This contradicts the choice of ξ as an element of smallest norm with this property.[11] $\square$

The integers φ, which are not associated to a natural number, and which yet are primes, occur if and only if $N(\varphi) = \varphi\bar{\varphi}$ is a prime number. For on the one hand a decomposition $\varphi = \xi\eta$ would imply a decomposition $N(\varphi) = N(\xi)N(\eta)$, and on the other, if φ is a prime then so is $\bar{\varphi}$, so by Proposition 6 $N(\varphi) = \varphi\bar{\varphi}$ must represent the uniquely determined decomposition of $N(\varphi)$ into prime factors in $\mathbb{Z}(i)$. Therefore $N(\varphi)$ has only 1 and itself as natural divisors.

If one considers a prime number p, one can start from the decomposition $p = \varphi_1 \ldots \varphi_R$ in $\mathbb{Z}(i)$. For the norm one obtains $N(p) = p^2 = N(\varphi_1) \ldots N(\varphi_R)$. The $N(\varphi_r)$ are distinct natural numbers, distinct also from 1, and there are only two possibilities. Either $R = 1$ – then p is associated to φ_1 and so is also prime in $\mathbb{Z}(i)$, or one has $R = 2$ – then $p^2 = N(\varphi_1)N(\varphi_2)$, implying that $N(\varphi_1) = N(\varphi_2) = p$. In this case φ_1 and φ_2 are conjugate prime elements $\varphi_1 = x + iy$, $\varphi_2 = x - iy$, and in $\mathbb{Z}(i)$ p decomposes as $p = \varphi_1\bar{\varphi}_1 = x^2 + y^2$. With this Proposition 6 implies the uniqueness asserted in Corollary 8; moreover one sees that

Proposition 7. *The prime elements from $\mathbb{Z}(i)$ are*
(1) $1 + i$,
(2) the prime numbers $p \equiv 3 (\mathrm{mod}\ 4)$ from $\mathbb{Z}$,
(3) pairs of elements $\varphi = x + iy$ and $\bar{\varphi} = x - iy$, for which $N(\varphi) = \varphi\bar{\varphi} = x^2 + y^2$, and which give a prime number $p \equiv 1 (\mathrm{mod}\ 4)$.
In addition there are all the associates of the elements above.

As an application of corollary 8 and proposition 6 one can answer the problem of the possible representations of a natural number n as a sum of integral squares

$$n = x^2 + y^2$$

and the number of these representations. In $\mathbb{Z}$ n has a prime factor decomposition of the form

$$n = 2^D \cdot \prod_h p_h^{E_h} \cdot \prod_k q_k^{F_k} \ .$$

The p_h denote prime numbers $\equiv 1 (\mathrm{mod}\ 4)$ and the q_k prime numbers $\equiv 3 (\mathrm{mod}\ 4)$. Since the p_h decompose in $\mathbb{Z}(i)$ as $p_h = \varphi_h\bar{\varphi}_h$, in $\mathbb{Z}(i)$ one can decompose n in the following way:

$$n = (1 + i)^D (1 - i)^D \prod_h \varphi_h^{E_h} \bar{\varphi}_h^{E_h} \prod_k q_k^{F_k} \ .$$

An expression $n = x^2 + y^2 = (x + iy)(x - iy)$ implies nothing more than the existence of some ξ from $\mathbb{Z}(i)$, with $n = N(\xi) = \xi\bar{\xi}$. Since each decomposition

$$\xi = \prod_j \xi_j$$

gives rise to a corresponding decomposition

$$\bar{\xi} = \prod_j \bar{\xi}_j$$

it follows that φ is a prime divisor of ξ if and only if $\bar{\varphi}$ is a prime divisor of $\bar{\xi}$. One can find a solution of the equation $n = \xi\bar{\xi}$ if and only if the prime factors of n in $\mathbb{Z}(i)$ arise as conjugate pairs. Because $\bar{q}_k = q_k$ this is at best possible when all the F_k are even.

How many possibilities are there – assuming that all the F_k are even – of constructing ξ in $\mathbb{Z}(i)$ so that $n = \xi\bar{\xi}$? For arbitrary integers d, e_h with $0 \le d \le D, 0 \le e_h \le E_h$, we write

$$\xi = (1+i)^d (1-i)^{D-d} \prod_h \varphi_h^{e_h} \bar{\varphi}_h^{E_h - e_h} \prod_k q_k^{F_k/2} \ .$$

Then we have

$$\bar{\xi} = (1+i)^{D-d} (1-i)^d \prod_h \varphi_h^{E_h - e_h} \bar{\varphi}_h^{e_h} \prod_k q_k^{F_k/2} \ .$$

Clearly in this way all possibilities are exhausted – we have only to look at the numbers associated to ξ. In addition we have only to consider that the mere alteration of the value of d only leads to one of the numbers associated to ξ, because $1+i$, $1-i$ are associated primes. Hence the number of non-associated ξ with $n = \xi\bar{\xi}$ corresponds to the number of possibilities of specifying the integers e_h. This equals

$$\prod_h (E_h + 1) \ .$$

If we now count the numbers associated to ξ, this number is multiplied up, implying

Proposition 8. *If*

$$n = 2^D \cdot \prod_h p_h^{E_h} \cdot \prod_k q_k^{F_k}$$

expresses the decomposition of the natural number n into prime factors, where the $p_h \equiv 1$ and the $q_k \equiv 3 \pmod 4$, then the number $r(n)$ of integral solutions of the diophantine equation $x^2 + y^2 = n$ is given by $r(n) = 0$ if one of the components F_k is odd, and by

$$r(n) = 4 \cdot \prod_h (E_h + 1)$$

if all the multiplicities F_k are even.

The decisive condition

$$\mathrm{vol}(K) > 2^N \cdot d(\mathfrak{G})$$

in the Minkowski theorem can be weakened by restriction to specific symmetric
and convex bodies.

Theorem 2: Blichfeldt's Theorem. *Let P be the mid-point of the n-dimensional ball K and let $\mathfrak{G}$ be a lattice with*

$$\operatorname{vol}(K) \geq \frac{N+2}{2} \cdot 2^{N/2} \cdot d(\mathfrak{G}) \ ,$$

then K contains at least one lattice point $R = P + \mathfrak{g}$, with $\mathfrak{g}$ in $\mathfrak{G}$ distinct from P from the point lattice $P + \mathfrak{G}$ based at P.

Proof. Arguing in a similar way to the proof of his Proposition 4 Blichfeldt uses the formula

$$\int_F \left(\sum_{\mathfrak{g} \in \mathfrak{G}} f(Q + \mathfrak{g}) \right) dQ = \int_{\mathbb{P}} f(Q) dQ \ ,$$

in which $\mathbb{P}$ denotes the point space and F a fundamental parallelepiped of the lattice. However Blichfeldt does not now simply choose $f = c_K$, but

$$f(Q) = \max \left(1 - \frac{2\|\overrightarrow{PQ}\|^2}{r^2}, 0 \right) \ ,$$

where r denotes the radius of the ball. Then one has

$$
\begin{aligned}
\int_{\mathbb{P}} f(Q) dQ &= \int \cdots \int_{x_1^2 + \ldots + x_N^2 \leq \frac{r^2}{2}} \left(1 - \frac{2\left(x_1^2 + \ldots + x_N^2\right)}{r^2} \right) dx_1 \ldots dx_N \\
&= \frac{2}{r^2} \cdot \int \cdots \int_{x_1^2 + \ldots + x_N^2 \leq \frac{r^2}{2}} \left(\frac{r^2}{2} - \left(x_1^2 + \ldots + x_N^2\right) \right) dx_1 \ldots dx_N \\
&= \frac{2}{r^2} \cdot \int \cdots \int_{x_0 + x_1^2 + \ldots + x_N^2 \leq \frac{r^2}{2}} \int_{0 \leq x_0 \leq \frac{r^2}{2}} dx_0 dx_1 \ldots dx_N \\
&= \frac{2}{r^2} \int_0^{r^2/2} \int \cdots \int_{x_1^2 + \ldots + x_N^2 \leq \frac{r^2}{2} - x_0} dx_1 \ldots dx_N dx_0 \\
&= \frac{2}{r^2} \cdot \int_0^{r^2/2} \frac{\pi^{N/2}}{\Gamma\left(\frac{N}{2} + 1\right)} \left(\frac{r^2}{2} - x_0 \right)^{N/2} dx_0 \\
&= -\frac{2}{r^2} \cdot \frac{\pi^{N/2}}{\Gamma\left(\frac{N}{2} + 1\right)} \cdot \frac{\left(\frac{r^2}{2} - x_0 \right)^{\frac{N+2}{2}}}{\frac{N+2}{2}} \Bigg|_{x_0=0}^{x_0=\frac{r^2}{2}} \\
&= \frac{1}{\frac{N+2}{2} \cdot 2^{N/2}} \cdot \frac{\pi^{N/2} r^N}{\Gamma\left(\frac{N}{2} + 1\right)} = \frac{1}{\frac{N+2}{2} \cdot 2^{N/2}} \cdot \operatorname{vol}(K) \ .
\end{aligned}
$$

By assumption this quantity is not less than $d(\mathfrak{G})$, i.e.

$$\int_F \left(\sum_{\mathfrak{g} \in \mathfrak{G}} f(Q + \mathfrak{g}) \right) dQ \geq d(\mathfrak{G}) \ .$$

As a result there exists some point Q with

$$\sum_{\mathfrak{g} \in \mathfrak{G}} \max \left(1 - \frac{2\|\overrightarrow{PQ} + \mathfrak{g}\|^2}{r^2}, 0 \right)$$

$$= \sum_{\substack{\mathfrak{g} \in \mathfrak{G} \\ \|\overrightarrow{PQ}+\mathfrak{g}\| \leq r/\sqrt{2}}} \left(1 - \frac{2\|\overrightarrow{PQ} + \mathfrak{g}\|^2}{r^2} \right) \geq 1 \ .$$

If A denotes the number of lattice vectors $\mathfrak{g}$ with $\|\overrightarrow{PQ} + \mathfrak{g}\| \leq r/\sqrt{2}$, then from the inequality above it follows that

$$A - \frac{2}{r^2} \cdot \sum_{\substack{\mathfrak{g} \in \mathfrak{G} \\ \|\overrightarrow{PQ}+\mathfrak{g}\| \leq r/\sqrt{2}}} \|\overrightarrow{PQ} + \mathfrak{g}\|^2 \geq 1 \ ,$$

i.e.

$$\sum_{\substack{\mathfrak{g} \in \mathfrak{G} \\ \|\overrightarrow{PQ}+\mathfrak{g}\| \leq r/\sqrt{2}}} \|\overrightarrow{PQ} + \mathfrak{g}\|^2 \leq \frac{r^2}{2}(A - 1) \ .$$

Lower estimates for sums of this kind are however easy to obtain: if $Q_1, \ldots, Q_A$ are A arbitrary points, then because

$$\sum_{\substack{a,b=1 \\ a \neq b}}^{A} \|\overrightarrow{Q_b Q_a}\|^2$$

$$= \sum_{\substack{a,b=1 \\ a \neq b}}^{A} \|\overrightarrow{PQ}_a - \overrightarrow{PQ}_b\|^2$$

$$= \sum_{\substack{a,b=1 \\ a \neq b}}^{A} \|\overrightarrow{PQ}_a\|^2 + \sum_{\substack{a,b=1 \\ a \neq b}}^{A} \|\overrightarrow{PQ}_b\|^2 - 2 \cdot \sum_{\substack{a,b=1 \\ a \neq b}}^{A} \overrightarrow{PQ}_a \cdot \overrightarrow{PQ}_b$$

$$= (A-1) \cdot \sum_{a=1}^{A} \|\overrightarrow{PQ}_a\|^2 + (A-1) \cdot \sum_{b=1}^{A} \|\overrightarrow{PQ}_b\|^2 - 2 \cdot \sum_{\substack{a,b=1 \\ a \neq b}}^{A} \overrightarrow{PQ}_a \cdot \overrightarrow{PQ}_b$$

$$= 2A \cdot \sum_{a=1}^{A} \|\overrightarrow{PQ}_a\|^2 - 2 \left(\sum_{a=1}^{A} \overrightarrow{PQ}_a \right) \cdot \left(\sum_{b=1}^{A} \overrightarrow{PQ}_b \right)$$

$$\leq 2A \cdot \sum_{a=1}^{A} \|\overrightarrow{PQ}_a\|^2$$

one has the inequality

$$\sum_{\substack{a,b=1\\a\neq b}}^{A} \|\overrightarrow{Q_bQ_a}\|^2 \leq 2A \cdot \sum_{a=1}^{A} \|\overrightarrow{PQ_a}\|^2 \ .$$

In particular, if $\mathfrak{g}(1),\ldots,\mathfrak{g}(A)$ denote the A lattice points with the property $\|\overrightarrow{PQ} + \mathfrak{g}(a)\| \leq r\sqrt{2}$, and Q_a is defined by $Q_a = Q + \mathfrak{g}(a)$, it follows that

$$\sum_{\substack{a,b=1\\a\neq b}}^{A} \|\mathfrak{g}(a) - \mathfrak{g}(b)\|^2 = \sum_{\substack{a,b=1\\a\neq b}}^{A} \|\overrightarrow{Q_aQ_b}\|^2$$

$$\leq 2A \cdot \sum_{a=1}^{A} \|\overrightarrow{PQ_a}\|^2 \leq 2A \cdot \frac{r^2}{2}(A-1)$$

$$= r^2 A(A-1) \ .$$

This shows that it is impossible to have $\|\mathfrak{g}(a) - \mathfrak{g}(b)\| > r$ for all indices a, b with $b \neq a$. For at least two lattice vectors $\mathfrak{g}(a)$ and $\mathfrak{g}(b)$ $\|\mathfrak{g}(a) - \mathfrak{g}(b)\| \leq r$ is satisfied. If we put $\mathfrak{g} = \mathfrak{g}(a) - \mathfrak{g}(b)$, $R = P + \mathfrak{g}$ has to be a lattice point inside the ball of radius a distinct from the mid-point P. The existence of such an R was to be shown. $\qquad\square$

The question, which Blichfeldt answered in his theorem for the ball, can be more sharply posed. How small can we choose a positive constant c, so that for the symmetric body K and each lattice $\mathfrak{G}$ with $\text{vol}(K) > c \cdot d(\mathfrak{G})$ there always exists some lattice point $Q = P + \mathfrak{g}$ $(\mathfrak{g} \in \mathfrak{G})$ in K other than the centre of symmetry P? (We restrict ourselves to bodies of a certain shape.) Minkowski stated as a conjecture that c cannot be chosen to be arbitrarily small. This conjecture gives rise to

Theorem 3: Theorem of Minkowski and Hlawka. *For each bounded symmetric set K one can construct a lattice $\mathfrak{G}$ with the property that apart from the centre of symmetry P, there exists no lattice point $Q = P + \mathfrak{g}$ $(\mathfrak{g} \in \mathfrak{G})$ from the point lattice $P + \mathfrak{G}$ based at P lying in K, so long as*

$$\text{vol}(K) < 2 \cdot d(\mathfrak{G}) \ .$$

Proof. The proof rests on the implicitly assumed Jordan measurability of K. Indeed theorem 3 follows from a more general result for Riemann integrable functions.

Lemma 3: Deformation Theorem. *If the non-negative Riemann integrable function f takes the constant value zero outside some bounded set, then for each $\varepsilon > 0$ one can find some lattice $\mathfrak{G}$ with $d(\mathfrak{G}) = 1$, which possesses the property*

$$\sum_{\mathfrak{g}\in\mathfrak{G},\mathfrak{g}\neq\mathfrak{o}} f(P+\mathfrak{g}) < \int_{\mathbb{P}} f(Q)dQ + \varepsilon$$

(where the point P can be chosen arbitrarily).

If one puts $f = c_K$, then theorem 3 follows:

$$A = \sum_{\mathfrak{g}\in\mathfrak{G},\mathfrak{g}\neq\mathfrak{o}} c_K(P+\mathfrak{g})$$

gives the number of lattice points in K distinct from P. If one chooses $\varepsilon = 1 - \mathrm{vol}(K)/2$, it follows from

$$A < \int_{\mathbb{P}} c_K(Q)dQ + \varepsilon = \mathrm{vol}(K) + \varepsilon < 2 \ ,$$

that there can exist at most one $Q = P + \mathfrak{g}$ with $\mathfrak{g} \neq \mathfrak{a}$ in K. Because K is symmetric with respect to P, $Q' = P - \mathfrak{g}$ would also have to lie in K. Hence in point of fact there exists no $P + \mathfrak{g} \neq P$ in K. The restriction $d(\mathfrak{G}) = 1$ is clearly unimportant for the validity of the theorem.

Proof of Lemma 3. Let p denote a prime number, $\mathfrak{A}(p)$ the set of all vectors

$$\mathfrak{a} = a_1\mathfrak{e}_1 + \ldots + a_N\mathfrak{e}_N = \mathfrak{e}_1 + a_2\mathfrak{e}_2 + \ldots + a_N\mathfrak{e}_N$$

with $a_1 = 1$, and $0 \leq a_n \leq p-1$ for $n = 2,\ldots,N$, where all the a_n are integral, so that $\mathfrak{A}(p)$ contains p^{N-1} vectors. The lattice $\mathfrak{B} = \mathfrak{B}(p,\mathfrak{a})$ is spanned by the basis $\mathfrak{b}_1 = \mathfrak{a}$, $\mathfrak{b}_2 = p\mathfrak{e}_2,\ldots,\mathfrak{b}_N = p\mathfrak{e}_N$. If one chooses $q = \sqrt[N]{p^{N-1}}$, then for the lattice $\mathfrak{G} = \mathfrak{G}(p,\mathfrak{a})$ spanned by $(1/q)\mathfrak{b}_1,\ldots,(1/q)\mathfrak{b}_N$ it follows from $d(\mathfrak{B}) = p^{N-1}$ that $d(\mathfrak{G}) = 1$.

Referred to the standard basis the lattice vectors $\mathfrak{g} = g_1\mathfrak{b}_1 + \ldots + g_N\mathfrak{b}_N = g_1'\mathfrak{e}_1 + \ldots + g_N'\mathfrak{e}_N$ possess components $g_1' = g_1, g_2' = g_1a_2 + g_2p, g_3' = g_1a_3 + g_3p,\ldots,g_N' = g_1a_N + g_Np$.

The set $\mathfrak{C}(p)$ consists of all vectors $\mathfrak{c} = c_1\mathfrak{e}_1 + \ldots + c_N\mathfrak{e}_N$ with integral components $c_1,\ldots,c_N$ and the single component c_1 not divisible by p. In the first step of the proof we claim that for each $\mathfrak{c}$ from $\mathfrak{C}(p)$ there exists one and only one vector $\mathfrak{a}$ from $\mathfrak{A}(p)$ so that $\mathfrak{c}$ belongs to the lattice $\mathfrak{B}(p,\mathfrak{a})$.

If r denotes the smallest residue modulo p of c_1, i.e. $c_1 = r(\mathrm{mod}\ p), 1 \leq r \leq p-1$, the integers $a_2, a_3,\ldots,a_N$ are chosen to be the unique solutions of the congruences

$$ra_n \equiv c_n(\mathrm{mod}\ p), n = 2,\ldots,N \ ,$$

for which $0 \leq a_n \leq p-1$ holds. Hence there exist integers k_n with

$$c_1 = r + k_1p, \ c_2 = ra_2 + k_2p, \ c_3 = ra_3 + k_3p, \ \ldots, \ c_N = ra_N + k_Np \ .$$

If one writes

$$g_1 = c_1, \ g_2 = k_2 - a_2k_1, \ g_3 = k_3 - a_3k_1, \ \ldots, \ g_N = k_N - a_Nk_1 \ ,$$

then because $r = c_1 - k_1 p$, this implies that

$$c_1 = g_1,$$
$$c_n = c_1 a_n + (k_n - a_n k_1) p = g_1 a_n + g_n p, \quad n = 2, \ldots, N \ .$$

This already shows that $\mathfrak{c}$ belongs to $\mathfrak{B}(p, \mathfrak{a})$.

The intersection $\mathfrak{D}(p, \mathfrak{a})$ formed from $\mathfrak{B}(p, \mathfrak{a})$ and $\mathfrak{C}(p)$ therefore consists of all $\mathfrak{g} = g_1 \mathfrak{b}_1 + \ldots + g_N \mathfrak{b}_N = g_1' \mathfrak{e}_1 + \ldots + g_N' \mathfrak{e}_N$, for which $g_1 = g_1' \not\equiv 0 (\mathrm{mod}\ p)$ holds. On account of the considerations above the $\mathfrak{D}(p, \mathfrak{a})$ are pairwise disjoint, as $\mathfrak{a}$ runs through $\mathfrak{A}(p)$. Their union generates all of $\mathfrak{C}(p)$.

In the second step of the proof the definition of the Riemann integral is suitably formulated: if $\mathfrak{C}$ consists of all $\mathfrak{c} = c_1 \mathfrak{e}_1 + \ldots + c_N \mathfrak{e}_N$ with integral c_n then

$$\sum_{\mathfrak{c} \in \mathfrak{C}} f\left(P + \frac{1}{q}\mathfrak{c}\right) \cdot \left(\frac{1}{q}\right)^N$$

is a Riemann approximating sum with the $P + (1/q)\mathfrak{c}$ as division points for the cube consisting of $X = P + x_1 \mathfrak{e}_1 + \ldots + x_N \mathfrak{e}_N$ with $c_n/q \leq x_n \leq (c_n+1)/q, \mathfrak{c} = c_1 \mathfrak{e}_1 + \ldots + c_N \mathfrak{e}_N$. From the choice of a sufficiently large prime number p it follows that the approximating sum is arbitrarily close to the Riemann integral. More precisely: for each positive ε we can find a sufficiently large prime number p to ensure that

$$-\varepsilon < \frac{1}{q^N} \sum_{\mathfrak{c} \in \mathfrak{C}} f\left(P + \frac{1}{q}\mathfrak{c}\right) - \int_{\mathbb{P}} f(Q) dQ < \varepsilon \ .$$

The function f is non-negative, $\mathfrak{C}(p)$ is a subset of $\mathfrak{C}$, and one has $q^N = p^{N-1}$. The result of the second step in the proof therefore reads:

$$\frac{1}{p^{N-1}} \sum_{\mathfrak{c} \in \mathfrak{C}(p)}{}' f\left(P + \frac{1}{q}\mathfrak{c}\right) < \int_{\mathbb{P}} f(Q) dQ + \varepsilon \ .$$

Put together these two steps give the following important formula:

$$\frac{1}{p^{N-1}} \sum_{\mathfrak{a} \in \mathfrak{A}(p)} \sum_{\mathfrak{c} \in \mathfrak{D}(p,\mathfrak{a})} f\left(P + \frac{1}{q}\mathfrak{c}\right) < \int_{\mathbb{P}} f(Q) dQ + \varepsilon \ .$$

It shows that for at least one vector $\mathfrak{a}$

$$\sum_{\mathfrak{c} \in \mathfrak{D}(p,\mathfrak{a})} f\left(P + \frac{1}{q}\mathfrak{c}\right) < \int_{\mathbb{P}} f(Q) dQ + \varepsilon$$

must hold. For in the contrary case one would have

$$\sum_{\mathfrak{a} \in \mathfrak{A}(p)} \sum_{\mathfrak{c} \in \mathfrak{D}(p,\mathfrak{a})} f\left(P + \frac{1}{q}\mathfrak{c}\right) \geq \left(\int_{\mathbb{P}} f(Q) dQ + \varepsilon\right) \cdot p^{N-1}$$

$(\mathfrak{A}(p)$ consisting of p^{N-1} elements) contradicting the formula above.

In the final step of the proof we show that for all sufficiently large p we have

$$\sum_{\mathfrak{c}\in\mathfrak{B}(p,\mathfrak{a}),\mathfrak{c}\neq\mathfrak{o}} f\left(P+\frac{1}{q}\mathfrak{c}\right) < \int_{\mathbb{P}} f(Q)dQ + \varepsilon \ .$$

Here we make essential use of the fact that f vanishes outside a sufficiently large cube consisting of $X = P + x_1\mathfrak{e}_1 + \ldots + x_N\mathfrak{e}_N$ with $|x_n| \leq M$. To the index set $\mathfrak{D}(p,\mathfrak{a})$ for the sum above one must add only lattice vectors $\mathfrak{c} = g_1\mathfrak{b}_1 + \ldots + g_N\mathfrak{b}_N$ with $g_1 \equiv 0(\text{mod } p)$. If p is chosen so large that $\sqrt[N]{p} > M$, then since $g_1 \neq 0$ it follows from the divisibility of g_1 by p that $\left|g_1/p^{1-1/N}\right| > M$, $f\left(P+\frac{1}{q}\mathfrak{c}\right) = 0$, i.e. the sum is not increased by these additional lattice vectors. If $g_1 = 0$ holds, then

$$\frac{1}{q}\mathfrak{c} = \sum_{n=2}^{N} \frac{g_n p}{p^{1-1/N}}\,\mathfrak{e}_n = \sum_{n=2}^{N} g_n p^{1/N}\mathfrak{e}_n \ .$$

Some g_n is certainly distinct from zero, which for $\sqrt[N]{p} > M$ again has $\left|g_n p^{1/N}\right| > M$, $f\left(P+\frac{1}{q}\mathfrak{c}\right) = 0$ as a consequence. Again no contribution is made to the sum above. Paying due regard to the definition of $\mathfrak{G} = \mathfrak{G}(p,\mathfrak{a})$, the final inequality implies the formula given in Lemma 3. In this way both Lemma 3 and Theorem 3 are proved using a simplified method of C.A. Rogers.

Exercises on Chapter 3

If $\mathfrak{M} \subseteq \mathbb{R}^n$, we denote by $\langle\mathfrak{M}\rangle$ the subspace generated by $\mathfrak{M}$. The *rank of* $\mathfrak{M}$ equals $\dim\langle\mathfrak{M}\rangle$. A set $\mathfrak{X} \subseteq \mathbb{R}^n$ is called *discrete*, if it has no accumulation points (in $\mathbb{R}^n$).

* 1. Let $\mathfrak{G}$ be a discrete subgroup of $\mathbb{R}^n$ of rank k. Let $\{\mathfrak{a}_1,\ldots,\mathfrak{a}_{k-1},\mathfrak{v}_k\} \subseteq \mathfrak{G}$ be linearly independent. Suppose that $\mathfrak{G} \cap (\mathbb{R}\mathfrak{a}_1 + \ldots + \mathbb{R}\mathfrak{a}_{k-1}) = \mathbb{Z}\mathfrak{a}_1 + \ldots + \mathbb{Z}\mathfrak{a}_{k-1}$. Put $\mathfrak{S} = \mathfrak{G} \cap ([0,1)\mathfrak{a}_1 + \ldots + [0,1)\mathfrak{a}_{k-1} + [0,1]\mathfrak{v}_k)$, $\alpha_k = \min\{\alpha > 0:$ there exist $\alpha_1,\ldots,\alpha_{k-1} \in \mathbb{R}$ with $\alpha_1\mathfrak{a}_1 + \ldots + \alpha_{k-1}\mathfrak{a}_{k-1} + \alpha\mathfrak{v}_k \in \mathfrak{S}\}$, and choose $\alpha_1,\ldots,\alpha_{k-1} \in \mathbb{R}$ so that $\mathfrak{a}_k := \alpha_1\mathfrak{a}_1 + \ldots + \alpha_{k-1}\mathfrak{a}_{k-1} + \alpha_k\mathfrak{v}_k \in \mathfrak{S}$. Show that $\mathfrak{G} = \mathbb{Z}\mathfrak{a}_1 + \ldots + \mathbb{Z}\mathfrak{a}_k$.

2. Let $\mathfrak{G}$ be a discrete subgroup of $\mathbb{R}^n$ of rank k, $0 \leq k \leq n$. Show that there exist linearly independent vectors $\mathfrak{a}_1,\ldots,\mathfrak{a}_k$ with $\mathfrak{G} = \mathbb{Z}\mathfrak{a}_1 + \ldots + \mathbb{Z}\mathfrak{a}_k$.

3. Let $\mathfrak{G}$ be a closed, non-discrete subgroup of $\mathbb{R}^n$, $(\mathfrak{x}_p)_{p\geq 1}$ a null-sequence in $\mathfrak{G}$, $0 < |\mathfrak{x}_p| \leq 1$ and $k_p = \max\{h \in \mathbb{N} : h|\mathfrak{x}_p| \leq 1\}$ for $p \geq 1$. Let $\mathfrak{a}$ be an accumulation point of $(k_p\mathfrak{x}_p)_{p\geq 1}$. Show that $\mathfrak{a} \neq \mathfrak{o}$ and $\mathbb{R}\mathfrak{a} \subseteq \mathfrak{G}$.

4. Let $\mathfrak{G}$ be a closed subgroup of $\mathbb{R}^n$ of rank r, $0 \leq r \leq n$. Show that $\mathfrak{G}$ contains a maximal subspace $\mathfrak{V}$ of $\mathbb{R}^n$. If $\mathfrak{W}$ is a subspace of $\mathbb{R}^n$ with $\mathfrak{V} + \mathfrak{W} = \mathbb{R}^n$, $\mathfrak{V} \cap \mathfrak{W} = \{0\}$, then $\mathfrak{G} \cap \mathfrak{W}$ is discrete and $\mathfrak{G} = \mathfrak{V} + (\mathfrak{W} \cap \mathfrak{G})$.

5. Let $\mathfrak{G}$ be a closed subgroup of $\mathbb{R}^n$ of rank r. Then there exists some p, $0 \leq p \leq r$ and some basis $\{\mathfrak{a}_1, \ldots, \mathfrak{a}_n\}$ of $\mathbb{R}^n$, so that

$$\mathfrak{G} = \mathbb{R}\mathfrak{a}_1 + \ldots + \mathbb{R}\mathfrak{a}_p + \mathbb{Z}\mathfrak{a}_{p+1} + \ldots + \mathbb{Z}\mathfrak{a}_r \ .$$

6. Let $\mathfrak{G}$ be a subgroup of $\mathbb{R}^n$ and $\mathfrak{G}^* = \{\mathfrak{u} \in \mathbb{R}^n : \langle \mathfrak{u}, \mathfrak{x} \rangle \in \mathbb{Z} \text{ for all } \mathfrak{x} \in \mathfrak{G}\}$. $\mathfrak{G}^*$ is called the *group associated with* $\mathfrak{G}$. Show that $\mathfrak{G}^*$ is a closed subgroup of $\mathbb{R}^n$ and that $\mathfrak{G}^* = \left(\bar{\mathfrak{G}}\right)^*$.

7. Let $\{\mathfrak{a}_1, \ldots, \mathfrak{a}_n\}$ be a basis of $\mathbb{R}^n$, $\mathfrak{G} = \mathbb{R}\mathfrak{a}_1 + \ldots + \mathbb{R}\mathfrak{a}_p + \mathbb{Z}\mathfrak{a}_{p+1} + \mathbb{Z}\mathfrak{a}_{p+q}$ and $\{\mathfrak{a}_1^*, \ldots, \mathfrak{a}_n^*\}$ the basis dual to $\{\mathfrak{a}_1, \ldots, \mathfrak{a}_n\}$. Show that

$$\mathfrak{G}^* = \mathbb{Z}\mathfrak{a}_{p+1}^* + \ldots + \mathbb{Z}\mathfrak{a}_{p+q}^* + \mathbb{R}\mathfrak{a}_{p+q}^* + \ldots + \mathbb{R}\mathfrak{a}_n^* \ .$$

8. Let $\mathfrak{G}$ be a subgroup of $\mathbb{R}^n$. Show that $\bar{\mathfrak{G}} = (\mathfrak{G}^*)^*$.

9. Let $L : \mathbb{R}^n \to \mathbb{R}^m$ be linear, $\mathfrak{b} \in \mathbb{R}^m$ and let $\{\mathfrak{e}_1, \ldots, \mathfrak{e}_n\}$ be the standard basis of $\mathbb{R}^n$. Show that the following statements are equivalent:
   (i)   For all $\varepsilon > 0$ there exists $(\mathfrak{p}, \mathfrak{q}) \in \mathbb{Z}^m \times \mathbb{Z}^n$ with $|L(\mathfrak{q}) - \mathfrak{p} - \mathfrak{b}| < \varepsilon$.
   (ii) If $\mathfrak{u} \in \mathbb{Z}^m$ is such that $\langle \mathfrak{u}, L(\mathfrak{e}_i) \rangle \in \mathbb{Z}$ for $1 \leq i \leq n$, then $\langle \mathfrak{u}, \mathfrak{b} \rangle \in \mathbb{Z}$.
   (Kronecker's Approximation Theorem.)

10. Let $L : \mathbb{R}^n \to \mathbb{R}^m$ be linear and let $\{\mathfrak{e}_1, \ldots, \mathfrak{e}_n\}$ be the standard basis of $\mathbb{R}^n$. Show that the following statements are equivalent:
   (i)   For all $\varepsilon > 0$ and all $\mathfrak{b} \in \mathbb{R}^m$ there exists $(\mathfrak{p}, \mathfrak{q}) \in \mathbb{Z}^m \times \mathbb{Z}^n$ with $|L(\mathfrak{q}) - \mathfrak{p} - \mathfrak{b}| < \varepsilon$.
   (ii) If $\mathfrak{u} \in \mathbb{Z}^m$ is such that $\langle \mathfrak{u}, L(\mathfrak{e}_i) \rangle = 0$ for $1 \leq i \leq n$, then $\mathfrak{u} = 0$.
   (Kronecker's Approximation Theorem).

* 11. Let $\mathfrak{G}$ be a lattice in $\mathbb{R}^n$, $\mathfrak{P}$ a fundamental parallelepiped of $\mathfrak{G}$, $\mathfrak{X} \subseteq \mathbb{R}^n$ discrete, $\mathfrak{M} \subseteq \mathbb{R}^n$ measurable and $\lambda(\mathfrak{M}) < \infty$. For all $\mathfrak{g} \in \mathfrak{G}$ suppose that $\mathfrak{X} + \mathfrak{g} = \mathfrak{X}$. Show that there exists some point $\mathfrak{x} \in \mathfrak{P}$ with

$$|(\mathfrak{M} + \mathfrak{x}) \cap \mathfrak{X}| \geq \lambda(\mathfrak{M})|\mathfrak{X} \cap \mathfrak{P}|/d(\mathfrak{G}) \ .$$

* 12. Let $\mathfrak{G}$ be a lattice in $\mathbb{R}^n$, $\mathfrak{P}$ a fundamental parallelepiped of $\mathfrak{G}$, $\mathfrak{X} \subseteq \mathbb{R}^n$ discrete, $\mathfrak{M} \subseteq \mathbb{R}^n$ compact. For all $\mathfrak{g} \in \mathfrak{G}$ suppose that $\mathfrak{X} + \mathfrak{g} = \mathfrak{X}$. Show that there exists some $\mathfrak{x} \in \mathfrak{P}$ with

$$|(\mathfrak{M} + \mathfrak{x}) \cap \mathfrak{X}| > \lambda(\mathfrak{M})|\mathfrak{X} \cap \mathfrak{P}|/d(\mathfrak{G})$$

(Blichfeldt's theorem).

* 13. Let $L : \mathbb{R}^n \to \mathbb{R}^m$ be linear and $N > 1$ real. Show that there exists some $(\mathfrak{x}, \mathfrak{y}) \in \mathbb{Z}^n \times \mathbb{Z}^m$ with $0 < \|\mathfrak{x}\| < N^{m/n}$ and $\|L(\mathfrak{x}) - \mathfrak{y}\| \leq \frac{1}{N}$. Here $\|\mathfrak{a}\| = \max_{1 \leq i \leq t} |a_i|$ for $\mathfrak{a} \in \mathbb{R}^t$. (Dirichlet's approximation theorem.)

14. Let $\mathfrak{o} \in \mathfrak{K} \subseteq \mathbb{R}^n$. Then the *lattice constant of* $\mathfrak{K}$, $\Delta(\mathfrak{K})$, equals $\inf \{d(\mathfrak{G}): \mathfrak{G} \text{ is a lattice in } \mathbb{R}^n, \mathfrak{G} \cap \mathfrak{K} = \{\mathfrak{o}\}\}$ (where $\inf \emptyset = \infty$). Show that for $t > 0$, $\Delta(t\mathfrak{K}) = t^n \Delta(\mathfrak{K})$.

15. Let $\mathfrak{o} \in \mathfrak{K} \subseteq \mathbb{R}^n$ be Jordan measurable. Show that $\Delta(\mathfrak{K}) \leq \lambda(\mathfrak{K})$ (Hlawka 1943).

16. Let $\mathfrak{K} \subseteq \mathbb{R}^n$ be compact, convex and contain $\mathfrak{o}$ as an interior point. Then $\mathfrak{K}$ is called a *convex body*. If $\mathfrak{K}$ is such, $1 \leq j \leq n$ and $I_j = \{\lambda > 0 : \lambda\mathfrak{K} \cap \mathbb{Z}^n \text{ has}$

rank $\geq j\}$, then I_j is a closed interval, unbounded from above. $\lambda_j(\mathfrak{K}) := \inf I_j$ is called the *jth successive minimum of $\mathfrak{K}$*. Show that $\lambda_1(\mathfrak{K}) \leq \ldots \leq \lambda_n(\mathfrak{K})$.

17. Let $a_1, \ldots, a_n$ be positive numbers, $\mathfrak{K} = \prod_{i=1}^{n} [-a_i, a_i]$, and σ a permutation with $a_{\sigma(1)} \geq \ldots \geq a_{\sigma(n)}$. Show that $\lambda_j(\mathfrak{K}) = a_{\sigma(j)}^{-n}$.

18. Let $\mathfrak{K}$ be a convex body, $1 \leq j \leq n$. Show:
    (i)   If $L : \mathbb{R}^n \to \mathbb{R}^n$ is linear and $L(\mathbb{Z}^n) = \mathbb{Z}^n$, then $\lambda_j\big(L(\mathfrak{K})\big) = \lambda_j(\mathfrak{K})$.
    (ii)   If $t > 0$, then $\lambda_j(t\mathfrak{K}) = \frac{1}{t}\lambda_j(\mathfrak{K})$.

19. If $\mathfrak{K}$ is a convex body, then there exists some basis $\{\mathfrak{g}_1, \ldots, \mathfrak{g}_n\}$ of $\mathbb{R}^n$, so that for $1 \leq j \leq n$, $\mathfrak{g}_j \in \lambda_j(\mathfrak{K})\mathfrak{K} \cap \mathbb{Z}^n$.

* 20. Let $d < 0$ be square-free, $d \not\equiv 1 \pmod 4$. Show that the following statements are equivalent:
    (i)   Prime decomposition is unique in $\mathcal{O}_d$
    (ii) $d \in \{-1, -2\}$.
    (For the definition of $\mathcal{O}_d$ see page 139 of our book on elementary number theory.)

Hints for the Exercises on Chapter 3

1. Let $\mathfrak{x} = \sum_{i=1}^{k-1} \beta_i \mathfrak{a}_i + \beta_k \mathfrak{v}_k \in \mathfrak{G}$. For $1 \leq i \leq k$ put $\gamma_i = \beta_i - \alpha_i \left[\frac{\beta_k}{\alpha_k}\right]$ and $\mathfrak{w} = \sum_{i=1}^{k-1} [\gamma_i]\,\mathfrak{a}_i$. Show that $\mathfrak{x} - \mathfrak{w} - \left[\frac{\beta_k}{\alpha_k}\right] \mathfrak{a}_k \in \mathfrak{G}$, then prove that $\frac{\beta_k}{\alpha_k} \in \mathbb{Z}$, and finally that $\gamma_i \in \mathbb{Z}$ for $1 \leq i < k$.

2. Induction on k. Apply (1).

11. If $\mathfrak{p} \in \mathfrak{X} \cap \mathfrak{P}$ and $\mathfrak{G} \subseteq \mathbb{R}^n$, put $\nu_\mathfrak{p}(\mathfrak{G}) := \sum_{\mathfrak{g} \in \mathfrak{G}} c_\mathfrak{G}(\mathfrak{p} + \mathfrak{g})$. Show that $\sum_{\mathfrak{p} \in \mathfrak{P} \cap \mathfrak{X}} \nu_\mathfrak{p}(\mathfrak{G}) = |\mathfrak{G} \cap \mathfrak{X}|$ and calculate $\int_\mathfrak{P} \nu_\mathfrak{p}(\mathfrak{M} + \mathfrak{x})d\mathfrak{x}$. Finally put $\mathfrak{G} = \mathfrak{M} + \mathfrak{x}$.

12. The case $\nu := \lambda(\mathfrak{M})|\mathfrak{X} \cap \mathfrak{P}|/d(\mathfrak{G}) \notin \mathbb{Z}$ is trivial. If $\nu \in \mathbb{Z}$, choose $\lambda_k = 1 + \frac{1}{k}$ and apply (11) to $\lambda_k \mathfrak{M}$.

13. Put $\mathfrak{A} = \{\mathfrak{x} \in \mathbb{Z}^n : 0 \leq x_i < N^{m/n}$ for $1 \leq i \leq n\}$, so that $|\mathfrak{A}| \geq N^m$. For $\mathfrak{y} \in \mathbb{R}^m$ put $\{\mathfrak{y}\} = (\{y_1\}, \ldots, \{y_m\})$. If $\mathfrak{x} \mapsto \{L(\mathfrak{x})\}$ is not injective on $\mathfrak{A}$, then one immediately finds $\mathfrak{x}, \mathfrak{y}$. If $\mathfrak{x} \mapsto \{L(\mathfrak{x})\}$ is injective, then in (12) put $\mathfrak{X} = \mathbb{Z}^m + L(\mathfrak{A})$, $\mathfrak{G} = \mathbb{Z}^m$, $\mathfrak{P} = [0, 1)^m$ and $\mathfrak{M} = \left[0, \frac{1}{N}\right]^m$. If $\mathfrak{a} \in \mathfrak{A}$, then $\{L(\mathfrak{a})\} \in \mathfrak{P} \cap \mathfrak{X}$, so that $|\mathfrak{P} \cap \mathfrak{X}| \geq |\mathfrak{A}|$.

15. Apply Lemma 3.

20. In order to prove (i) $\Rightarrow$ (ii) note that for $d < -2$, 2 is irreducible in $\mathcal{O}_d$, but not prime.

4. Number Theoretic Functions

Until now the approximation of individual real numbers has been studied with the help of integers, and for this the methods of Analysis and Geometry have been seen to be helpful. From now on the reverse problem will be attacked: number theoretic functions, the calculation of which for large values is severely limited because of the irregularity of their behaviour, will be represented approximately by known functions from differential and integral calculus with easily described behaviour. Moreover the methods of analysis are also available for these approximation problems. For example in Chapter 5 we prove the prime number theorem, which has for its subject the approximation of the number theoretic function $\pi(x)$, which counts all prime numbers $p \leq x$ by means of the function $x/\log x$. In this chapter we develop the notation and necessary techniques, discuss how good an approximation is, and present some simple examples.

First Tool of the Trade: The Landau Symbol

The symbols O, o named after Landau, although used earlier by Bachmann, describe similarities in the growth of functions.

$$f(x) = g(x) + O\big(h(x)\big)$$

implies that as $x \to \infty$,

$$\frac{f(x) - g(x)}{h(x)}$$

remains bounded, i.e. that $f(x) - g(x)$ increases at most as fast as $h(x)$. If on the other hand $f(x) - g(x)$ increases much more slowly than $h(x)$, more precisely if

$$\lim_{x \to \infty} \frac{f(x) - g(x)}{h(x)} = 0 \ ,$$

then Landau writes $f(x) = g(x) + o\big(h(x)\big)$.

Of course in these definitions the common domain of definition of f, g and h must not be bounded above. If it is clear from the context that instead of $x \to \infty$ we are interested in the convergence $x \to \xi$, another accumulation point of the functions' common domain of definition, then we shall use the symbols in the same way, implicitly supposing that $x \to \xi$. If

$$\lim_{x \to \infty} \frac{f(x)}{g(x)} = 1 \ ,$$

i.e.

$$f(x) = g(x) + o\big(g(x)\big)$$

which is equivalent to

$$f(x) = g(x) + o\big(f(x)\big)$$

then the two functions are said to be *asymptotically equal*, and we write

$$f(x) \sim g(x) \ .$$

From differential calculus one has the familiar examples

$$x^{\Omega} = 0 + o(e^{\varepsilon x}) = o(e^{\varepsilon x}) \ ,$$
$$(\log x)^{\Omega} = 0 + o(x^{\varepsilon}) = o(x^{\varepsilon})$$

for arbitrary $\varepsilon > 0$ and Ω, or

$$\sinh x \sim \frac{e^x}{2} \ ,$$
$$\sin x = O(1) \ ,$$
$$\sin x = x + O(x^3), \qquad (x \to 0) \ .$$

By direct application of the definitions one also sees the validity of the following statements: if g is non-negative and $f(x) = O(g(x))$, then (by continuity of the functions for $x \geq a$)

$$\int_a^x f(t)dt = O\left(\int_a^x g(t)dt \right) \ .$$

Furthermore if $f(x) = o\big(g(x)\big)$ and $\int_a^\infty g(t)dt$ diverges, then

$$\int_a^x f(t)dt = o\left(\int_a^x g(t)dt \right) \ .$$

Applications:

(1) One says that a number theoretic function, that is a function f defined on $\mathbb{N}$, is multiplicative, if it is not identically zero, and for all coprime pairs (m,n) we have

$$f(mn) = f(m)f(n) \ .$$

If we know the value of f on all prime powers then f is completely determined by its multiplicity. This is also the basis for

Lemma 1. *If for the multiplicative function f one has*

$$\lim_{p^m \to \infty} f(p^m) = 0 \ ,$$

as p runs through all prime powers, then

$$\lim_{n \to \infty} f(n) = 0 \ .$$

Proof. By assumption $f(p^m)$ remains bounded for all prime powers p^m with $m \geq 0$,

$$|f(p^m)| \leq A, \qquad (A \geq 1) \ .$$

Further there exists some positive B, so that $p^m \geq B$ implies that

$$|f(p^m)| \leq 1 \ .$$

For each arbitrarily chosen $\varepsilon > 0$ we can find some $C = C(\varepsilon)$ with

$$|f(p^m)| \leq \frac{\varepsilon}{A^B}$$

for $p^m \geq C$. If in the prime factorisation

$$n = \prod_p p^{\nu(p,n)}$$

all prime powers $p^{\nu(p,n)} < C$, then one would have

$$n \leq \prod_{p:p^{\nu(p,n)}<C} p^{\nu(p,n)} < C^C \ ,$$

that is for all $n \geq C^C$ for at least one prime factor p one would have to have

$$p^{\nu(p,n)} \geq C \ .$$

Thus in the relations

$$|f(n)| = \prod_p \left| f\left(p^{\nu(p,n)} \right) \right|$$

$$= \prod_{p:p^{\nu(p,n)}<B} \left| f\left(p^{\nu(p,n)} \right) \right| \times$$

$$\times \prod_{p:B \leq p^{\nu(p,n)}<C} \left| f\left(p^{\nu(p,n)} \right) \right| \times$$

$$\times \prod_{p:C \leq p^{\nu(p,n)}} \left| f\left(p^{\nu(p,n)} \right) \right|$$

the first factor on the right-hand side is bounded by A^B, the second by 1 and the third (which for $n \geq C^C$ does not reduce to the empty product) by ε/A^B. This shows that for $n \geq C^C$

$$|f(n)| \leq \varepsilon \ ,$$

concluding the proof. $\square$

This lemma contributes to the proofs of the following results:

Proposition 1. *If $\tau(n)$ counts the natural divisors of n, then for each positive ε,*

$$\tau(n) = o(n^\varepsilon) \ .$$

Proof. Elementary number theory shows the multiplicativity of the divisor function $\tau(n)$, hence if

$$f(n) = \frac{\tau(n)}{n^\varepsilon}$$

f is also multiplicative. By Lemma 1 we have to show that

$$\lim_{p^m \to \infty} \frac{\tau(p^m)}{p^{\varepsilon m}} = \lim_{p^m \to \infty} \frac{m+1}{p^{\varepsilon m}} = 0$$

which follows from

$$\frac{m+1}{p^{\varepsilon m}} \le \frac{2m}{p^{\varepsilon m}} = \frac{2 \log p^m}{p^{\varepsilon m} \log p} \le \frac{2}{\log 2} \cdot \frac{\log p^m}{(p^m)^\varepsilon} \ . \qquad \Box$$

(2) The Euler function $\varphi(n)$ counts the prime residue classes modulo n, i.e. the number of all natural numbers $k \le n$ with the highest common factor $(k, n) = 1$. This is also multiplicative, and the formula

$$\frac{\varphi(p^m)}{p^m} = \frac{p^m - p^{m-1}}{p^m} = 1 - \frac{1}{p}$$

together with $\varphi(n)/n \le 1$ shows that

$$\overline{\lim}_{n \to \infty} \frac{\varphi(n)}{n} = 1 \ .$$

In the reverse direction

$$f(n) = \frac{n^{1-\varepsilon}}{\varphi(n)}$$

is multiplicative, and for $\varepsilon > 0$ the convergence of $f(p^m)$ to 0 follows from

$$f(p^m) = \frac{p^{m(1-\varepsilon)}}{p^m - p^{m-1}} = \frac{p^{-m\varepsilon}}{1 - \dfrac{1}{p}} \le 2p^{-m\varepsilon} \ .$$

By Lemma 1 this implies that

$$\lim_{n \to \infty} \frac{n^{1-\varepsilon}}{\varphi(n)} = 0 \ ,$$

with which the following statement is proved:

Proposition 2. *If $\varphi(n)$ counts the prime residues modulo n, then on the one hand $\varphi(n) = O(n)$, and on the other for each positive ε one has $n^{1-\varepsilon} = o(\varphi(n))$.*

(3) The function $\sigma(n)$ the sum of the natural divisors of n is also multiplicative. From $\sigma(n) \geq n+1 > n$ and $\sigma(p) = p+1$ one obtains

$$\varlimsup_{n \to \infty} \frac{n}{\sigma(n)} = 1 \ .$$

The function

$$f(n) = \frac{\sigma(n)}{n^{1+\varepsilon}}$$

is multiplicative, and from the convergence of

$$f(p^m) = \frac{p^{m+1} - 1}{p - 1} \cdot \frac{1}{p^{m(1+\varepsilon)}}$$
$$= \frac{1}{p^{\varepsilon m}} \left(\frac{1 - 1/p^{m+1}}{1 - 1/p} \right)$$

to zero for all $\varepsilon > 0$ follows

Proposition 3. *If $\sigma(n)$ sums the natural divisors of n, then on the one hand $n = O(\sigma(n))$, and on the other, for each positive ε, one has $\sigma(n) = O\left(n^{1+\varepsilon}\right)$.*

The result stated here is so weak that it ought to be ashamed of the title "proposition". A simple estimate gives

$$\sigma(n) = \sum_{m|n} \frac{n}{m} = n \cdot \sum_{m|n} \frac{1}{m} \leq n \cdot \sum_{m \leq n} \frac{1}{m}$$
$$= n + n \cdot \sum_{m=2}^{n} \int_{m-1}^{m} \frac{dt}{m} \leq n + n \cdot \sum_{m=2}^{n} \int_{m-1}^{m} \frac{dt}{t}$$
$$= n + n \log n$$

and with it the more precise statement

$$\sigma(n) = O(n \log n) \ .$$

The results stated in the propositions above give only scanty information on the behaviour of number theoretic functions. The last calculation shows that clever replacement of a sum by an integral yields deeper insight into the structure of these functions. Thus Dirichlet proposed, rather than describe the behaviour of number theoretic functions directly by estimates of $f(n)$, to use the means

$$\frac{1}{N} \sum_{n=1}^{N} f(n) \ .$$

This is because taking the sum of isolated extreme values of f, which perturb the average behaviour, has a smoothing effect. Dirichlet contrasts the sum above with the integral

$$\frac{1}{T}\int_0^T f(t)dt \ .$$

As was indicated in the last calculation performed, his idea rests on the replacement of the sums above by appropriate integrals, and then draws on the Analyst's trusted rules for the asymptotic calculation of the sums. This introduces the second calculatory technique to be considered in this chapter.

Second Tool of the Trade: Analogy Between Sum and Integral

(1) There is a *qualitative analogy* (i.e. an existential analogy between the sum and integral) because the sum like the integral is a positive linear functional – in this connection recall the discussion of the Weyl criterion carried out in the second chapter.

(2) The qualitative analogy is not limited to positivity and linearity; the rule

$$\int_{x=a}^{x=b} f(x)d(g(x)) = f(b)g(b) - f(a)g(a) - \int_{x=a}^{x=b} g(x)d(f(x))$$

for integrating by parts corresponds via the so-called *Abel transformation* to the following sum

$$\sum_{n=P}^{Q} f(n)\big(g(n) - g(n-1)\big) = f(Q+1)g(Q)$$
$$- f(P)g(P-1)$$
$$- \sum_{n=P}^{Q} g(n)\big(f(n+1) - f(n)\big) \ .$$

One shows this by direct calculation:

$$\sum_{n=P}^{Q} f(n)(g(n) - g(n-1))$$

$$= \sum_{n=P}^{Q} f(n)g(n) - \sum_{n=P}^{Q} f(n)g(n-1)$$

$$= \sum_{n=P}^{Q} f(n)g(n) - \sum_{n=P-1}^{Q-1} f(n+1)g(n)$$

$$= \sum_{n=P}^{Q-1} (f(n) - f(n+1))g(n) + f(Q)g(Q) - f(P)g(P-1)$$

$$= \sum_{n=P}^{Q} (f(n) - f(n+1))g(n) - (f(Q) - f(Q+1))g(Q)$$

$$+ f(Q)g(Q) - f(P)g(P-1)$$

$$= f(Q+1)g(Q) - f(P)g(P-1)$$

$$- \sum_{n=P}^{Q} g(n)(f(n+1) - f(n)) \ .$$

If $g(n)$ is itself a sum

$$g(n) = \sum_{m=P}^{n} h(m) \ ,$$

then the Abel transformation reduces to the following formula:

$$\sum_{n=P}^{Q} f(n)h(n) = f(Q+1) \cdot \sum_{n=P}^{Q} h(n)$$

$$+ \sum_{n=P}^{Q} (f(n) - f(n+1)) \cdot \sum_{m=P}^{n} h(m) \ .$$

Finally, taking $P = 1$ for simplicity, let f be a continuously differentiable function defined on $[1, \infty[$. If for $x \geq 1$ one defines the sum function by

$$g(x) = \sum_{m=1}^{[x]} h(m) \ ,$$

then because of

$$f(n) - f(n+1) = - \int_{n}^{n+1} f'(t)dt$$

and the transformations

$$\sum_{n=1}^{[x]} (f(n) - f(n+1)) \cdot \sum_{m=1}^{n} h(m)$$

$$= -\sum_{n=1}^{[x]} \int_{n}^{n+1} f'(t)dt \cdot g(n)$$

$$= -\sum_{n=1}^{[x]} \int_{n}^{n+1} f'(t)g(t)dt$$

$$= -\int_{1}^{[x]+1} g(t)f'(t)dt$$

$$= -\int_{1}^{x} g(t)f'(t)dt - \int_{x}^{[x]+1} g(t)f'(t)dt$$

$$= -\int_{1}^{x} g(t)f'(t)dt - g([x]) \int_{x}^{[x]+1} f'(t)dt$$

$$= -\int_{1}^{x} g(t)f'(t)dt - g(x)f([x]+1) + g(x)f(x)$$

for $Q = [x]$ the right hand side gives the following statements.

Proposition 4: Abel Transformation. *Let f be continuously differentiable on $[1, \infty[$, h be a number theoretic function, and g be defined by*

$$g(x) = \sum_{m=1}^{[x]} h(m)$$

then

$$\sum_{n=1}^{[x]} f(n)h(n) = f(x)g(x) - \int_{1}^{x} g(t)f'(t)dt \ .$$

This version of Abel transformation appears to allow meaningful *extension of number theoretic functions g* from $\mathbb{N}$ to $[1, \infty[$ via $g(x) = g([x])$. Moreover instead of $\sum_{n=1}^{[x]}$ one willingly writes $\sum_{n \leq x}$ or $\sum_{1 \leq n \leq x}$. With this Abel transformation is elegantly expressed as

$$\sum_{n \leq x} f(n)h(n) = f(x)g(x) - \int_{1}^{x} g(t)f'(t)dt$$

for

$$g(x) = \sum_{m \leq x} h(m) \ .$$

By means of a simple substitution one easily sees that for an arbitrary monotone increasing, unbounded sequence of real numbers $\lambda_1 < \lambda_2 < \ldots <$

$\lambda_n < \ldots$ and a function f continuously differentiable on $[\lambda_1, \infty[$ the Abel transformation holds in the form

$$\sum_{\lambda_1 \leq \lambda_n \leq x} f(\lambda_n) h(n) = f(x) g(x) - \int_{\lambda_1}^{x} g(t) f'(t) dt$$

with

$$g(x) = \sum_{\lambda_1 \leq \lambda_m \leq x} h(m) \ .$$

In the initially proved "sum version" of the Abel transformation one does not need to assume that the number theoretic functions take complex values - it suffices to assume that these values lie in some (not necessarily commutative) ring. All such generalisations of the Abel transformation are useful; however for the purpose of this book the connection between sum and integral stated in and after Proposition 4 suffices.

(3) The qualitative analogy between sum and integral is rounded out by a *quantitative analogy.*

Proposition 5: The Euler Sum Formula. *Let a be a natural number and f a continuously differentiable function defined on $[a, \infty[$. Then we have*

$$\sum_{a \leq n \leq x} f(n) = \int_{a}^{x} f(t) dt + R$$

with the remainder term

$$R = \int_{a}^{x} \{t\} f'(t) dt + f(a) - \{x\} f(x) \ .$$

In particular, if f increases (decreases) monotonically, then $R = O(f(x))$ (resp. $R = O(f(a))$); we assume $f \geq 0$.

Proof. As usual let $\{t\} = t - [t]$ denote the fractional part of the real quantity t. If in the formula for the Abel transformation one substitutes the arithmetic function h with $h(m) = 0$ for $m = 1, 2, \ldots, a - 1$ and $h(m) = 1$ for $m \geq a$, then one obtains

$$\sum_{a \leq n \leq x} f(n) = f(x)([x] - a + 1) - \int_{a}^{x} ([t] - a + 1) f'(t) dt$$

$$= [x] f(x) - (a - 1) f(x) - \int_{a}^{x} [t] f'(t) dt$$

$$+ (a - 1)(f(x) - f(a))$$

$$= [x] f(x) - (a - 1) f(a) - \int_{a}^{x} [t] f'(t) dt \ .$$

If from this relation one subtracts the formula

$$\int_a^x f(t)dt = xf(x) - af(a) - \int_a^x tf'(t)dt$$

obtained by partial integration, then one is left with the remainder term of the Euler sum formula. Since from the monotonicity of f the sign of f' remains unchanged, from

$$\left| \int_a^x \{t\} f'(t)dt \right|$$

$$\leq \int_a^x |\{t\} f'(t)|dt \leq \int_a^x |f'(t)|dt$$

$$= \left| \int_a^x f'(t)dt \right| = |f(x) - f(a)|$$

one deduces in general that

$$R = O(|f(x)| + |f(a)|) \ ,$$

from which the remainder of Proposition 5 follows. $\square$

If for example for some constant $s > 1$ one writes $f(t) = t^{-s}$, then, because of the decreasing monotonicity of the function, for natural numbers P, Q with $P < Q$ one has

$$\sum_{n=P}^{Q} \frac{1}{n^s} = \int_P^Q \frac{dt}{t^s} + O\left(\frac{1}{P^s}\right)$$

$$= \frac{1}{s-1}\left(\frac{1}{P^{s-1}} - \frac{1}{Q^{s-1}}\right) + O\left(\frac{1}{P^s}\right) \ .$$

This justifies the following statement:

Proposition 6. *For natural numbers P, Q with $P < Q$ and real $s > 1$ one has*

$$\sum_{n=P}^{Q} \frac{1}{n^s} = \frac{1}{s-1}\left(\frac{1}{P^{s-1}} - \frac{1}{Q^{s-1}}\right) + O\left(\frac{1}{P^s}\right) \ .$$

In particular

$$\sum_{n=1}^{\infty} \frac{1}{n^s} = \frac{1}{s-1} + O(1) \ ,$$

and

$$\sum_{n=N}^{\infty} \frac{1}{n^s} = \frac{1}{s-1} \cdot \frac{1}{N^{s-1}} + O\left(\frac{1}{N^s}\right) \ .$$

If one puts $f(t) = 1/t$ one obtains the relation

$$\sum_{n=1}^{N} \frac{1}{n} = \int_{1}^{N} \frac{dt}{t} + 1 - \int_{1}^{N} \frac{\{t\}dt}{t^2}$$

$$= \log N + 1 - \int_{1}^{\infty} \frac{\{t\}dt}{t^2} + \int_{N}^{\infty} \frac{\{t\}dt}{t^2} \quad,$$

where the first integral certainly converges, and one estimates the second by

$$\int_{N}^{\infty} \frac{\{t\}dt}{t^2} \leq \int_{N}^{\infty} \frac{dt}{t^2} = \frac{1}{N} \quad.$$

The number

$$C = 1 - \int_{1}^{\infty} \frac{\{t\}dt}{t^2}$$

is called the *Euler-Mascheroni* constant, and in turn this calculation justifies

Proposition 7. *One has*

$$\sum_{n=1}^{N} \frac{1}{n} = \log N + C + O\left(\frac{1}{N}\right) \quad,$$

where

$$C = 1 - \int_{1}^{\infty} \frac{\{t\}dt}{t^2} = \lim_{N \to \infty} \left(\sum_{n=1}^{N} \frac{1}{n} - \log N\right)$$

denotes the Euler-Mascheroni constant.

Applications:

(1) In the Dirichlet divisor problem the averaged behaviour of the divisor function

$$\frac{1}{N} \sum_{n=1}^{N} \tau(n)$$

is to be asymptotically calculated. The sum above calculates all natural numbers m, which divide one of the natural numbers $n \leq N$; in other words it counts all pairs (m, k) of natural numbers for which $mk = n \leq N$ holds. An analytic formulation of the sum above, which is easier to handle, reads

$$\frac{1}{N}\sum_{n=1}^{N}\tau(n) = \frac{1}{N}\sum_{n\leq N}\sum_{m\leq N/n}1 = \frac{1}{N}\sum_{n\leq N}\left[\frac{N}{n}\right]$$

$$= \frac{1}{N}\sum_{n\leq N}\frac{N}{n} - \frac{1}{N}\sum_{n\leq N}\left\{\frac{N}{n}\right\} = \sum_{n\leq N}\frac{1}{n}$$

$$+ O\left(\frac{1}{N}\sum_{n\leq N}1\right) = \log N + C + O\left(\frac{1}{N}\right) + O(1)$$

$$= \log N + O(1) \ .$$

Although with this

$$\frac{1}{N}\sum_{n=1}^{N}\tau(n) \sim \log N \ ,$$

or more exactly

$$\frac{1}{N}\sum_{n=1}^{N}\tau(n) = \log N + O(1)$$

is proved, the result is not yet satisfactory, since the crude estimate

$$\frac{1}{N}\sum_{n=1}^{N}\left\{\frac{N}{n}\right\} = O(1)$$

dominates the more subtle calculation of the harmonic sum in Proposition 7. Dirichlet solved this mismatch by bringing the symmetry between divisors and complementary divisors into play. If into A we collect all pairs (m,n) of natural numbers with $nm \leq N$ and $n \leq \sqrt{N}$, and into B all pairs (n,m) with $nm \leq N$ and $m \leq \sqrt{N}$, then the union of A and B describes all the divisors counted in

$$\sum_{n=1}^{N}\tau(n) \ .$$

Only the pairs with $n \leq \sqrt{N}$ and $m \leq \sqrt{N}$ are counted twice in a common listing of A and B; the number of these doubly counted pairs comes to $\left[\sqrt{N}\right]^{2}$, from which it follows that

$$\frac{1}{N}\sum_{n=1}^{N}\tau(n) = \frac{1}{N}\left(2\sum_{n\leq\sqrt{N}}\sum_{m\leq N/n}1 - \left[\sqrt{N}\right]^{2}\right)$$

$$= \frac{2}{N}\sum_{n\leq\sqrt{N}}\left[\frac{N}{n}\right] - \frac{1}{N}\left[\sqrt{N}\right]^{2}$$

$$= \frac{2}{N}\sum_{n\leq\sqrt{N}}\left(\frac{N}{n} - \left\{\frac{N}{n}\right\}\right) - \frac{1}{N}\left(\sqrt{N} - \left\{\sqrt{N}\right\}\right)^{2}$$

$$= 2\sum_{n\leq\sqrt{N}}\frac{1}{n} - 1 - \frac{2}{N}\sum_{n\leq\sqrt{N}}\left\{\frac{N}{n}\right\} + \frac{2}{\sqrt{N}}\left\{\sqrt{N}\right\}$$

$$\quad - \frac{1}{N}\left\{\sqrt{N}\right\}^{2}$$

$$= 2\log\left[\sqrt{N}\right] + 2C + O\left(\frac{1}{\sqrt{N}}\right) - 1 - \frac{2}{N}O\left(\sqrt{N}\right)$$

$$\quad + \frac{2}{\sqrt{N}}O(1) - \frac{1}{N}O(1)$$

$$= 2\log\left(\sqrt{N} - \left\{\sqrt{N}\right\}\right) + 2C - 1 + O\left(\frac{1}{\sqrt{N}}\right)$$

$$= 2\log\sqrt{N} + O\left(\frac{2}{\sqrt{N} - \left\{\sqrt{N}\right\}}\cdot\left\{\sqrt{N}\right\}\right)$$

$$\quad + 2C - 1 + O\left(\frac{1}{\sqrt{N}}\right)$$

$$= \log N + 2C - 1 + O\left(\frac{1}{\sqrt{N}}\right)\ .$$

(In the last line but one the mean value theorem is applied.)   □

Proposition 8. *The averaged behaviour of the divisor function $\tau(n)$ is described asymptotically by*

$$\frac{1}{N}\sum_{n=1}^{N}\tau(n) = \log N + 2C - 1 + O\left(\frac{1}{\sqrt{N}}\right)\ .^{12}$$

(2) In the third chapter we introduced the number theoretic function $r(n)$, which counts the representation of n as the sum of two integral squares. Although here also we are concerned with a typical number theoretic function with unusually jumpy behaviour, asymptotic calculation of the mean

$$\frac{1}{N}\sum_{n=1}^{N}r(n)$$

shows itself to be extraordinarily simple. Thus let

$$\sum_{n=1}^{N} r(n) = \sum_{n=1}^{N} \sum_{x} \sum_{y:x^2+y^2=n} 1$$

$$= \sum_{x} \sum_{y:1\leq x^2+y^2\leq N} 1$$

give the number of points (x,y) distinct from the origin $(0,0)$ with integral coordinates inside the circle centred at $(0,0)$ with radius $\sqrt{N}$. If with each (x,y) with integral coordinates one associates the square $Q(x,y) = [x, x+1[\times[y,y+1[$ and lets K_N denote the disc, i.e. the set of all (ξ,η) with $\xi^2 + \eta^2 \leq N$, then

$$1 + \sum_{n=1}^{N} r(n) = \sum_{(x,y)\in K_N} 1$$

$$= \sum_{(x,y)\in K_N} \iint_{Q(x,y)} d\xi\, d\eta \ .$$

It follows that it is only necessary to estimate the total area of the square $Q(x,y)$ with (x,y) from K_N. Since the diameter of the square is $\sqrt{2}$, all squares under consideration must lie inside the concentric circle with radius $\sqrt{N}+\sqrt{2}$ and also must cover the concentric disc with radius $\sqrt{N}-\sqrt{2}$. Thus

$$\pi\cdot\left(\sqrt{N}-\sqrt{2}\right)^2 \leq 1+\sum_{n=1}^{N} r(n) \leq \pi\cdot\left(\sqrt{N}+\sqrt{2}\right)^2 \ .$$

Hence on the one hand

$$\sum_{n=1}^{N} r(n) \leq \pi N + 2\pi\sqrt{2}\sqrt{N} + 2\pi - 1$$

$$= \pi N + O\left(\sqrt{N}\right) \ ,$$

and on the other

$$\sum_{n=1}^{N} r(n) \geq \pi N - 2\pi\sqrt{2}\sqrt{N} + 2\pi - 1$$

$$= \pi N + O\left(\sqrt{N}\right) \ .$$

This simple calculation was first carried out by Gauss, for which reason the asymptotic averaging of

$$\frac{1}{N}\sum_{n=1}^{N} r(n)$$

is, with hindsight of its geometric significance, called the *Gauss circle problem*. The following theorem ties the result together:

Proposition 9. *The behaviour of the function $r(n)$, which counts the representation of n as the sum of two squares is described asymptotically by*

$$\frac{1}{N} \sum_{n=1}^{N} r(n) = \pi + O\left(\frac{1}{\sqrt{N}}\right) \quad .^{13}$$

(3) For the calculation of the averaged behaviour of the Euler φ-function it is most convenient to use the well-known representation in terms of the Möbius μ-function from elementary number theory,

$$\varphi(n) = \sum_{m|n} \frac{n}{m} \cdot \mu(m) \ .$$

Thus

$$\frac{1}{N} \sum_{n=1}^{N} \varphi(n) = \frac{1}{N} \sum_{n=1}^{N} \sum_{m|n} \frac{n}{m} \cdot \mu(m)$$

$$= \frac{1}{N} \sum_{k} \sum_{m:km=n\leq N} k \cdot \mu(m)$$

$$= \frac{1}{N} \sum_{m=1}^{N} \mu(m) \cdot \sum_{k:km\leq N} k$$

$$= \frac{1}{N} \sum_{m=1}^{N} \mu(m) \cdot \sum_{k\leq N/m} k$$

$$= \frac{1}{N} \sum_{m=1}^{N} \mu(m) \frac{[N/m]([N/m]+1)}{2}$$

$$= \frac{1}{2N} \sum_{m=1}^{N} \mu(m) \left(\frac{N}{m} - \left\{\frac{N}{m}\right\}\right) \left(\frac{N}{m} - \left\{\frac{N}{m}\right\} + 1\right)$$

$$= \frac{N}{2} \sum_{m=1}^{N} \frac{\mu(m)}{m^2} + \frac{1}{2} \sum_{m=1}^{N} \frac{\mu(m)(1 - 2\{N/m\})}{m}$$

$$+ \frac{1}{2N} \sum_{m=1}^{N} \mu(m) \left\{\frac{N}{m}\right\} \left(1 - \left\{\frac{N}{m}\right\}\right) \ .$$

The orders of magnitude of the second and third summand are

$$O\left(\frac{1}{2}\sum_{m=1}^{N}\frac{1}{m}\right) = O(\log N) \ ,$$

$$O\left(\frac{1}{2N}\sum_{m=1}^{N}1\right) = O(1) \ .$$

For the estimation of the first summand one applies the identity

$$\sum_{k=1}^{\infty}\frac{1}{k^s}\sum_{m=1}^{\infty}\frac{\mu(m)}{m^s} = \sum_{k\geq1}\sum_{m\geq1}\frac{\mu(m)}{(km)^s}$$
$$= \sum_{n=1}^{\infty}\sum_{m|n}\frac{\mu(m)}{n^s}$$
$$= \sum_{n=1}^{\infty}\frac{1}{n^s}\left(\sum_{m|n}\mu(m)\right) = 1 \ ,$$

which on the one hand depends on the Möbius inversion formula, and on the other on the interchangeability of series members, given the absolute convergence for $s > 1$. In short one has a special case of a more general allowable procedure. From this it follows that

$$\sum_{m=1}^{\infty}\frac{\mu(m)}{m^s} = \left(\sum_{k=1}^{\infty}\frac{1}{k^s}\right)^{-1} = \frac{1}{\zeta(s)}, \quad s > 1 \ .$$

In particular by Proposition 6

$$\sum_{m=1}^{N}\frac{\mu(m)}{m^2} = \sum_{m=1}^{\infty}\frac{\mu(m)}{m^2} - \sum_{m=N+1}^{\infty}\frac{\mu(m)}{m^2}$$
$$= \frac{1}{\zeta(2)} + O\left(\frac{1}{2-1}\cdot\frac{1}{(N+1)^{2-1}}\right)$$
$$= \frac{6}{\pi^2} + O\left(\frac{1}{N}\right) \ .$$

Proposition 10. *The averaged behaviour of the Euler function $\varphi(n)$ is described asymptotically by*

$$\frac{1}{N}\sum_{n=1}^{N}\varphi(n) = \frac{3}{\pi^2}\cdot N + O(\log N) \ .$$

Since this proposition indicates linear behaviour for the mean of the Euler function, the quotient $\varphi(n)/n$ is also of interest, since in mean it must be asymptotically equal to a constant. In the calculation of

$$\sum_{n=1}^{N} \frac{\varphi(n)}{n}$$

we use the Abel transformation (Proposition 4) together with the expressions $f(n) = 1/n$, $h(n) = \varphi(n)$ and

$$g(n) = \sum_{m=1}^{n} h(m) = \frac{3}{\pi^2} \cdot n^2 + O(n \log n) \ .$$

Thus

$$\sum_{n=1}^{N} \frac{\varphi(n)}{n} = \frac{1}{N} \cdot \frac{3N^2}{\pi^2} + \frac{1}{N} O(N \log N)$$

$$- \int_{1}^{N} \frac{3t^2}{\pi^2} \cdot \frac{-1}{t^2} dt + O\left(\int_{1}^{N} \frac{t \log t}{t^2} dt \right)$$

$$= \frac{3N}{\pi^2} + O(\log N) + \frac{3}{\pi^2} \cdot N + O\left(\int_{t=1}^{t=N} \log t \cdot d(\log t) \right)$$

$$= \frac{6N}{\pi^2} + O(\log^2 N) \ .$$

Proposition 11. *The averaged asymptotic behaviour of the quotient $\varphi(n)/n$ is described by*

$$\frac{1}{N} \sum_{n=1}^{N} \frac{\varphi(n)}{n} = \frac{6}{\pi^2} + O\left(\frac{\log^2 N}{N} \right) \ .$$

(4) A natural number n is called *k-free* (if $k = 2$, *square-free*), if it is divisible by no integral kth power other than 1^k. The number theoretic function μ_k determines whether n is k-free or not, since for k-free n it has the value $\mu_k(n) = 1$ and for the remaining n $\mu_k(n) = 0$. If one writes $n = h^k \cdot l$ with l k-free, then from $m^k \mid n$ it follows necessarily that $m^k \mid h^k$, that is $m \mid h$. In the same way one passes from $m \mid h$ to $m^k \mid n$. Therefore

$$\sum_{m:m^k \mid n} \mu(m) = \sum_{m \mid h} \mu(m)$$

holds. On the right stands the sum function of the Möbius function for h; obviously it only equals 1 for $h = 1$, and is otherwise always zero. This shows that

$$\mu_k(n) = \sum_{m:m^k \mid n} \mu(m)$$

and helps to justify

Proposition 12. *The averaged behaviour of the function $\mu_k(n)$, which determines if n is k-free (if yes then $\mu_k(n) = 1$, if no then $\mu_k(n) = 0$) is described asymptotically by*

$$\frac{1}{N} \sum_{n=1}^{N} \mu_k(n) = \frac{1}{\zeta(k)} + O\left(\frac{\sqrt[k]{N}}{N}\right) .$$

Proof. Use the relations

$$\frac{1}{N} \sum_{n=1}^{N} \mu_k(n) = \frac{1}{N} \sum_{n \leq N} \sum_{m:m^k|n} \mu(m)$$

$$= \frac{1}{N} \sum_{m,l:m^k l \leq N} \mu(m)$$

$$= \frac{1}{N} \sum_{m \leq \sqrt[k]{N}} \mu(m) \sum_{l \leq N/m^k} 1$$

$$= \frac{1}{N} \sum_{m \leq \sqrt[k]{N}} \mu(m) \left[\frac{N}{m^k}\right]$$

$$= \frac{1}{N} \sum_{m \leq \sqrt[k]{N}} \mu(m)\frac{N}{m^k} - \frac{1}{N} \sum_{m \leq \sqrt[k]{N}} \mu(m)\left\{\frac{N}{m^k}\right\}$$

$$= \sum_{m=1}^{\infty} \frac{\mu(m)}{m^k} - \sum_{m > \sqrt[k]{N}} \frac{\mu(m)}{m^k} + \frac{1}{N}O\left(\sum_{m \leq \sqrt[k]{N}} 1\right)$$

$$= \frac{1}{\zeta(k)} + O\left(\sum_{m > \sqrt[k]{N}} \frac{1}{m^k}\right) + O\left(\frac{\sqrt[k]{N}}{N}\right)$$

$$= \frac{1}{\zeta(k)} + O\left(\frac{1}{\sqrt[k]{N^{k-1}}}\right) + O\left(\frac{\sqrt[k]{N}}{N}\right)$$

$$= \frac{1}{\zeta(k)} + O\left(\frac{\sqrt[k]{N}}{N}\right) .$$

If $A(x)$ counts the natural numbers $n \leq x$ which satisfy a certain property A, then one calls

$$\underline{\lim}_{x \to \infty} \frac{A(x)}{x}$$

the *density* of the natural numbers with property A. In the sense of this convention it follows from Proposition 12 that

Corollary 1. *The density of the k-free natural numbers equals $1/\zeta(k)$, in particular the density of the square-free natural numbers equals $6/\pi^2$.*

(5) As a final illustrative example we study the probability that a pair (m,n) of natural numbers is coprime. As a technical tool we use a basic property of the Möbius function.

Lemma 2: Vinogradov's Lemma. *Let S denote a finite set, G a commutative group (written additively), f and g two functions defined on S with values in $\mathbb{N}$ (respectively G). Then the formula*

$$\sum_{s\in S: f(s)=1} g(s) = \sum_{m=1}^{\infty} \mu(m) \cdot \sum_{s\in S: m|f(s)} g(s)$$

holds.

Proof. Since the sum function of the Möbius function $\sum_{m|f(s)} \mu(m)$ only takes the value 1 for $f(s) = 1$, otherwise always taking the value 0, we must have

$$\sum_{s\in S: f(s)=1} g(s) = \sum_{s\in S} g(s) \cdot \sum_{m|f(s)} \mu(m)$$

$$= \sum_{m=1}^{\infty} \mu(m) \cdot \sum_{s\in S: m|f(s)} g(s) \ ,$$

(where the final infinite sequence is actually terminating). $\square$

As an application of the Vinogradov lemma put $G = \mathbb{C}$, S equal to the set of all natural divisors of n and $f(s) = n/s$. The formula

$$g(n) = \sum_{m=1}^{\infty} \mu(m) \cdot \sum_{s: m|n/s} g(s)$$

$$= \sum_{m|n} \mu(m) \sum_{s|n/m} g(s)$$

obtained in this way proves the Möbius inversion formula.

A second application concerns the exercise mentioned at the beginning, which we can state more generally: how large is the number $A_k(N)$ of k-tuples $(n_1,\ldots,n_k)$ of natural numbers $n_j \leq N$, which are relatively prime to each other, that is have highest common factor $g.c.d.(n_1,\ldots,n_k) = 1$? If in the Vinogradov lemma one sets S equal to the set of k-tuples under consideration, takes $g(s) = g(n_1,\ldots,n_k) = 1$ for all $s = (n_1,\ldots,n_k)$ from S, $G = \mathbb{C}$, and $f(s) = f(n_1,\ldots,n_k) = g.c.d(n_1,\ldots,n_k)$, then Lemma 2 implies that

$$A_k(N) = \sum_{m=1}^{\infty} \mu(m) \cdot \sum_{(n_1,\ldots,n_k)\in S:\, m|n_1,\ldots,m|n_k} 1$$

$$= \sum_{m=1}^{\infty} \mu(m) \cdot \sum_{l_1,\ldots,l_k:\, l_1 \le N/m,\ldots,l_k \le N/m} 1$$

$$= \sum_{m=1}^{\infty} \mu(m) \left[\frac{N}{m}\right]^k$$

$$= \sum_{m=1}^{\infty} \mu(m)\frac{N^k}{m^k} + O\left(\sum_{m=1}^{N} |\mu(m)| \left(\frac{N}{m}\right)^{k-1}\right)$$

$$= \frac{N^k}{\zeta(k)} + O\left(N^{k-1} \sum_{m=1}^{N} m^{1-k}\right) \ .$$

If we distinguish between the cases $k = 2$ and $k > 2$ then we obtain the following answer

Proposition 13. *The number $A_k(N)$ of k-tuples $(n_1,\ldots,n_k)$ of natural numbers $n_j \le N$, which are relatively prime to each other, for $k > 2$ satisfy*

$$A_k(N) = N^k/\zeta(k) + O(N^{k-1})$$

and for $k = 2$

$$A_2(N) = 6N^2/\pi^2 + O(N \log N) \ .$$

In particular the probability that an arbitrarily chosen fraction $a = n/m$ cannot be reduced tends to $6/\pi^2$.

Third Tool of the Trade: Dirichlet Series

The formula

$$\sum_{n=1}^{\infty} \frac{\mu(n)}{n^s} = \frac{1}{\zeta(s)} = \left(\sum_{n=1}^{\infty} \frac{1}{n^s}\right)^{-1}$$

introduced in connection with the third application for $s > 1$ is only an example of the general concept of so-called Dirichlet series: if f is a number theoretic function we call

$$\sum_{n=1}^{\infty} \frac{f(n)}{n^s} = F(s)$$

the *Dirichlet series* formed from this function. In order to understand the importance of these functions for number theory, let us first put on one side the study of their convergence. Neglecting convergence suppose first that Dirichlet series are considered formally as finite sums – if we multiply $F(s)$ by the series

$$\sum_{n=1}^{\infty} \frac{g(n)}{n^s} = G(s) \ ,$$

with the same exponentiated power in the denominator one obtains

$$F(s) \cdot G(s) = \sum_{m=1}^{\infty} \frac{f(m)}{m^s} \sum_{k=1}^{\infty} \frac{g(k)}{k^s}$$

$$= \sum_{m=1}^{\infty} \sum_{k=1}^{\infty} \frac{f(m)g(k)}{(mk)^s}$$

$$= \sum_{n=1}^{\infty} \frac{1}{n^s} \sum_{m,k:mk=n} f(m)g(k)$$

$$= \sum_{n=1}^{\infty} \frac{1}{n^s} \sum_{m|n} f(m)g\left(\frac{n}{m}\right) \ .$$

Thus the result is again a Dirichlet series forming the convolution $f * g$ of the two number theoretic functions f and g. We define

$$f * g(n) = \sum_{m|n} f(m)g\left(\frac{n}{m}\right) \ .$$

In particular if $G(s) = \zeta(s)$, i.e. $g(n) = 1$ for all n, one obtains

$$F(s)\zeta(s) = \sum_{n=1}^{\infty} \frac{1}{n^s} \sum_{m|n} f(m) \ ,$$

and hence this Dirichlet series is generated by the sum function of f. Thus the formula

$$\sum_{n=1}^{\infty} \frac{\mu(m)}{n^s} = \frac{1}{\zeta(s)}$$

only mirrors

$$\sum_{m|n} \mu(m) \cdot \sum_{k|n/m} f(k) = f(n) \ ,$$

i.e. once more the Möbius inversion formula.

Simple examples show the connection between Dirichlet series and number theoretic functions. Because

$$\tau(n) = \sum_{m|n} 1$$

holds, we also have

$$\sum_{n=1}^{\infty} \frac{\tau(n)}{n^s} = \zeta(s)^2 \ .$$

For

$$\sigma(n) = \sum_{m|n} m$$

the representation

$$\sum_{n=1}^{\infty} \frac{\sigma(n)}{n^s} = \zeta(s)\zeta(s-1)$$

follows from

$$\zeta(s)\zeta(s-1) = \sum_{k=1}^{\infty} \frac{1}{k^s} \sum_{m=1}^{\infty} \frac{m}{m^s}$$

$$= \sum_{n=1}^{\infty} \frac{1}{n^s} \sum_{m|n} m \ .$$

Since it is clear that

$$\sum_{m|n} \varphi(m) = n$$

and the associated Dirichlet series agree with

$$\sum_{n=1}^{\infty} \frac{n}{n^s} = \zeta(s-1)$$

it follows for the Euler φ-function from the Möbius inversion formula that

$$\sum_{n=1}^{\infty} \frac{\varphi(n)}{n^s} = \frac{\zeta(s-1)}{\zeta(s)} \ .$$

For prime powers $n = p^\nu$ let the *Mangoldt function* Λ be given by $\Lambda(n) = \Lambda(p^\nu) = \log p$. On the other hand if n is not a pure prime power, set $\Lambda(n) = 0$. Since in the sum function

$$\sum_{m|n} \Lambda(m) \ ,$$

for each prime divisor p $\log p$ is counted according to the multiplicity $\nu(p,n)$ arising in the prime factorisation of n, and the sum of the logarithms can be expressed as the logarithm of the product, the sum function above reads

$$\sum_{m|n} \Lambda(m) = \log n \ .$$

For the formal differentiation of the Riemann ζ-function and the Möbius inversion formula it follows that

$$-\zeta'(s) = \sum_{n=1}^{\infty} \frac{\log n}{n^s}$$

$$= \sum_{n=1}^{\infty} \frac{1}{n^s} \sum_{m|n} \Lambda(m)$$

$$= \zeta(s) \cdot \sum_{n=1}^{\infty} \frac{\Lambda(n)}{n^s} \ ,$$

which implies

$$\sum_{n=1}^{\infty} \frac{\Lambda(n)}{n^s} = -\frac{\zeta'(s)}{\zeta(s)} \ .$$

For multiplicative number theoretic functions f the Dirichlet series can be transformed into product form, since, starting from the unique product decomposition

$$n = p_1^{\nu_1} p_2^{\nu_2} \dots p_N^{\nu_N} \dots$$

of each natural number as a product of prime powers, we can carry through the formal manipulation

$$F(s) = \sum_{n=1}^{\infty} \frac{f(n)}{n^s}$$

$$= \sum_{\nu_1 \geq 0} \sum_{\nu_2 \geq 0} \cdots \sum_{\nu_N \geq 0} \cdots \frac{f\left(p_1^{\nu_1}\right) f\left(p_2^{\nu_2}\right) \dots f\left(p_N^{\nu_N}\right) \dots}{p_1^{\nu_1 s} p_2^{\nu_2 s} \dots p_N^{\nu_N s} \dots}$$

$$= \prod_p \sum_{\nu=0}^{\infty} \frac{f(p^\nu)}{p^{\nu s}} \ ,$$

where the product is taken over all prime numbers p. Furthermore, if f is strongly multiplicative – i.e. in each case $f(mn) = f(m)f(n)$ holds – this shows, taking into consideration the geometric series

$$\sum_{\nu=0}^{\infty} \frac{f(p)^\nu}{p^{\nu s}} = \frac{1}{1 - \dfrac{f(p)}{p^s}}$$

that

$$F(s) = \sum_{n=1}^{\infty} \frac{f(n)}{n^s} = \prod_p \frac{1}{1 - \dfrac{f(p)}{p^s}} \ .$$

A last example shows how this formula serves to represent number theoretic functions.

The *Liouville function* λ is defined for prime powers by $\lambda(p^\nu) = (-1)^\nu$, and is extended to all natural numbers so as to be multiplicative. Hence $\lambda(n)$ takes the value $+1$ or -1, depending on whether n possesses an even or odd number of prime factors, due regard being paid to multiplicities. Since λ is even strongly multiplicative, it follows that

$$\sum_{n=1}^{\infty} \frac{\lambda(n)}{n^s} = \prod_p \frac{1}{1 + \dfrac{1}{p^s}}$$

$$= \prod_p \frac{1 - \dfrac{1}{p^s}}{1 - \dfrac{1}{p^{2s}}} = \frac{\zeta(2s)}{\zeta(s)} \ .$$

In the formula

$$\zeta(2s) = \sum_{m=1}^{\infty} \frac{1}{m^{2s}} = \sum_{n=1}^{\infty} \frac{q(n)}{n^s}$$

we introduce the number theoretic function $q(n)$, which for a natural square $n = m^2$ takes the value $q(n) = q(m^2) = 1$, and otherwise is equal to zero. It follows

$$\sum_{n=1}^{\infty} \frac{\lambda(n)}{n^s} = \frac{\zeta(2s)}{\zeta(s)} = \sum_{n=1}^{\infty} \frac{1}{n^s} \sum_{m|n} q(m)\mu\left(\frac{n}{m}\right)$$

$$= \sum_{n=1}^{\infty} \frac{1}{n^s} \sum_{m^2|n} \mu\left(\frac{n}{m^2}\right) ,$$

which gives

$$\lambda(n) = \sum_{m^2|n} \mu\left(\frac{n}{m^2}\right) .$$

These formal manipulations can be put on a sound foundation, since calculations with Dirichlet series, such as differentiation, transformation to infinite products, multiplication with arbitrary rearrangement of the component sub-products, etc. can be considered as formal operations taking place inside the framework of formal Dirichlet series. For the analyst the setting up of the framework is uninteresting – he justifies all the operations arising here by the demonstration of absolute (resp. uniform) convergence for the series (resp. products) which occur. This is achieved by means of

Proposition 14. *If*

$$F(s) = \sum_{n=1}^{\infty} \frac{f(n)}{n^s}$$

is the Dirichlet series formed from the number theoretic function f, and if $F(s)$ converges for the complex number $s = s_0$, then it converges uniformly for all complex s in the corner region $-\alpha \leq \arg(s - s_0) \leq \alpha$ for an arbitrary constant $\alpha < \pi/2$.

The fact that s is considered as a complex variable, $s = \sigma + it$, should be particularly clearly emphasised; it allows us to use complex variable techniques for Dirichlet series. The theorem visibly asserts that the uniform convergence of Dirichlet series is guaranteed in a region of the form drawn in Fig. 14. In particular the function $F(s)$ converges for $\mathrm{Re}(s) > \mathrm{Re}(s_0)$.

Proof. Passing from f to the number theoretic function $f(n)/n^{s_0}$ one recognises that without loss of generality one can take $s_0 = 0$, i.e. one can assume the convergence of

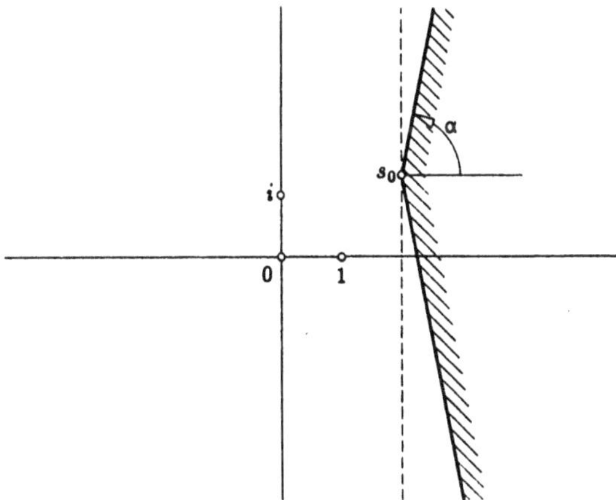

Fig. 14. Region of convergence for Dirichlet series

$$F(0) = \sum_{n=1}^{\infty} f(n) \; .$$

This implies that for each arbitrarily preassigned $\varepsilon > 0$, one can find some natural number K, such that for all N, M with $N \geq M \geq K$,

$$\left| \sum_{n=M}^{N} f(n) \right| < \varepsilon \cos \alpha$$

holds. Thus for s with $|\arg s| \leq \alpha$, $|\cos \arg s| = \mathrm{Re}(s)/|s| \geq \cos \alpha$, because of

$$\left| \frac{1}{n^s} - \frac{1}{(n+1)^s} \right| = \left| \int_n^{n+1} t^{-s-1} dt \right| |s|$$

$$\leq |s| \cdot \int_n^{n+1} t^{-\mathrm{Re}(s)-1} dt$$

$$= \frac{|s|}{\mathrm{Re}(s)} \left(n^{-\mathrm{Re}(s)} - (n+1)^{-\mathrm{Re}(s)} \right)$$

$$\leq \frac{1}{\cos \alpha} \left(n^{-\mathrm{Re}(s)} - (n+1)^{-\mathrm{Re}(s)} \right) \; ,$$

one can apply the Abel transformation, obtaining

$$\left| \sum_{n=M}^{N} f(n)n^{-s} \right| = \left| (N+1)^{-s} \sum_{n=M}^{N} f(n) \right.$$

$$+ \left. \sum_{n=M}^{N} \left(n^{-s} - (n+1)^{-s}\right) \sum_{m=M}^{n} f(m) \right|$$

$$\leq (N+1)^{-\mathrm{Re}(s)} \left| \sum_{n=M}^{N} f(n) \right|$$

$$+ \sum_{n=M}^{N} \frac{1}{\cos \alpha} \left(n^{-\mathrm{Re}(s)} - (n+1)^{-\mathrm{Re}(s)} \right) \left| \sum_{m=M}^{n} f(m) \right|$$

$$\leq (N+1)^{-\mathrm{Re}(s)} \cdot \varepsilon \cdot \cos \alpha$$

$$+ \left(M^{-\mathrm{Re}(s)} - (N+1)^{-\mathrm{Re}(s)} \right) \cdot \varepsilon$$

$$\leq M^{-\mathrm{Re}(s)} \cdot \varepsilon < \varepsilon \ .$$

The assertion in the proposition now follows.   □

The infimum σ_0 of the real parts of all s for which $F(s)$ converges is called the *convergence abscissa* of the Dirichlet series, since for $\mathrm{Re}(s) > \sigma_0$ the Dirichlet series converges, and in each compact subregion of the half-plane $\mathrm{Re}(s) > \sigma_0$ is even uniformly convergent. Hence by the complex variable convergence theorem of Weierstrass it follows that the Dirichlet series

$$F(s) = \sum_{n=1}^{\infty} \frac{f(n)}{n^s}$$

is holomorphic and may be differentiated term by term,

$$F'(s) = \sum_{n=1}^{\infty} \frac{-f(n)\log n}{n^s} \ .$$

The convergence abscissa σ_0' of the Dirichlet series

$$\sum_{n=1}^{\infty} \frac{|f(n)|}{n^s}$$

is called the *abscissa of absolute convergence* of $F(s)$. The example of the Riemann ζ-function with convergence abscissa $\sigma_0' = 1$ and of the Dirichlet series

$$\sum_{n=1}^{\infty} \frac{(-1)^{n-1}}{n^s}$$

with convergence abscissa $\sigma_0' = 0$ (by the Leibniz convergence test for alternating series) shows that the abscissa of absolute convergence, which for the series given below coincides with that of the Riemann ζ-function, may in general be greater than the convergence abscissa itself. However the difference is at most 1: indeed if σ_0 (resp. σ_0') denotes the convergence abscissa (resp. the abscissa of absolute convergence) of

$$\sum_{n=1}^{\infty} \frac{f(n)}{n^s} \, ,$$

then, because of the boundedness of the terms for $\sigma > \sigma_0$, there must exist some K with

$$\left| f(n) \cdot n^{-\sigma} \right| < K$$

for all n. Hence for an arbitrary positive ε

$$\sum_{n=1}^{\infty} \left| \frac{f(n)}{n^{\sigma+1+\varepsilon}} \right| \leq K \cdot \sum_{n=1}^{\infty} \frac{1}{n^{1+\varepsilon}} < \infty \, ,$$

so that it is certainly true that $\sigma_0' \leq \sigma + 1 + \varepsilon$. Because of the free choice of $\sigma > \sigma_0$, $\varepsilon > 0$, the inequality $\sigma_0' \leq \sigma_0 + 1$ has to hold, as asserted.

Proposition 15. *Let the Dirichlet series*

$$F(s) = \sum_{n=1}^{\infty} \frac{f(n)}{n^s}, \quad G(s) = \sum_{n=1}^{\infty} \frac{g(n)}{n^s}$$

be convergent for $\mathrm{Re}(s) > \sigma_0$ *and be absolutely convergent for* $\mathrm{Re}(s) > \sigma_0'$. *If for all s with* $\mathrm{Re}(s) > \sigma_0$ *the equation $F(s) = G(s)$ holds, then the same also holds for the underlying number theoretic functions, i.e. for all natural numbers n $f(n) = g(n)$. For all s with* $\mathrm{Re}(s) > \sigma_0'$ *the product may be expressed as*

$$F(s) \cdot G(s) = \sum_{n=1}^{\infty} \frac{f * g(n)}{n^s}$$

with

$$f * g(n) = \sum_{m,k:mk=n} f(m)g(k) = \sum_{m|n} f(m)g\left(\frac{n}{m}\right) \, .$$

If f is a multiplicative function, then for all s with $\mathrm{Re}(s) > \sigma_0'$ *there exists the product representation*

$$F(s) = \prod_{p} \sum_{\nu \geq 0} \frac{f(p^\nu)}{p^{\nu s}} \, ,$$

which for a strongly multiplicative function simplifies to

$$F(s) = \prod_{p} \frac{1}{1 - \dfrac{f(p)}{p^s}} \, .$$

(As always the index p runs through all prime numbers.)

Proof. For the verification of the first statement it must follow from

$$H(s) = F(s) - G(s) = \sum_{n=1}^{\infty} \frac{f(n) - g(n)}{n^s} = \sum_{n=1}^{\infty} \frac{h(n)}{n^s} = 0$$

that $h(n) = 0$ for all n. If N were the smallest natural number with $h(N) \neq 0$, then from the uniform convergence of

$$\sum_{n=N}^{\infty} \frac{h(n) \cdot N^s}{n^s}$$

for $s \geq \sigma_0 + 1$, it would follow that for all $\varepsilon > 0$ we could find some M, depending only on ε, with

$$\left| \sum_{n=M+1}^{\infty} \frac{h(n) \cdot N^s}{n^s} \right| < \frac{\varepsilon}{2} \ .$$

Since for a sufficiently large s one can also suppose that

$$\left| \sum_{n=N+1}^{M} h(n) \cdot \left(\frac{N}{n} \right)^s \right| < \frac{\varepsilon}{2} \ ,$$

one concludes that

$$|h(N)| \leq \left| \sum_{n=N+1}^{M} h(n) \cdot \left(\frac{N}{n} \right)^s \right| + \left| \sum_{n=M+1}^{\infty} \frac{h(n) \cdot N^s}{n^s} \right| < \varepsilon \ ,$$

which is arbitrarily small. Thus in spite of everything $h(N) = 0$.

The second statement of the proposition is clear from the known rearrangement theorem, given the assumption of absolute convergence. The simplest way to derive the third is the following: on the one hand

$$\sum_{\nu \geq 0} \frac{f(p^\nu)}{p^{\nu s}}$$

is absolutely convergent for all prime numbers p and for $\mathrm{Re}(s) > \sigma_0'$. If on the other hand $\mathbb{N}_k$ denotes the set of all natural numbers, which only have prime factors $p \leq k$, then $\lim_{k \to \infty} \mathbb{N}_k$ agrees with the union of all the $\mathbb{N}_k$, i.e. with all of $\mathbb{N}$. The conclusion follows from

$$\prod_{p \leq k} \sum_{\nu \geq 0} \frac{f(p^\nu)}{p^{\nu s}} = \sum_{n \in \mathbb{N}_k} \frac{f(n)}{n^s}$$

by taking the limit as $k \to \infty$. $\qquad\qquad\qquad\qquad\qquad\qquad\qquad\square$

With this the formal operations in connection with the introduction of Dirichlet series are analytically justified. In particular we have

Proposition 16. *The Möbius function $\mu(n)$, which is only distinct from zero for square-free n and in this case takes the value ± 1 depending on whether n*

has an even or odd number of distinct prime factors, and the number theoretic functions

$$\tau(n) = \sum_{m|n} 1 \; ,$$

which counts the divisors,

$$\sigma(n) = \sum_{m|n} m \; ,$$

which adds the divisors,

$$\varphi(n) = \sum_{m \le n: g.c.d.(m,n)=1} 1 = \sum_{m|n} \mu(m) \cdot \frac{n}{m} \; ,$$

which counts the prime residue classes,

$$\lambda(n) = \sum_{m^2|n} \mu\left(\frac{n}{m^2}\right) \; ,$$

which determines by ± 1 whether the number of prime factors of n (due regard being paid to their multiplicities) is even or odd,

$$\Lambda(n) = \sum_{m|n} \mu(m) \log\left(\frac{n}{m}\right) \; ,$$

which for prime powers $p^\nu = n$ takes the value $\log p$ and is otherwise zero, give rise to the Dirichlet series

$$\sum_{n=1}^{\infty} \frac{\mu(n)}{n^s} = \frac{1}{\zeta(s)} \; , \quad \mathrm{Re}(s) > 1 \; ,$$

(in particular this shows that $\zeta(s)$ is never zero for $\mathrm{Re}(s) > 1$),

$$\sum_{n=1}^{\infty} \frac{\tau(n)}{n^s} = \zeta(s)^2 \; , \quad \mathrm{Re}(s) > 1 \; ,$$

$$\sum_{n=1}^{\infty} \frac{\sigma(n)}{n^s} = \zeta(s)\zeta(s-1) \; , \quad \mathrm{Re}(s) > 2 \; ,$$

$$\sum_{n=1}^{\infty} \frac{\varphi(n)}{n^s} = \frac{\zeta(s-1)}{\zeta(s)} \; , \quad \mathrm{Re}(s) > 2 \; ,$$

$$\sum_{n=1}^{\infty} \frac{\lambda(n)}{n^s} = \frac{\zeta(2s)}{\zeta(s)} \; , \quad \mathrm{Re}(s) > 1 \; ,$$

$$\sum_{n=1}^{\infty} \frac{\Lambda(n)}{n^s} = -\frac{\zeta'(s)}{\zeta(s)} \; , \quad \mathrm{Re}(s) > 1 \; .$$

The holomorphic nature of the Dirichlet series to the right of the convergence abscissa raises the question whether holomorphic functions, which can be represented to the right of the convergence abscissa by Dirichlet series, admit holomorphic extensions to larger regions of the complex plane. Already the relation

$$(1 - 2^{1-s}) \cdot \zeta(s)$$

$$= 1 + \frac{1}{2^s} + \frac{1}{3^s} + \ldots - 2 \cdot \left(\frac{1}{2^s} + \frac{1}{4^s} + \frac{1}{6^s} + \ldots \right)$$

$$= \sum_{n=1}^{\infty} \frac{(-1)^{n-1}}{n^s} \,,$$

which holds for $\mathrm{Re}(s) > 1$, while the right hand side involves a holomorphic Dirichlet series for $\mathrm{Re}(s) > 0$, raises the possibility of extending $\zeta(s)$ for $\mathrm{Re}(s) > 0$. We make this more rigorous in the following manner: on the one hand one has the integral representation

$$\zeta(s) = \sum_{n=1}^{\infty} \frac{n}{n^s} - \sum_{n=1}^{\infty} \frac{n-1}{n^s}$$

$$= \sum_{n=1}^{\infty} \frac{n}{n^s} - \sum_{n=1}^{\infty} \frac{n}{(n+1)^s}$$

$$= \sum_{n=1}^{\infty} n(n^{-s} - (n+1)^{-s} \,,$$

$$= \sum_{n=1}^{\infty} ns \int_{n}^{n+1} x^{-s-1} dx$$

$$= s \cdot \sum_{n=1}^{\infty} \int_{n}^{n+1} [x] x^{-s-1} dx$$

$$= s \cdot \int_{1}^{\infty} [x] x^{-s-1} dx \,, \quad \mathrm{Re}(s) > 1 \,.$$

On the other hand there is the integral closely related to the one above

$$s \cdot \int_{1}^{\infty} x \cdot x^{-s-1} dx = \frac{s}{s-1} = 1 + \frac{1}{s-1} \,.$$

On the right hand side the difference

$$\zeta(s) - \frac{1}{s-1} = 1 + s \cdot \int_{1}^{\infty} ([x] - x) x^{-s-1} dx$$

$$= 1 - s \cdot \int_{1}^{\infty} \frac{\{x\}}{x^{s+1}} dx$$

describes a function holomorphic for $\mathrm{Re}(s) > 0$, so that analytic continuation

$$\zeta(s) = \frac{1}{s-1} + 1 - s \cdot \int_{1}^{\infty} \frac{\{x\}}{x^{s+1}} dx$$

yields a function holomorphic for all s with $\mathrm{Re}(s) > 0$ away from the pole at $s = 1$.

Proposition 17. *Because of the integral formula*

$$\zeta(s) = s \cdot \int_1^\infty [x] x^{-s-1} dx$$

and its consequence

$$\zeta(s) - \frac{1}{s-1} = 1 - s \cdot \int_1^\infty \frac{\{x\}}{x^{s+1}} dx$$

the Riemann ζ-function

$$\zeta(s) = \sum_{n=1}^\infty \frac{1}{n^s}, \quad \mathrm{Re}(s) > 1 ,$$

can be analytically extended as a holomorphic function in $\mathrm{Re}(s) > 0$ *with the exception of the simple pole at* $s = 1$ *with residue 1.* [14]

The fact that the analytic continuation of the Riemann zeta function cannot be achieved without the price of a singularity on the convergence abscissa of the Dirichlet series is based on a general theorem, which holds for all Dirichlet series formed from positive number theoretic functions.

Proposition 18: Landau's Theorem. *If for all natural numbers n the number theoretic function $f(n)$ takes only non-negative real values, and if the Dirichlet series*

$$F(s) = \sum_{n=1}^\infty \frac{f(n)}{n^s}$$

formed from it has the finite convergence abscissa σ_0, then the function $F(s)$ holomorphic in $\mathrm{Re}(s) > \sigma_0$ *cannot be analytically continued past* $\mathrm{Re}(s) = \sigma_0$ *to a region enclosing the point $s = \sigma_0$.*

Proof. For $\sigma = \sigma_0 + 1$ because $F(s)$ is holomorphic one has

$$F(s) = \sum_{k=0}^\infty \frac{F^{(k)}(\sigma)}{k!}(s - \sigma)^k$$

$$= \sum_{k=0}^\infty \frac{(-1)^k}{k!} \sum_{n=1}^\infty \frac{f(n)\log^k n}{n^\sigma}(s - \sigma)^k$$

$$= \sum_{k=0}^\infty \sum_{n=1}^\infty \frac{(\sigma - s)^k}{k!} \cdot \frac{f(n)\log^k n}{n^\sigma} .$$

If $F(s)$ were analytically continuable past $\mathrm{Re}(s) = \sigma_0$, the radius of convergence of the power series would have to exceed $\sigma - \sigma_0 = 1$. Hence the series would have to converge for some real $s < \sigma_0$, and because of the positivity of all summands this convergence would be absolute. Because of the permitted rearrangement of the members

$$F(s) = \sum_{n=1}^{\infty} \frac{f(n)}{n^{\sigma}} \sum_{k=0}^{\infty} \frac{\left((\sigma - s)\log n\right)^{k}}{k!}$$

$$= \sum_{n=1}^{\infty} \frac{f(n)}{n^{\sigma}} \cdot e^{(\sigma-s)\log n}$$

$$= \sum_{n=1}^{\infty} \frac{f(n)}{n^{s}}$$

and the convergence of the last series, it would follow that σ_0 would not be the convergence abscissa of $F(s)$.

Abel transformation shows that not only the Riemann ζ-function, but also other Dirichlet series possess integral representations. From

$$\sum_{n \leq x} \frac{f(n)}{n^{s}} = \frac{1}{x^{s}} \cdot g(x) - \int_{1}^{x} g(t) \cdot \frac{-s}{t^{s+1}} \, dt$$

with

$$g(x) = \sum_{n \leq x} f(n)$$

it follows that for a sufficiently large real part of s (chosen to ensure that $g(x)/x^{s}$ tends to zero as $x \to \infty$ and the convergence of the integral)

$$\sum_{n=1}^{\infty} \frac{f(n)}{n^{s}} = s \cdot \int_{1}^{\infty} g(x) x^{-s-1} dx$$

with

$$g(x) = \sum_{n \leq x} f(n) \ .$$

For example, if the ψ-function is defined by

$$\psi(x) = \sum_{n \leq x} \Lambda(n) = \sum\sum_{p,\nu \geq 1 : p^{\nu} \leq x} \log p$$

and one considers the trivial inequality $\psi(x) \leq x \log x$, then a consequence of representation as an integral as above is

Proposition 19. *For all s with* $\mathrm{Re}(s) > 1$ *one has the integral representation*

$$-\frac{\zeta'(s)}{\zeta(s)} = s \cdot \int_{1}^{\infty} \psi(x) x^{-s-1} dx \ ,$$

where $\psi(x)$ adds the values of $\Lambda(n)$ for all $n \leq x$.

Exercises on Chapter 4

1. Let Z denote the set of all complex-valued number theoretic functions. For $f, g \in Z$ and $n \in \mathbb{N}$ put $(f + g)(n) = f(n) + g(n)$ and $f * g(n) = \sum_{d|n} f(d) g\left(\frac{n}{d}\right)$. Show that $(Z, +, *)$ is an integral domain.

2. (i)   Show that the unitgroup Z^* of Z equals $\{f \in Z : f(1) \neq 0\}$.
   (ii)   For each prime number p let

$$g_p(n) = \begin{cases} 0 & n \neq p \\ 1 & n = p \end{cases} .$$

   Show that g_p is a prime element in Z.

* 3. Let $\mathfrak{K} \subseteq \mathbb{R}^n$ be measurable, and suppose either that $|\mathfrak{K} \cap \mathbb{Z}^n| < \infty$ or $\lambda(\mathfrak{K}) < \infty$. For $\mathfrak{g} \in \mathbb{Z}^n$ define $\mathfrak{Q}_{\mathfrak{g}} := \{\mathfrak{x} \in \mathbb{R}^n : 0 \leq x_i - g_i < 1 \text{ for } 1 \leq i \leq n\}$. Show that

$$\left| |\mathfrak{K} \cap \mathbb{Z}^n| - \lambda(\mathfrak{K}) \right| \leq \sum_{\mathfrak{g} \in \mathbb{Z}^n, \, \mathfrak{Q}_{\mathfrak{g}} \cap \partial \mathfrak{K} \neq \emptyset} 1 .$$

4. For $n \in \mathbb{N}$ let V_n be the volume of the unit ball in $\mathbb{R}^n$. Show that as $R \to \infty$

$$\sum_{\substack{\mathfrak{g} \in \mathbb{Z}^n \\ g_1^2 + \cdots + g_n^2 \leq R^2}} 1 = R^n V_n + O(R^{n-1}) .$$

5. For $n \in \mathbb{N}$ let

$$\varphi_n(x) = \prod_{\substack{k=0 \\ gcd(k,n)=1}}^{n-1} \left(x - e^{2\pi i k/n}\right)$$

be the nth *cyclotomic polynomial*. Show that $\varphi_n(x) = \prod_{d|n} \left(x^{n/d} - 1\right)^{\mu(d)} \in \mathbb{Z}[x]$.

6. Show that as $n \to \infty$ $\sum_{k=n}^{\infty} \frac{\varphi(k)}{k^3} = \frac{6}{\pi^2 n} + O\left(\frac{\log n}{n^2}\right)$.

7. For $n \geq 1$ let

$$A_n = \sum_{\substack{k=1 \\ gcd(k,n)=1}}^{n} \frac{1}{k+n} .$$

Show that $A_n = \frac{1}{n} \sum_{d|n} \sum_{k=1}^{d} \frac{d}{k+d} \mu\left(\frac{n}{d}\right)$.

8. Show that as $n \to \infty$ $A_n = \frac{\varphi(n)}{n} \log 2 + O\left(\frac{\tau(n)}{n}\right)$.

* 9. For $n \geq 1$ write

$$S_n = \sum_{\substack{k=1 \\ k+\ell>n, \, gcd(k,\ell)=1}}^{n} \sum_{\ell=1}^{n} \frac{1}{k\ell(k+\ell)} .$$

Show that $S_n = 2 \sum_{k=n+1}^{\infty} \frac{A_k}{k^2}$.

10. Show that as $n \to \infty$ $S_n = \frac{12 \log 2}{\pi^2 n} + O\left(\frac{\log n}{n^2}\right)$. (Lehner and Newman, 1968).

11. Show that for $\mathrm{Re}(s) = 1$ $\sum_{n=1}^{\infty} n^{-s}$ diverges.

12. Show that the Dirichlet series $\sum_{n=1}^{\infty} (-1)^n n^{-s}$ has infinitely many zeros on the line $\mathrm{Re}(s) = 1$.

13. (i)   The power series $\sum_{n=1}^{\infty} \frac{z^n}{n} = -\log(1 - z)$ converges compactly in $\{z \in \mathbb{C} : |z| \le 1, z \ne 1\}$.
   (ii)   For all $x \in \mathbb{R}$, $x \notin \mathbb{Z}$, $\sum_{n=1}^{\infty} \frac{\sin 2\pi n x}{n} = -\pi(\{x\} - 1/2)$.
   (iii) For all $N \in \mathbb{N}$ and all $x \in \mathbb{R}$, $\left|\sum_{n=1}^{N} \frac{\sin 2\pi n x}{n}\right| \le 2\pi$.

14. Show:
   (i)   For all $x \in \mathbb{C}$ the power series $\sum_{n=0}^{\infty} \frac{B_n(x)}{n!} z^n = \frac{z e^{xz}}{e^z - 1}$ converges for $|z| < 2\pi$.
   (ii)   For $n \ge 0$, $B_n(x)$ is a polynomial, $B_n(x) = \sum_{k=0}^{n} \binom{n}{k} B_k(0) x^{n-k}$. $B_n(x)$ is called the nth *Bernoulli polynomial* and $B_n := B_n(0)$ the nth *Bernoulli number*.
   (iii) For n odd, $n > 1$, $B_n = 0$.
   (iv)   For $n \ge 0$, $n \ne 1$, $B_n(1) = B_n$. We have $B_0(x) = 1$, $B_1(x) = x - \frac{1}{2}$.
   (v)   For $n > 1$, $\sum_{k=0}^{n-1} \binom{n}{k} B_k = 0$.

15. (i)   For $n \ge 0$, $B_n \in \mathbb{Q}$.
   (ii)   For $n \ge 1$, $B_n'(x) = n B_{n-1}(x)$.

16. Show that for $m \in \mathbb{N}$ and all $a, b \in \mathbb{R}$

$$\int_a^b B_m(\{x\})\,dx = \frac{1}{m+1}\left(B_{m+1}(\{b\}) - B_{m+1}(\{a\})\right)$$

and from this deduce that, for $m \ge 2$, $x \mapsto B_m(\{x\})$ is $(m-2)$fold continuously differentiable.

17. Let $m \ge 2$, or $m = 1$ and $x \notin \mathbb{Z}$. Show:
   (i) $B_m(\{x\}) = -\frac{m!}{(2\pi i)^m} \sum_{\substack{k \in \mathbb{Z} \\ k \ne 0}} k^{-m} e^{2\pi i k x}$.

   (ii)   For m even $B_m(\{x\}) = \frac{2 m! (-1)^{\frac{m}{2}+1}}{(2\pi)^m} \cdot \sum_{k=1}^{\infty} \frac{\cos 2\pi k x}{k^m}$.

   (iii) For m odd $B_m(\{x\}) = \frac{2 m! (-1)^{(m+1)/2}}{(2\pi)^m} \cdot \sum_{k=1}^{\infty} \frac{\sin 2\pi k x}{k^m}$.

   (iv) $\zeta(2m) = (-1)^{m+1} \frac{(2\pi)^{2m}}{2(2m)!} B_{2m}$.

18. Let $f : [a, b] \to \mathbb{C}$ be q times differentiable, $\int_a^b \left|f^{(q)}(x)\right| dx < \infty$. Show that for $1 \le m \le q$:

$$\sum_{a < n \le b} f(n) = \int_a^b f(x)\,dx$$

$$+ \sum_{k=1}^{m} \frac{(-1)^k}{k!} \left(B_k(\{b\}) f^{(k-1)}(b) - B_k(\{a\}) f^{(k-1)}(a)\right)$$

$$+ \frac{(-1)^{m+1}}{m!} \int_a^b B_m(\{x\}) f^{(m)}(x)\,dx \ .$$

(Euler-McLaurin sum formula)

19. Let $a, b \in \mathbb{Z}$, $f : [a, b] \to \mathbb{C}$ q times differentiable, $\int_a^b \left| f^{(q)}(x) \right| dx < \infty$. Show that for $1 \leq m \leq q$:

$$\sum_{a < n \leq b} f(n) = \int_a^b f(x)dx + \sum_{k=1}^m \frac{(-1)^k}{k!} B_k \left(f^{(k-1)}(b) - f^{(k-1)}(a) \right)$$
$$+ \frac{(-1)^{m+1}}{m!} \int_a^b B_m(\{x\}) f^{(m)}(x) dx$$

* 20. For $\omega > 0$ and $\mathrm{Re}(s) > 1$ define

$$\zeta(s, \omega) = \sum_{n=0}^{\infty} (n + \omega)^{-s} \ .$$

The map $s \mapsto \zeta(s, \omega)$ is called the *Hurwitz zeta function*. Show that it can be meromorphically continued on all of $\mathbb{C}$. It has a pole only at $s = 1$, which is simple, and whose residue equals 1. In particular the same holds for the Riemann zeta function.

Hints for the Exercises on Chapter 4

3. Put $\mathfrak{A} = \{\mathfrak{g} \in \mathbb{Z}^n : \mathfrak{Q}_\mathfrak{g} \subseteq \mathfrak{K}\}$, $\mathfrak{A}' = \{\mathfrak{g} \in \mathbb{Z}^n : \mathfrak{Q}_\mathfrak{g} \cap \partial\mathfrak{K} \neq \emptyset\}$. Show that $|\mathfrak{A}| \leq \lambda(\mathfrak{K})$, $|\mathfrak{A}| \leq |\mathfrak{K} \cap \mathbb{Z}^n|$ and that from $\mathfrak{K} \cap \mathfrak{Q}_\mathfrak{g} \neq \emptyset$ and $\mathfrak{g} \notin \mathfrak{A}$ it follows that $\mathfrak{Q}_\mathfrak{g} \cap \partial\mathfrak{K} \neq \emptyset$. Therefore $|\mathfrak{K} \cap \mathbb{Z}^n| \leq |\mathfrak{A}| + |\mathfrak{A}'|$ and $\lambda(\mathfrak{K}) \leq |\mathfrak{A}| + |\mathfrak{A}'|$.

9. Show that $S_n - S_{n-1} = -2n^{-2} A_n$ and that $\lim_{n \to \infty} S_n = 0$.

17. Induction on m. Apply (16), (13ii) and (13iii). In order to determine the constant of integration, integrate a second time.

20. Apply (19) with m large to the function $f : [0, \infty) \to \mathbb{C}$, $f(x) = (x + \omega)^{-s}$.

5. The Prime Number Theorem

Whereas in the previous chapter we presented asymptotic calculations for several number theoretic functions, in this chapter we consider essentially the asymptotic description of a single number theoretic function, namely the function $\pi(n)$, which counts all prime numbers [15] between 1 and n, or extended to $\mathbb{R}$:

$$\pi(x) = \sum_{p \leq x} 1 \ .$$

It is well-known that Euclid showed

$$\lim_{x \to \infty} \pi(x) = \infty$$

in that he proved that there exist infinitely many prime numbers. If there were only finitely many prime numbers $2, 3, \ldots, P$, then one of them would have to divide

$$2 \cdot 3 \cdot 5 \cdot \ldots \cdot P - 1$$

which is clearly false. If one orders the prime numbers $p_1 = 2$, $p_2 = 3, \ldots, p_n, \ldots$ according to size, then one sees from Euclid's proof that

$$p_n \leq 2^{2^{n-1}} \ ,$$

since there must exist some p_m with $m > n$ dividing the number $p_1 p_2 \ldots p_n - 1$, and not exceeding it in size. It follows by induction that

$$p_{n+1} \leq 2^{2^0} \cdot 2^{2^1} \cdot \ldots \cdot 2^{2^{n-1}}$$
$$= 2^{2^0 + 2^1 + \ldots + 2^{n-1}} \leq 2^{2^n} \ ,$$

as asserted. If for some $x \geq 2$ $\quad$ n denotes the largest natural number with

$$2^{2^{n-1}} \leq x \ ,$$

then

$$\pi(x) \geq n \geq 1 + \left[\frac{1}{\log 2} \cdot \log \left(\frac{\log x}{\log 2} \right) \right] \ ,$$

where the right-hand side increases at least proportionately to $\log \log x$. Hence Euclid implicitly proved that

$$\pi(x) \geq c \cdot \log \log x, \quad c > 0 \ .$$

This result is a long way from the observed growth rate of $\pi(x)$. Using a method of Erdös Dressler argued in a more subtle way: since each square-free integer $n \leq x$ can only be divided by $p_1, \ldots, p_K$ with $K = \pi(x)$, and since n can be written uniquely as

$$n = \prod_{k=1}^{\pi(x)} p_k^{v_k}$$

where v_k takes only the values 0 or 1, there are at most $2^{\pi(x)}$ square-free integers $n \leq x$ available. We know that the density of the square-free integers tends to $6/\pi^2$, i.e. the number of square-free numbers $n \leq x$ grows asymptotically as $6x/\pi^2$. This means that for some positive constant $c_1 < 6/\pi^2$

$$c_1 \cdot x \leq 2^{\pi(x)}$$

for sufficiently large x. Hence

$$\pi(x) \geq c \cdot \log x$$

for some $c > 0$.

But neither does this result describe the asymptotic growth of $\pi(x)$. Numerical considerations come closer to the goal: in the following table we let the values for x be increasing powers of ten, and next to the value for $\pi(x)$ we enter the quantity of $x/\pi(x)$:

x	$\pi(x)$	$x/\pi(x)$
10	4	2.5
10^2	25	4.0
10^3	168	6.0
10^4	1 229	8.1
10^5	9 592	10.4
10^6	78 498	12.7
10^7	664 579	15.0
10^8	5 761 455	17.4
10^9	50 847 534	19.7
10^{10}	455 052 511	22.0

If x is multiplied by 10, the value of $x/\pi(x)$ in the table increases approximately by $2.3 \simeq \log 10$; for this reason Legendre and Gauss conjectured the asymptotic growth $x/\pi(x) \sim \log x$, i.e. $\pi(x) \sim x/\log x$. The truth of this assertion is the content of the prime number theorem. If the *integral logarithm* is defined as the Cauchy principal value integral

$$\mathrm{li}\, x = \int_0^x \frac{dt}{\log t}$$

$$= \lim_{\varepsilon \to 0} \left(\int_0^{1-\varepsilon} + \int_{1+\varepsilon}^x \right) \frac{dt}{\log t} \; ,$$

then by de l'Hopital's rule,

$$\lim_{x \to \infty} \frac{\operatorname{li} x}{\dfrac{x}{\log x}} = \lim_{x \to \infty} \frac{\dfrac{1}{\log x}}{\dfrac{1}{\log x} - \dfrac{1}{\log^2 x}} = 1$$

we obtain the asymptotic equation

$$\frac{x}{\log x} \sim \operatorname{li} x \ .$$

Hence $\pi(x) \sim \operatorname{li} x$ can also be called the prime number theorem. Gauss conjectured that $\operatorname{li}(x)$ describes $\pi(x)$ even better than $x/\log x$, and the table below seems to justify this:

x	$\pi(x)$	$\operatorname{li} x$	$x/\log x$
10^3	168	177. ...	144. ...
10^4	1 229	1 246. ...	1 085. ...
10^5	9 592	9 629. ...	8 685. ...
10^6	78 498	78 627. ...	72 382. ...
10^7	664 579	664 918. ...	620 420. ...
10^8	5 761 455	5 762 209. ...	5 428 681. ...
10^9	50 847 534	50 849 234. ...	48 254 942. ...
10^{10}	455 052 511	455 055 614. ...	434 294 481. ... .

However in this book we aim only towards proving that $\pi(x) \sim x/\log x \sim \operatorname{li} x$, and neglect more delicate remainder term estimates. [16] Chebyshev almost arrived at the statement of the prime number theorem in the following result:

Theorem 1: Chebyshev's Theorem. *There exist two positive constants c_1 and c_2, so that for sufficiently large x we have*

$$c_1 \cdot \frac{x}{\log x} \leq \pi(x) \leq c_2 \cdot \frac{x}{\log x} \ .$$

Proof. Here for the first time one sees the importance of the Mangoldt function, since the proof rests on the following technical result:

Lemma 1. *The following estimates hold for the Mangoldt function:*

$$\sum_{n \leq x} \Lambda(n) \left[\frac{x}{n}\right] = x \log x - x + O(\log x) \ ,$$

$$\sum_{n \leq x} \Lambda(n) \left(\left[\frac{x}{n}\right] - 2 \left[\frac{x}{2n}\right] \right) = x \log 2 + O(\log x) \ .$$

Proof of the lemma. It follows from the definition of the Mangoldt function that

$$\sum_{n \leq x} \log n = \sum_{n \leq x} \sum_{m \mid n} \Lambda(m)$$

$$= \sum_{m \leq x} \Lambda(m) \sum_{n \leq x : m \mid n} 1 \ .$$

Since the inner sum counts the multiples n of m with $n \leq x$, it equals $[x/m]$, so that from the Euler sum formula we obtain

$$\sum_{n \leq x} \Lambda(n) \left[\frac{x}{n}\right] = \sum_{n \leq x} \log n$$

$$= \int_1^x \log t \, dt + O(\log x)$$

$$= x \log x - x + O(\log x)$$

which yields the first estimate. The second is a consequence of the first, because in

$$\sum_{n \leq x} \Lambda(n) \left(\left[\frac{x}{n}\right] - 2\left[\frac{x}{2n}\right]\right)$$

$$= \sum_{n \leq x} \Lambda(n) \left[\frac{x}{n}\right] - 2 \cdot \sum_{n \leq x/2} \Lambda(n) \left[\frac{x}{2n}\right]$$

$$- 2 \cdot \sum_{x/2 < n \leq x} \Lambda(n) \left[\frac{x}{2n}\right]$$

the final sum equals zero, and from the equations above

$$= x \log x - x - 2\left(\frac{x}{2} \log \frac{x}{2} - \frac{x}{2}\right) + O(\log x)$$

$$= x \log 2 + O(\log x) \ .$$

Continuation of the proof of Theorem 1. Since in general

$$[a] - 2\left[\frac{a}{2}\right] < a - 2\left(\frac{a}{2} - 1\right) = 2$$

and the left-hand side is integral, one has

$$[a] - 2\left[\frac{a}{2}\right] \leq 1 \ .$$

Therefore from the second formula of Lemma 1

$$x \log 2 + O(\log x) = \sum_{n \leq x} \Lambda(n) \left(\left[\frac{x}{n}\right] - 2\left[\frac{x}{2n}\right]\right)$$

$$\leq \sum_{n \leq x} \Lambda(n) = \sum_{p \leq x} \left[\frac{\log x}{\log p}\right] \log p$$

$$\leq \log x \cdot \sum_{p \leq x} 1 = \pi(x) \log x \ ,$$

which after division by $\log x$ implies the left-hand inequality in Chebyshev's theorem. In order to verify the right-hand inequality one proceeds analogously, starting from the general estimate

$$[a] - 2\left[\frac{a}{2}\right] > a - 1 - 2\frac{a}{2} = -1 \ .$$

Because the left-hand side is again integral this can be sharpened to

$$[a] - 2\left[\frac{a}{2}\right] \geq 0 \ ,$$

and now one seeks to bring the second inequality of Lemma 1 into play. This turns out to be somewhat more complicated than before – the calculation leading to the goal reads

$$
\begin{aligned}
\pi(x)\log x &- \pi\left(\frac{x}{2}\right)\log\frac{x}{2} \\
&= \log\frac{x}{2}\cdot\left(\pi(x) - \pi\left(\frac{x}{2}\right)\right) + \pi(x)\log 2 \\
&= \log\frac{x}{2}\cdot\left(\pi(x) - \pi\left(\frac{x}{2}\right)\right) + O(x) \\
&= O\left(\sum_{x/2 < p \leq x}\log p + x\right) \\
&= O\left(\sum_{x/2 < n \leq x}\Lambda(n)\cdot(1-0) + x\right) \\
&= O\left(\sum_{x/2 < n \leq x}\Lambda(n)\left(\left[\frac{x}{n}\right] - 2\left[\frac{x}{2n}\right]\right) + x\right) \\
&= O\left(\sum_{n \leq x}\Lambda(n)\left(\left[\frac{x}{n}\right] - 2\left[\frac{x}{2n}\right]\right) + x\right) \\
&= O(x) \ .
\end{aligned}
$$

Since more generally this gives the estimate

$$\pi\left(\frac{x}{2^k}\right)\log\frac{x}{2^k} - \pi\left(\frac{x}{2^{k+1}}\right)\log\frac{x}{2^{k+1}} = O\left(\frac{x}{2^k}\right) \ ,$$

for K with $2^K \leq x < 2^{K+1}$ we obtain

$$
\begin{aligned}
\pi(x)\log x \\
&= \sum_{k=0}^{K}\left(\pi\left(\frac{x}{2^k}\right)\log\frac{x}{2^k} - \pi\left(\frac{x}{2^{k+1}}\right)\log\frac{x}{2^{k+1}}\right) \\
&= O\left(\sum_{k=0}^{K}\frac{x}{2^k}\right) = O(x) \ .
\end{aligned}
$$

After division by $\log x$ this leads to the right-hand inequality in Chebyshev's theorem. $\square$

Chebyshev tried a further attempt to force the at that time unproved prime number theorem, and introduced the following equivalence.

Proposition 1. *The statement*

$$\pi(x) \sim \frac{x}{\log x}$$

is equivalent to

$$\psi(x) \sim x$$

where

$$\psi(x) = \sum_{n \leq x} \Lambda(n) = \sum \sum_{p,v \geq 1 : p^v \leq x} \log p$$

denotes the ψ-function (introduced by Chebyshev).

Proof. On the one hand by definition

$$\psi(x) = \sum_{p \leq x} \left[\frac{\log x}{\log p} \right] \log p$$

$$\leq \log x \cdot \sum_{p \leq x} 1 = \pi(x) \log x \ .$$

On the other hand for any y in $1 < y < x$, we have

$$\pi(x) = \pi(y) + \sum_{y < p \leq x} 1$$

$$\leq \pi(y) + \sum_{y < p \leq x} \frac{\log p}{\log y}$$

$$\leq c_2 \cdot \frac{y}{\log y} + \frac{\psi(x)}{\log y}$$

from which the inequality

$$\pi(x) \cdot \frac{\log x}{x} \leq c_2 \cdot \frac{y \log x}{x \log y} + \frac{\psi(x)}{x} \cdot \frac{\log x}{\log y}$$

follows. In the special case of $y = x / \log x$, this implies

$$\pi(x) \cdot \frac{\log x}{x} \leq \frac{c_2}{\log x - \log \log x}$$

$$+ \frac{\psi(x)}{x} \cdot \frac{1}{1 - (\log \log x)/\log x} \ .$$

Taken together the two inequalities show that the statements

$$\lim_{x \to \infty} \pi(x) \cdot \frac{\log x}{x} = 1,$$

$$\lim_{x \to \infty} \frac{\psi(x)}{x} = 1$$

are equivalent, as claimed by Chebyshev in Proposition 1. $\square$

After the fruitless efforts of Chebyshev to find a proof of the prime number theorem in the years 1848 and 1850, Mertens in 1874 produced asymptotic calculations for means of functions of prime numbers.

Theorem 2: Mertens' Theorem. *If the variable p runs through all prime numbers, the following asymptotic approximations hold:*

$$\sum_{p \le x} \frac{\log p}{p} = \log x + O(1),$$

$$\sum_{p \le x} \frac{1}{p} = \log \log x + c + O\left(\frac{1}{\log x}\right) .$$

Here c denotes a constant, and

$$\prod_{p \le x} \left(1 - \frac{1}{p}\right) = \frac{c'}{\log x} \cdot \left(1 + O\left(\frac{1}{\log x}\right)\right)$$

with another constant c'.

Proof. Because of Lemma 1 we have

$$x \log x - x + O(\log x)$$

$$= \sum_{n \le x} \Lambda(n) \left[\frac{x}{n}\right]$$

$$= \sum_{p \le x} \left[\frac{x}{p}\right] \log p + \sum_{p \le \sqrt{x}} \sum_{v \ge 2 : p^v \le x} \left[\frac{x}{p^v}\right] \log p$$

$$= \sum_{p \le x} \frac{\log p}{p} \cdot x - \sum_{p \le x} \left\{\frac{x}{p}\right\} \log p + O\left(\sum_{p \le \sqrt{x}} \sum_{2 \le v \le \log x / \log p} x p^{-v} \log p\right)$$

$$= x \cdot \sum_{p \le x} \frac{\log p}{p} + O\left(\sum_{p \le x} \log p\right) + O\left(\sum_{n=1}^{\infty} x \cdot \frac{\log n}{n^2}\right)$$

$$= x \cdot \sum_{p \le x} \frac{\log p}{p} + O\left(\log x \cdot c_2 \cdot \frac{x}{\log x}\right) + O\left(x \cdot \sum_{n=1}^{\infty} \frac{\log n}{n^2}\right)$$

$$= x \cdot \sum_{p \le x} \frac{\log p}{p} + O(x) ,$$

and on division by x we obtain the first formula. The second formula is derived from the first by Abel transformation

$$\sum_{p \le x} \frac{1}{p} = \sum_{p \le x} \frac{\log p}{p} \cdot \frac{1}{\log p}$$

$$= \frac{1}{\log x} \cdot \sum_{p \le x} \frac{\log p}{p} + \int_2^x \sum_{p \le t} \frac{\log p}{p} \cdot \frac{dt}{t \cdot \log^2 t}$$

$$= 1 + O\left(\frac{1}{\log x}\right) + \int_2^x \frac{dt}{t \log t} + \int_2^x \left(\sum_{p \le t} \frac{\log p}{p} - \log t\right) \frac{dt}{t \cdot \log^2 t} \ .$$

Since

$$w(t) = \sum_{p \le t} \frac{\log p}{p} - \log t$$

is bounded, the integral

$$\int_2^\infty w(t) \frac{dt}{t \cdot \log^2 t}$$

converges, and furthermore we have

$$\int_2^x \frac{dt}{t \log t} = \log \log x - \log \log 2 \ .$$

Therefore

$$\sum_{p \le x} \frac{1}{p} = \log \log x + \left(1 - \log \log 2 + \int_2^\infty w(t) \frac{dt}{t \cdot \log^2 t}\right)$$

$$+ O\left(\frac{1}{\log x} + \int_x^\infty |w(t)| \frac{dt}{t \cdot \log^2 t}\right)$$

completing the proof.

If finally we define the constant c'' by

$$c'' = \sum_{n=2}^\infty \frac{1}{n} \sum_p \frac{1}{p^n} \ ,$$

then we obtain the third formula from the second by means of the calculation

$$\log \left(\prod_{p \le x} \left(1 - \frac{1}{p}\right)\right) = \sum_{p \le x} \log \left(1 - \frac{1}{p}\right)$$

$$= -\sum_{p \le x} \sum_{n=1}^\infty \frac{(1/p)^n}{n}$$

$$= -\sum_{p \le x} \frac{1}{p} - \sum_{n \ge 2} \frac{1}{n} \sum_{p \le x} \frac{1}{p^n}$$

$$= -\sum_{p \le x} \frac{1}{p} - c'' + O\left(\sum_{n \ge 2} \frac{1}{n} \sum_{p > x} \frac{1}{p^n} \right)$$

$$= -\sum_{p \le x} \frac{1}{p} - c'' + O\left(\sum_{n \ge 2} \frac{1}{n} \sum_{m > x} \frac{1}{m^n} \right)$$

$$= -\sum_{p \le x} \frac{1}{p} - c'' + O\left(\sum_{n \ge 2} \frac{1}{n} \cdot \frac{1}{(n-1)x^{n-1}} \right)$$

$$= -\sum_{p \le x} \frac{1}{p} - c'' + O\left(\frac{1}{x} \right) . \qquad \Box$$

However, even these calculations of Mertens did not lead him to a proof of the prime number theorem. Even though the elementary methods of Chebyshev and Mertens were rich in tricks, they were not powerful enough for the asymptotic calculation of $\pi(x)$.

The door to the prime number theorem was opened not by Euclid's but by *Euler's proof of the infinity of the set of all primes.* If there were only finitely many prime numbers,

$$\zeta(s) = \prod_{p} \frac{1}{1 - \dfrac{1}{p^s}}$$

could not diverge for $s = 1$. Here Riemann joined in with the study of the ζ-function named after him, and in 1896 with the help of techniques from complex variable theory Hadamard and de la Vallée Poussin were fortunate enough to find the first, very complicated, demonstration of the prime number theorem. Since then the methods of proof have become more and more refined; above all simplification has come through a theorem of Ikehara and Wiener from harmonic analysis. Although Wiener's work from the year 1932 deals with Tauberian theorems for Fourier integrals, i.e. is concerned with a relatively difficult area of analysis far from number theory, for a long time people were of the opinion that the proof of the prime number theorem via Ikehara's and Wiener's Tauberian theorem would turn out to be the simplest.[20] This conviction did not change, even though in 1947 and 1948 Selberg and Erdös found an elementary proof of the prime number theorem, that is a proof using only the simplest methods of real analysis, without recourse to complex variable theory or harmonic analysis. As the price of limiting itself to elementary analysis the proof of Selberg and Erdös demands manipulative skill of high order, rich in tricks, together with a far from transparent thought process. In 1980 Newman found an elegant complex variable theoretic proof of a weaker version of the Tauberian theorem of Wiener and Ikehara, and this version just suffices for the demonstration of the prime number theorem. In what follows we shall describe this cornerstone of the prime number theorem discovered by Newman, since in character it has great simplicity and clarity.

Basic for the following considerations is

Lemma 2: Tauberian Theorem of Ingham and Newman. *Let $F(t)$ be a bounded complex valued function, defined for $0 < t < \infty$ and integrable over every compact subset of the domain of definition. Let a holomorphic function $G(z)$ be defined in a region containing the closed half-plane $\mathrm{Re}(z) \geq 0$, and for all z in $\mathrm{Re}(z) > 0$ let G agree with the Laplace transform of $F(t)$, that is*

$$G(z) = \int_0^\infty F(t)e^{-zt}dt, \quad \mathrm{Re}(z) > 0 \ .$$

Then the improper integral

$$\int_0^\infty F(t)dt$$

converges.

Proof. Without loss of generality assume for all $t > 0$ that $|F(t)| \leq 1$, and for an arbitrary positive λ write

$$G_\lambda(z) = \int_0^\lambda F(t)e^{-zt}dt \ .$$

$G_\lambda(z)$ is analytic for all complex z. The lemma is proved if we show that

$$\lim_{\lambda \to \infty} G_\lambda(0) = \lim_{\lambda \to \infty} \int_0^\lambda F(t)dt = G(0) \ .$$

To this end Newman estimates the difference $G(0) - G_\lambda(0)$ and for this applies complex integration. First of all, given the Cauchy integral formula, we have the relation

$$G(0) - G_\lambda(0) = \frac{1}{2\pi i} \int_\gamma (G(z) - G_\lambda(z)) \frac{dz}{z} \ ,$$

where γ denotes a suitable, simple closed, positively oriented curve about the origin. The estimates for $x = \mathrm{Re}(z) > 0$

$$|G(z) - G_\lambda(z)| = \left| \int_\lambda^\infty F(t)e^{-zt}dt \right|$$

$$\leq \int_\lambda^\infty e^{-xt}dt = \frac{1}{x} \cdot e^{-\lambda x}$$

and for $x = \mathrm{Re}(z) < 0$

$$|G_\lambda(z)| = \left| \int_0^\lambda F(t)e^{-zt}dt \right|$$

$$\leq \int_0^\lambda e^{-xt}dt = \frac{1}{|x|} \cdot e^{-\lambda x}$$

lead to the relation

$$G(0) - G_\lambda(0) = \frac{1}{2\pi i} \int_\gamma \left(G(z) - G_\lambda(z)\right) e^{\lambda z} \cdot \left(\frac{1}{z} + \frac{z}{R^2}\right) dz \ ,$$

which is more suited to our purpose. Here the constant R is taken to be the distance of the curve from the origin in the neighbourhood of the intersection points of γ with the imaginary axis, i.e. in the neighbourhood of $x = 0$. For on the one hand by the Cauchy theorem this equation is just as valid as the relation originally stated. On the other, given the additional factor $e^{\lambda z}$, both estimates have the common bound $e^{-\lambda x}/|x|$ removed. And by means of the additional factor $(1/z + z/R^2)$, which agrees with $2x/R^2$ in the neighbourhood of the intersection points of γ with the imaginary axis, one also removes the for z awkward limit $1/|x|$ near the imaginary axis. The following calculations show this more precisely. Choose some arbitrarily small $\varepsilon > 0$; let the curve γ be specially chosen so that for $\mathrm{Re}(z) \geq -\delta$ it traces out the circle $|z| = R = 3/\varepsilon$ in a positive direction, and then connects the end-points on the circular arc by the line $\mathrm{Re}(z) = -\delta$.

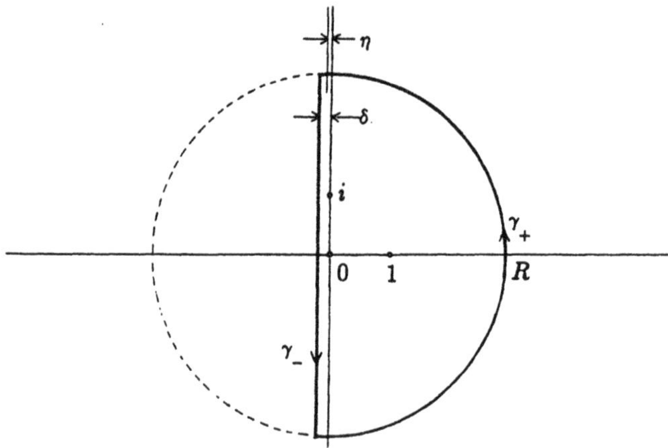

Fig. 15. Path of integration γ

It would indeed be better to choose γ to be the positively oriented circle $|z| = R$; however since $G(z)$ is only holomorphic in a neighbourhood of each point on the imaginary axis, γ cannot stray too far to the left. One must choose $\delta = \delta(\varepsilon)$ so small that $G(z)$ remains analytic on the compact circular segment bounded by γ. If γ_+ denotes the part of γ lying in $\mathrm{Re}(z) > 0$, it follows from

$$1/z + z/R^2 = 2x/R^2$$

that

$$\left| (G(z) - G_\lambda(z)) \, e^{\lambda z} \left(\frac{1}{z} + \frac{z}{R^2} \right) \right|$$

$$\leq \frac{1}{x} \cdot e^{-\lambda x} e^{\lambda x} \cdot \frac{2x}{R^2} = \frac{2}{R^2} \ ,$$

$$\left| \frac{1}{2\pi i} \int_{\gamma_+} (G(z) - G_\lambda(z)) \, e^{\lambda z} \left(\frac{1}{z} + \frac{z}{R^2} \right) dz \right|$$

$$\leq \frac{1}{2\pi} \cdot \frac{2}{R^2} \cdot \pi R = \frac{1}{R} \ .$$

If γ_- denotes the part of γ lying in $\mathrm{Re}(z) < 0$ then for the estimate of

$$\frac{1}{2\pi i} \int_{\gamma_-} G_\lambda(z) e^{\lambda z} \left(\frac{1}{z} + \frac{z}{R^2} \right) dz$$

one can deform it to the half-circle $|z| = R$ lying in $\mathrm{Re}(z) \leq 0$, given that $G_\lambda(z)$ is holomorphic in the entire complex plane. Here, bearing in mind that for $\mathrm{Re}(z) \leq 0 \ |G_\lambda(z)| \leq (1/|x|)e^{-\lambda x}$, a similar calculation shows that

$$\left| \frac{1}{2\pi i} \int_{\gamma_-} G_\lambda(z) e^{\lambda z} \left(\frac{1}{z} + \frac{z}{R^2} \right) dz \right| \leq \frac{1}{R} \ .$$

For the remaining integral

$$\frac{1}{2\pi i} \int_{\gamma_-} G(z) e^{\lambda z} \left(\frac{1}{z} + \frac{z}{R^2} \right) dz$$

we note from the holomorphic nature of $G(z) \left(\frac{1}{z} + \frac{z}{R^2} \right)$ on the path of γ_-, that the boundedness of

$$\left| G(z) \left(\frac{1}{z} + \frac{z}{R^2} \right) \right| \leq K$$

follows, with a constant $K = K(R, \delta) = K(\varepsilon)$. Because

$$\left| G(z) e^{\lambda z} \left(\frac{1}{z} + \frac{z}{R^2} \right) \right| \leq K e^{\lambda x}$$

on each part of γ_- for which $\mathrm{Re}(z) = x \leq -\eta < 0$ (with some $\eta < \delta$ yet to be determined), it follows that the integrand tends uniformly to zero as $\lambda \to \infty$. In the remaining parts of γ_- the integrand is bounded by K; η can be fixed so small, that integration over these remaining parts of γ_- undershoots the preassigned quantity $\varepsilon/3$. With this, because of our arbitrary choice of ε, it follows from

$$\overline{\lim}_{\lambda \to \infty} \left| \frac{1}{2\pi i} \int_{\gamma} (G(z) - G_\lambda(z)) \, e^{\lambda z} \left(\frac{1}{z} + \frac{z}{R^2} \right) dz \right|$$

$$\leq \frac{1}{R} + \frac{1}{R} + \frac{\varepsilon}{3} = \varepsilon$$

that the limiting value must equal zero, proving Lemma 2. $\square$

Corollary 1: Simplified Version of the Theorem of Wiener and Ikehara. *Let $f(x)$ be a monotone non-decreasing function defined for $x \geq 1$ with $f(x) = O(x)$. In some region containing the closed half-plane $\mathrm{Re}(s) \geq 1$ let the function $g(s)$ be defined, holomorphic except for a simple pole at $s = 1$ with residue c. (This means that the map given by*

$$g(s) - \frac{c}{s-1}$$

is holomorphic in the region containing the closed half-plane $\mathrm{Re}(s) \geq 1$.) Here for all s with $\mathrm{Re}(s) > 1$ $g(s)$ coincides with the Mellin transform of $f(x)$, i.e.

$$g(s) = s \cdot \int_1^\infty f(x) x^{-s-1} dx, \quad \mathrm{Re}(s) > 1 \ .$$

Then we have the asymptotic expression $f(x) \sim c\, x$.

Proof. If $F(t)$ is defined by

$$F(t) = e^{-t} \cdot f(e^t) - c$$

we are concerned with a bounded function defined on $0 < t < \infty$ and integrable on each bounded subinterval of the domain of definition. Its Laplace transform

$$\begin{aligned}
G(z) &= \int_0^\infty \left(e^{-t} f(e^t) - c \right) e^{-zt} dt \\
&= \int_1^\infty f(x) x^{-z-2} dz - \frac{c}{z} \\
&= \frac{1}{z+1} \left(g(z+1) - \frac{c}{z} - c \right)
\end{aligned}$$

is well-defined in $\mathrm{Re}(z) > 0$, and by assumption the right-hand side represents a holomorphic function in some region containing the closed half-plane $\mathrm{Re}(z) \geq 0$. Since by Lemma 2 for $t = \log x$ the improper integral

$$\int_0^\infty \left(e^{-t} f(e^t) - c \right) dt = \int_1^\infty \frac{f(x) - cx}{x^2} \, dx$$

converges, the conclusion that $f(x) \sim c \cdot x$ follows very easily from the non-decreasing monotonicity of f.

Indeed were $\overline{\lim}_{x \to \infty} f(x)/x > c$, there would exist some positive δ, so that for infinitely many, arbitrarily large y one would have $f(y) > (c + 2\delta)y$. Hence for all x with

$$y < x < y \cdot \frac{c+2\delta}{c+\delta} : \qquad f(x) > (c+2\delta)y > (c+\delta)x \ .$$

Given that

$$\int_{y}^{y \cdot \frac{c+2\delta}{c+\delta}} \frac{f(x) - cx}{x^2} dx > \int_{y}^{y \cdot \frac{c+2\delta}{c+\delta}} \frac{\delta}{x} dx = \delta \cdot \log \left(\frac{c + 2\delta}{c + \delta} \right)$$

and the fact that there exist infinitely many and arbitrarily large y of this kind, we obtain a contradiction. As a consequence

$$\overline{\lim}_{x \to \infty} \frac{f(x)}{x} \leq c .$$

Furthermore were $\underline{\lim}_{x \to \infty} f(x)/x < c$ there would exist some positive δ, so that for infinitely many, arbitrarily large y, one would have $f(y) < (c - 2\delta)y$ (with $c - 2\delta > 0$). Hence for all x with

$$y \cdot \frac{c - 2\delta}{c - \delta} < x < y ,$$

it would follow that $f(x) < (c - 2\delta)y < (c - \delta)x$. In this way we derive the analogous contradiction, with which Corollary 1 is proved. $\qquad\square$

With Corollary 1 one achieves breakthrough to the proof of the prime number theorem – by Proposition 1 we have only to show the equivalent assertion that

$$\psi(x) \sim x$$

and Proposition 19 of the previous chapter calculates the Mellin transform of $\psi(x)$:

$$-\frac{\zeta'(s)}{\zeta(s)} = s \cdot \int_{1}^{\infty} \psi(x) x^{-s-1} dx .$$

Because of Chebyshev's theorem we have

$$\psi(x) = \sum_{p \leq x} \left[\frac{\log x}{\log p} \right] \log p \leq \log x \cdot \sum_{p \leq x} 1$$
$$= \pi(x) \log x = O(x) ,$$

and in addition $\psi(x)$ is monotone non-decreasing. Hence by Corollary 1 we have only to show that the function $-\zeta'(s)/\zeta(s)$, holomorphic for $\mathrm{Re}(s) > 1$, is holomorphically extendable out past $\mathrm{Re}(s) = 1$, $s \neq 1$. More precisely:

$$-\frac{\zeta'(s)}{\zeta(s)} - \frac{1}{s - 1}$$

can be considered as a holomorphic function in some region containing the closed half-plane $\mathrm{Re}(s) \geq 1$. For $s = 1$ there is no problem, since by Proposition 17 of the previous chapter

$$\zeta(s) = \frac{1}{s - 1}(1 + h(s))$$

where the holomorphic function $h(s)$ remains bounded in some suitable neighbourhood of $s = 1$. More precisely we may require that $|h(s)| < 1$. The holomorphic nature of

$$-\frac{\zeta'(s)}{\zeta(s)} - \frac{1}{s-1}$$

in this neighbourhood of $s = 1$ follows from

$$-\frac{\zeta'(s)}{\zeta(s)} = \frac{1}{s-1} - \frac{h'(s)}{1+h(s)} \; .$$

For all other points s with $\mathrm{Re}(s) = 1$, because of the analytic continuability of the Riemann ζ-function (even for $\mathrm{Re}(s) > 0$), we have only now to clarify the role of $\zeta(s)$ in the denominator. Put another way we must prove the following statement:

Lemma 3. *For* $\mathrm{Re}(s) = 1$, $s \neq 1$, $\zeta(s) \neq 0$.

Proof. Already in 1898 Mertens found the following brilliant argument, which applies the inequality

$$3 + 4\cos\varphi + \cos 2\varphi = 2(1 + \cos\varphi)^2 \geq 0$$

for real φ. If with $t \neq 0 \qquad \zeta(1 + it)$ were ever zero, then

$$\Theta(s) = \zeta(s)^3 \cdot \zeta(s + it)^4 \cdot \zeta(s + 2it)$$

would also have to possess a zero for $s = 1$, since the four-fold zero of $\zeta(s+it)^4$ removes the three-fold pole of $\zeta(s)^3$. From this one has

$$\lim_{s \to 1} \log |\Theta(s)| = -\infty \; ,$$

and in particular for $s = \sigma > 1$, using the product representation of the ζ-function

$$\log |\zeta(\sigma + it)| = -\mathrm{Re}\left(\sum_p \log\left(1 - p^{-\sigma-it}\right)\right)$$

$$= \mathrm{Re}\left(\sum_p \left(p^{-\sigma-it} + \frac{1}{2}\left(p^2\right)^{-\sigma-it} + \frac{1}{3}\left(p^3\right)^{-\sigma-it} + \dots\right)\right)$$

$$= \mathrm{Re}\left(\sum_n a_n n^{-\sigma-it}\right)$$

with non-negative coefficients a_n. This leads to the contradiction

$$\log |\Theta(\sigma)| = \mathrm{Re}\left(\sum_{n=1}^{\infty} a_n n^{-\sigma}\left(3 + 4n^{-it} + n^{-2it}\right)\right)$$

$$= \sum_{n=1}^{\infty} a_n n^{-\sigma}\left(3 + 4\cos(t \log n) + \cos(2t \log n)\right) \geq 0 \; .$$

Given this we have set aside all obstructions to applying Corollary 1, and the goal is achieved.

Theorem 3: Prime Number Theorem *If $\pi(x)$ counts the prime numbers $p \leq x$, then one has $\pi(x) \sim x/\log x$.* [14]

The method used in the proof of the prime number theorem applies also to a larger class of Dirichlet series.

Corollary 2. *Let $f(x)$ denote a number theoretic function with non-negative values and with*

$$\sum_{n \leq x} f(n) = O(x) \ ,$$

and let the Dirichlet series formed from it

$$F(s) = \sum_{n=1}^{\infty} \frac{f(n)}{n^s}$$

be holomorphic for $\mathrm{Re}(s) > 1$ in the sense that the map given by

$$F(s) - \frac{c}{s-1}$$

for some fixed constant c, is defined to be holomorphic in some region containing the closed half-plane $\mathrm{Re}(s) \geq 1$. Then

$$\sum_{n \leq x} f(n) \sim c \cdot x \ .$$

Proof. Starting from the integral representation

$$F(s) = \sum_{n=1}^{\infty} \frac{f(n)}{n^s} = s \cdot \int_1^{\infty} \left(\sum_{n \leq x} f(n) \right) x^{-s-1} dx$$

at the end of the previous chapter, we see that $F(s)$ plays the role of $g(s)$ and $\sum_{n \leq x} f(n)$ the role of $f(x)$ in Corollary 1. $\qquad\square$

Corollary 3. *Let $f(n)$ and $g(n)$ be two number theoretic functions, where $f(n)$ takes only non-negative values and the formulas*

$$g(n) = O\big(f(n)\big)$$

and

$$\sum_{n \leq x} f(n) = O(x)$$

hold. If the Dirichlet series which they form

$$F(s) = \sum_{n=1}^{\infty} \frac{f(n)}{n^s} \; ,$$

$$G(s) = \sum_{n=1}^{\infty} \frac{g(n)}{n^s}$$

are holomorphic for $\mathrm{Re}(s) > 1$, *in the sense that the maps given by*

$$F(s) - \frac{c}{s-1}, \quad G(s) - \frac{\gamma}{s-1}$$

for specific constants c *and* γ *are defined to be holomorphic in some region containing the closed half-plane* $\mathrm{Re}(s) \geq 1$, *then*

$$\sum_{n \leq x} g(n) \sim \gamma \cdot x \; .$$

Proof. If K is chosen so large that for all n $|g(n)| \leq K f(n)$, then for the real-valued $g(n)$ one can apply Corollary 2 to the Dirichlet series formed by the number theoretic function $h(n) = K f(n) + g(n)$, denoted by

$$H(s) = \sum_{n=1}^{\infty} \frac{h(n)}{n^s} = KF(s) + G(s) \; .$$

Hence

$$\sum_{n \leq x} h(n) = K \cdot \sum_{n \leq x} f(n) + \sum_{n \leq x} g(n) \sim Kcx + \sum_{n \leq x} g(n)$$

and

$$\sum_{n \leq x} h(n) \sim Kcx + \gamma x$$

which leads to the conclusion. If $g(n)$ is complex-valued and $G^*(s) = \overline{G(\bar{s})}$, one carries out the calculation separately for

$$G_1(s) = \frac{1}{2}\left(G(s) + G^*(s)\right) = \sum_{n=1}^{\infty} \frac{\mathrm{Re}\big(g(n)\big)}{n^s}$$

and for

$$G_2(s) = \frac{1}{2i}\left(G(s) - G^*(s)\right) = \sum_{n=1}^{\infty} \frac{\mathrm{Im}\big(g(n)\big)}{n^s}$$

proceeding in the same way on this path to the conclusion. $\square$

Essentially Corollary 3 says that the assumption made in Corollary 2 that the number theoretic function be non-negative, can be set aside in the case that the complex-valued number theoretic function under consideration is dominated by a suitable non-negative number theoretic function. In the last chapter this generalisation will be unavoidable – here we give three examples as a foretaste of its importance.

Corollary 4. *For the Möbius function $\mu(n)$ and the Liouville function $\lambda(n)$ the following formulae hold,* [18,21]

$$\sum_{n \leq x} \mu(n) = o(x) \ ,$$

$$\sum_{n \leq x} \lambda(n) = o(x) \ .$$

Proof. In both these applications of Corollary 3 the associated Dirichlet series $G(s) = 1/\zeta(s)$ and $G(s) = \zeta(2s)/\zeta(s)$ are analytically continuable across $\mathrm{Re}(s) = 1$, without singularity, i.e. in Corollary 3 we can take $\gamma = 0$. $\square$

For a third example we examine more closely the Dirichlet series

$$\zeta_i(s) = \sum_{n=1}^{\infty} \frac{r(n)}{n^s} \ ,$$

which is formed from the number theoretic function $r(n)$ counting the representations of n as the sum of two squares. In Chapter 3 we explained how $r(n)$ can also be considered as the number of representations $n = \gamma\bar{\gamma}$, where $\gamma = x + iy$ runs through the elements of the ring $\mathbb{Z}(i)$. This leads us to make the following transformation of $\zeta_i(s)$, namely

$$\zeta_i(s) = \sum_{\gamma \neq 0} \frac{1}{|\gamma|^{2s}} = \sum_{\gamma \neq 0} \frac{1}{(\gamma\bar{\gamma})^s} \ ,$$

where $\zeta_i(s)$ now denotes the *ζ-function for the number theory of the ring $\mathbb{Z}(i)$* – analogous to the Riemann ζ-function

$$\zeta(s) = \sum_{n=1}^{\infty} \frac{1}{n^s} \ ,$$

for the ring $\mathbb{Z}$ of ordinary integers. $\gamma \neq 0$ under the summation sign signifies that γ runs through all elements of $\mathbb{Z}(i)$ with the exception of 0. However in $\zeta_i(s)$ only the magnitude of the element γ comes into play. In order to keep track of the arguments of these complex numbers, for integers h Hecke defined the following Dirichlet series

$$\Xi(h, s) = \sum_{\gamma \neq 0} \frac{1}{|\gamma|^{2s}} \cdot e^{4ih \arg \gamma} \ .$$

For $h = 0$ one clearly obtains $\Xi(0, s) = \zeta_i(s)$, and the transformation

$$\Xi(h,s) = \sum_{n=1}^{\infty} \frac{1}{n^s} \left(\sum_{|\gamma|^2=n} e^{4ih \arg \gamma} \right),$$

which for $\mathrm{Re}(s) > 1$ follows from the absolute convergence, i.e. the convergence of $\zeta_i(s)$, shows that we are indeed dealing with a Dirichlet series. The convergence of $\zeta_i(s)$ for $\mathrm{Re}(s) > 1$ follows by applying Abel's transformation and Proposition 9 of the previous chapter

$$\sum_{x \le n \le y} \frac{r(n)}{n^s} = \frac{1}{y^s} O(y-x) + s \cdot \int_x^y O(t-x) t^{-s-1} dt$$

$$= O\left(\frac{1}{x^{s-1}} \right).$$

The argument function $\arg(\gamma)$ in $\Xi(h,s)$ is uniquely defined by the principal value $-\pi < \arg(\gamma) \le \pi$. The factor 4 in the exponent serves to suppress the role of associated elements, i.e. for each of the numbers $1, i, -1, -i$ $4\arg(\varepsilon) \equiv 0 (\mathrm{mod}\ 2\pi)$ holds, i.e.

$$e^{4ih \arg \varepsilon\gamma} = e^{4ih \arg \gamma}.$$

Finally the function

$$f(\gamma) = e^{4ih \arg \gamma}$$

is *strongly multiplicative* over $\mathbb{Z}(i)$, i.e. $f(\gamma_1 \gamma_2) = f(\gamma_1)f(\gamma_2)$. If for a number theoretic function $f(\gamma)$ over $\mathbb{Z}(i)$ the relation $f(\gamma_1 \gamma_2) = f(\gamma_1)f(\gamma_2)$ only holds if γ_1 and γ_2 have no prime factor in common, then f is said to be only *multiplicative*. If in addition for each unit ε we have $f(\varepsilon) = 1$, then in the region of absolute convergence of

$$F(s) = \sum_{\gamma \ne 0} \frac{1}{|\gamma|^{2s}} f(\gamma)$$

we have the product representation

$$F(s) = 4 \cdot \prod_{\omega} \left(\sum_{v=0}^{\infty} \frac{f(\omega^v)}{|\omega|^{2vs}} \right),$$

where the index ω runs through all prime elements of $\mathbb{Z}(i)$ with $0 \le \arg(\omega) < \pi/2$.

The proof follows the lines of Proposition 15 in the previous chapter: on the one hand

$$\sum_{v=0}^{\infty} \frac{f(\omega^v)}{|\omega|^{2vs}}$$

is absolutely convergent, where $F(s)$ is absolutely convergent. If on the other hand $\mathbb{Z}_k(i)$ denotes the subset of all integral elements from $\mathbb{Z}(i)$, which only have prime factors $\omega = \omega_1, \omega_2, \ldots, \omega_j$ with $|\omega|^2 \le k$, then $\lim_{k \to \infty} \mathbb{Z}_k(i)$ agrees with the union of all the $\mathbf{Z}_k(i)$, i.e. with $\mathbf{Z}(i)$ with 0 removed. Since there are four units ε, the conclusion follows from

$$4 \cdot \prod_{|\omega|^2 \leq k} \sum_{v=0}^{\infty} \frac{f(\omega^v)}{|\omega|^{2vs}}$$

$$= \sum_{\varepsilon} \sum_{v_1=0}^{\infty} \cdots \sum_{v_j=0}^{\infty} \frac{f\left(\omega_1^{v_1} \ldots \omega_j^{v_j}\right)}{\left|\omega_1^{v_1} \ldots \omega_j^{v_j}\right|^{2s}}$$

$$= \sum_{\gamma \in \mathbf{Z}_k(i)} \frac{f(\gamma)}{|\gamma|^{2s}}$$

by taking the limit as $k \to \infty$. If for the strongly multiplicative function

$$f(\gamma) = e^{4ih \arg \gamma}$$

one applies the sum formula for the geometric series in the product representation, one obtains

$$\Xi(h, s) = 4 \cdot \prod_{\omega} \frac{1}{1 - \dfrac{e^{4ih \arg \omega}}{|\omega|^{2s}}}, \quad \operatorname{Re}(s) > 1 .$$

In particular

$$\zeta_i(s) = 4 \cdot \prod_{\omega} \frac{1}{1 - \dfrac{1}{|\omega|^{2s}}}, \quad \operatorname{Re}(s) > 1 .$$

The path to a proof of an analogue of the prime number theorem in $\mathbf{Z}(i)$ is now clearly signposted: for all γ we define $\Lambda_i(\varepsilon \omega^v) = \log |\omega|$ and $\Lambda_i(\gamma) = 0$ for all γ which cannot be decomposed as $\varepsilon \omega^v$ with ε a unit, ω a prime element and v a natural numerical exponent, analogously to the Mangoldt function. The connection between it and

$$-\frac{\Xi'(h, s)}{\Xi(h, s)}$$

is easy to show, in particular this quotient will play an important part in the Mellin transform of

$$\psi_i(x) = \sum_{|\gamma|^2 \leq x} \Lambda_i(\gamma) .$$

On the one hand the analytic continuability of $\Xi(h, s)$ and the fact that $\Xi(h, s)$ may be written in the denominator, i.e. that for $\operatorname{Re}(s) = 1$, $\Xi(h, s) \neq 0$, leads to the application of Corollary 1 (resp. Corollary 3) and hence to the desired result. Since the individual steps of the proof, in particular the one for the analytic continuation of $\Xi(h, s)$, are rather tedious, the sketch of the following considerations ought to be of help.

Lemma 4. *For* $\mathrm{Re}(s) > 1$ *and integers* h *one has*

$$-\frac{\Xi'(h,s)}{\Xi(h,s)} = \frac{1}{2} \sum_{\gamma \neq 0} \frac{1}{|\gamma|^{2s}} \Lambda_i(\gamma) e^{4ih \arg \gamma} \ .$$

Proof. Formally one obtains this result from the transformation below, using the product representation of $\Xi(h,s)$:

$$-\frac{\Xi'(h,s)}{\Xi(h,s)} = -\frac{d}{ds}\left(\log \Xi(h,s)\right)$$

$$= -\frac{d}{ds}\left(\log 4 - \sum_{\omega} \log\left(1 - \frac{e^{4ih \arg \omega}}{|\omega|^{2s}}\right)\right)$$

$$= \sum_{\omega} \frac{1}{1 - \frac{e^{4ih \arg \omega}}{|\omega|^{2s}}} \cdot \frac{e^{4ih \arg \omega} \cdot \log |\omega|^2}{|\omega|^{2s}}$$

$$= 2 \cdot \sum_{\omega} \frac{\log |\omega| \cdot e^{4ih \arg \omega}}{|\omega|^{2s}} \cdot \sum_{v=0}^{\infty} \frac{e^{4ih \arg(\omega^v)}}{|\omega^v|^{2s}}$$

$$= 2 \cdot \sum_{\omega} \sum_{m=1}^{\infty} \frac{\log |\omega| \cdot e^{4ih \arg(\omega^m)}}{|\omega^m|^{2s}}$$

$$= \frac{1}{2} \cdot \sum_{\varepsilon} \sum_{\omega} \sum_{m=1}^{\infty} \frac{\log |\varepsilon\omega| \cdot e^{4ih \arg((\varepsilon\omega)^m)}}{|(\varepsilon\omega)^m|^{2s}}$$

$$= \frac{1}{2} \cdot \sum_{\gamma \neq 0} \frac{1}{|\gamma|^{2s}} \Lambda_i(\gamma) e^{4ih \arg \gamma} \ .$$

One can however object to this line of argument, that writing $\Xi(h,s)$ in the denominator and too freely handling the logarithm and its functional equation are not justified. Accepting this one may proceed ad hoc with the following correction of the difficulties. Because

$$\log\left(1 - e^{4ih \arg \omega} \cdot |\omega|^{-2s}\right) = O\left(|\omega|^{-2\mathrm{Re}(s)}\right)$$

the series

$$H(s) = \log 4 - \sum_{\omega} \log\left(1 - \frac{e^{4ih \arg \omega}}{|\omega|^{2s}}\right)$$

converges uniformly in each compact subset inside the half-plane $\mathrm{Re}(s) > 1$, and thus represents a holomorphic function. Since the uniform convergence of holomorphic functions f_n in a compact subset to a limit function f implies the uniform convergence of e^{f_n} to e^f in the same compact subset, the formula

$$e^{H(s)} = \Xi(h,s)$$

follows. Consequences are the non-vanishing of $\Xi(h,s)$ and the equation

$$H'(s) \cdot \Xi(h,s) = \Xi'(h,s) \ ,$$

giving the conclusion of the lemma. $\square$

In particular one has

$$-\frac{\zeta_i'(s)}{\zeta_i(s)} = \frac{1}{2} \sum_{\gamma \neq 0} \frac{\Lambda_i(\gamma)}{|\gamma|^{2s}} \; .$$

For the analogue of the Chebyshev function $\psi(x)$ one defines

$$\psi_i(x) = \sum_{|\gamma|^2 \leq x} \Lambda_i(\gamma)$$

and by Abel transformation obtains

$$\sum_{|\gamma|^2 \leq x} \frac{\Lambda_i(\gamma)}{|\gamma|^{2s}} = \frac{1}{x^s} \cdot \psi_i(x) - \int_1^x \psi_i(t) \frac{-s}{t^{s+1}} dt \; .$$

By definition

$$\begin{aligned}
\psi_i(x) &= \sum_{|\gamma|^2 \leq x} \Lambda_i(\gamma) \\
&= 4 \cdot \sum_{|\omega|^2 \leq x} \left[\frac{\log x}{2 \log |\omega|} \right] \log |\omega| \\
&= O\left(\sum_{|\omega|^2 \leq x} \log x \right) = O\left(\log x \cdot \sum_{|\omega|^2 \leq x} 1 \right) \; ;
\end{aligned}$$

where the last sum describes the number of prime elements ω not associated to each other with $|\omega|^2 = \omega\bar\omega \leq x$. Since by Proposition 7 from Chapter 3 these can only be the numbers $1 + i$, $\omega = p$ for prime numbers $p \equiv 3 \pmod 4$, and prime elements ω with $\omega\bar\omega = p \equiv 1 \pmod 4$, p prime, this number can only increase according to the magnitude of the number of prime *numbers* smaller than or equal to x. Hence it follows from Chebyshev's theorem that

$$\psi_i(x) = O(x) \; ,$$

so that in the formula obtained by Abel transformation passage to the limit $x \to \infty$ may be carried out for $\mathrm{Re}(s) > 1$.

Lemma 5. *For all s with $\mathrm{Re}(s) > 1$ one has the integral representation*

$$-\frac{\zeta_i'(s)}{\zeta_i(s)} = \frac{s}{2} \cdot \int_1^\infty \psi_i(s) \cdot x^{-s-1} dx$$

where $\psi_i(x)$ adds the values of $\Lambda_i(\gamma)$ for all $|\gamma|^2 \leq x$.

This lemma is in the analogue of Proposition 19 in the previous chapter.

For the analytic continuability of $\Xi(h, s)$ it is important to know that the estimate

$$\sum_{1\le|\gamma|^2\le x} e^{4ih\,\arg\gamma} = O(\sqrt{x})$$

where the integer h is different from zero. This rests on the following consideration: since only the sum over non-associated elements is of interest, one has

$$\sum_{1\le|\gamma|^2\le x} e^{4ih\,\arg\gamma} = 4\cdot\sum_{a>0}\ \sum_{b\ge0:a^2+b^2\le x} e^{4ih\,\arg(a+ib)}\ ,$$

and since $\arg(a+ib) = \pi/2 - \arg(b+ia)$ holds, the sum simplifies to

$$\sum_{1\le|\gamma|^2\le x} e^{4ih\,\arg\gamma}$$

$$= 8\cdot\sum_{a>0}\ \sum_{b\ge a:a^2+b^2\le x} \cos\big(4h\arg(a+ib)\big) + O(\sqrt{x})$$

$$= 8\cdot\sum_{0<a\le\sqrt{x/2}}\ \sum_{a\le b\le\sqrt{x-a^2}} \cos\left(4h\arctan\frac{b}{a}\right) + O(\sqrt{x})\ .$$

Apply the Euler formula

$$= 8\cdot\sum_{1\le a\le\sqrt{x/2}} \left(\int_a^{\sqrt{x-a^2}} \cos\left(4h\arctan\frac{y}{a}\right) dy\right.$$

$$\left. + O\left(1+\int_a^{\sqrt{x-a^2}} \frac{1}{a}\cdot\frac{dy}{1+\frac{y^2}{a^2}}\right)\right) + O(\sqrt{x})\ .$$

The order of magnitude of the last integral is

$$\arctan\frac{\sqrt{x-a^2}}{a} - \arctan 1 = O(1)\ ,$$

and hence

$$\sum_{1\le|\gamma|^2\le x} e^{4ih\,\arg\gamma}$$

$$= 8\cdot\sum_{1\le a\le\sqrt{x/2}} \int_a^{\sqrt{x-a^2}} \cos\left(4h\arctan\frac{y}{a}\right) dy + O(\sqrt{x})\ .$$

Apply the Euler sum formula once again

$$= 8\cdot\int_1^{\sqrt{x/2}}\int_t^{\sqrt{x-t^2}} \cos\left(4h\arctan\frac{y}{t}\right) dy\, dt$$

$$+ O\left(\sqrt{x} + \int_1^{\sqrt{x/2}} \left|\frac{d}{dt}\int_t^{\sqrt{x-t^2}} \cos\left(4h\arctan\frac{y}{t}\right) dy\right| dt\right)$$

$$+ O(\sqrt{x})\ .$$

Because

$$\frac{d}{dt} \int_t^{\sqrt{x-t^2}} \cos\left(4h \arctan \frac{y}{t}\right) dy$$

$$= \frac{1}{t^2} \int_t^{\sqrt{x-t^2}} y \frac{4h \sin\left(4h \arctan \frac{y}{t}\right)}{1 + \frac{y^2}{t^2}} dy$$

$$- \frac{t}{\sqrt{x-t^2}} \cos\left(4h \arctan \frac{\sqrt{x-t^2}}{t}\right) - \cos h\pi$$

$$= O\left(\int_t^{\sqrt{x-t^2}} y \frac{dy}{t^2 + y^2} + \frac{t}{\sqrt{x-t^2}}\right)$$

$$= O\left(\log \frac{x}{2t^2} + 1\right)$$

and

$$\int_0^1 \int_t^{\sqrt{x-t^2}} \cos\left(4h \arctan \frac{y}{t}\right) dy\, dt = O(\sqrt{x})$$

one finally obtains

$$\sum_{1 \le |\gamma|^2 \le x} e^{4ih \arg \gamma}$$

$$= 8 \cdot \int_0^{\sqrt{x/2}} \int_t^{\sqrt{x-t^2}} \cos\left(4h \arctan \frac{y}{t}\right) dy\, dt + O(\sqrt{x}) \ .$$

Introducing polar coordinates $t = r \cos\varphi$, $y = r \sin\varphi$, $0 < r \le \sqrt{x}$, $\pi/4 \le \varphi \le \pi/2$:

$$= 8 \cdot \int_0^{\sqrt{x}} \int_{\pi/4}^{\pi/2} \cos(4h\varphi) d\varphi\, r\, dr + O(\sqrt{x})$$

$$= O(\sqrt{x}) \ ,$$

since the integral vanishes for $h \ne 0$. In this way one obtains the given estimate. Furthermore in discussing the convergence of

$$\lim_{N \to \infty} \frac{1}{N} \sum_{n=1}^{N} f(n) = c$$

the following application of Abel transformation is important:

$$\sum_{n=1}^{N}\left(\frac{1}{n^s}-\frac{1}{(n+1)^s}\right)\left(\sum_{m=1}^{n}f(m)-nc\right)$$

$$=\sum_{n=1}^{N}\frac{1}{n^s}\left(\left(\sum_{m=1}^{n}f(m)-nc\right)-\left(\sum_{m=1}^{n-1}f(m)-(n-1)c\right)\right)$$

$$-\frac{1}{(N+1)^s}\left(\sum_{m=1}^{N}f(m)-Nc\right)+1\cdot 0$$

$$=\sum_{n=1}^{N}\frac{1}{n^s}\left(f(n)-c\right)+\frac{Nc}{(N+1)^s}$$

$$-\frac{1}{(N+1)^s}\sum_{m=1}^{N}f(m)$$

$$=\sum_{n=1}^{N}\frac{f(n)}{n^s}-c\cdot\sum_{n=1}^{N}\frac{1}{n^s}+c\cdot\frac{N}{N+1}\cdot(N+1)^{1-s}$$

$$-(N+1)^{1-s}\cdot\frac{N}{N+1}\cdot\frac{1}{N}\sum_{n=1}^{N}f(n)$$

$$=\sum_{n=1}^{N}\frac{f(n)}{n^s}-c\cdot\sum_{n=1}^{N}\frac{1}{n^s}$$

$$+\frac{N}{N+1}\left(c-\frac{1}{N}\sum_{n=1}^{N}f(n)\right)(N+1)^{1-s}$$

$$=\sum_{n=1}^{N}\frac{f(n)}{n^s}-c\cdot\sum_{n=1}^{N}\frac{1}{n^s}+O\left((N+1)^{1-\operatorname{Re}(s)}\right)\ .$$

Since for $\operatorname{Re}(s)>1$ $(N+1)^{1-\operatorname{Re}(s)}$ tends to zero as $N\to\infty$, and also

$$\left|\sum_{n=1}^{N}\left(\frac{1}{n^s}-\frac{1}{(n+1)^s}\right)\left(\sum_{m=1}^{n}f(m)-nc\right)\right|$$

$$=\left|\sum_{n=1}^{N}s\cdot\int_{n}^{n+1}x^{-s-1}dx\cdot n\left(c-\frac{1}{n}\sum_{m=1}^{n}f(m)\right)\right|$$

$$\leq\sum_{n=1}^{N}|s|\cdot n^{-1-\operatorname{Re}(s)}n\cdot O(1)$$

$$=O\left(\sum_{n=1}^{N}n^{-\operatorname{Re}(s)}\cdot|s|\right)$$

has a convergent majorising series, for $\operatorname{Re}(s)>1$ one obtains the following important formula

$$\sum_{n=1}^{\infty} \frac{f(n)}{n^s} = c \cdot \zeta(s)$$

$$+ \sum_{n=1}^{\infty} \left(\frac{1}{n^s} - \frac{1}{(n+1)^s} \right) \left(\sum_{m=1}^{n} f(m) - nc \right) \ .$$

Lemma 6. *For integral $h \neq 0$ $\Xi(h,s)$ can be analytically continued on the half-plane of all s with $\operatorname{Re}(s) > \frac{1}{2}$. In the same way the function*

$$\zeta_i(s) - \frac{\pi}{s-1}$$

can be analytically continued on $\operatorname{Re}(s) > 1/2$ in the sense that $\zeta_i(s)$ is a holomorphic function on $\operatorname{Re}(s) > 1/2$ away from a simple pole at $s = 1$ with residue π.

Proof. Since for $h \neq 0$ the relation

$$c = \lim_{N \to \infty} \frac{1}{N} \sum_{n=1}^{N} f(n) = 0$$

holds for

$$f(n) = \sum_{|\gamma|^2 = n} e^{4ih \arg \gamma}$$

it follows from the formula above that

$$\Xi(h,s) = \sum_{n=1}^{\infty} \left(\frac{1}{n^s} - \frac{1}{(n+1)^s} \right) \sum_{1 \leq |\gamma|^2 \leq n} e^{4ih \arg \gamma} \ .$$

Since the expression

$$\sum_{n=M}^{N} \left(\frac{1}{n^s} - \frac{1}{(n+1)^s} \right) \sum_{1 \leq |\gamma|^2 \leq n} e^{4ih \arg \gamma}$$

$$= O \left(|s| \sum_{n=M}^{N} \sqrt{n} \left| \int_{n}^{n+1} x^{-s-1} dx \right| \right)$$

$$= O \left(|s| \sum_{n=M}^{N} n^{-1/2 - \operatorname{Re}(s)} \right)$$

converges uniformly to zero as $M \to \infty$ in each compact subset of the half-plane $\operatorname{Re}(s) > 1/2$, we have already shown the analytic continuability of $\Xi(h,s)$. Similarly the formula above helps with

$$\zeta_i(s) = \sum_{n=1}^{\infty} \frac{r(n)}{n^s}$$

$$= \pi \cdot \zeta(s) + \sum_{n=1}^{\infty} \left(\frac{1}{n^s} - \frac{1}{(n+1)^s} \right) \left(\sum_{m=1}^{n} r(m) - n\pi \right) ,$$

because, given the conclusion of Proposition 9 in the previous chapter, we may put $c = \pi$ and observe that

$$\sum_{m=1}^{n} r(m) - n\pi = O(\sqrt{n}) .$$

Lemma 7. *For all integers* h $\Xi(h,s) \neq 0$ *holds on the line* $\mathrm{Re}(s) = 1$.

Proof. This is obvious in the case $h = 0$, $s = 1$, because here a pole occurs. For the other cases we can use a modified version of Mertens' technique from Lemma 3. If $\Xi(s, 1 + it)$ were equal to zero, then

$$\Theta(s) = \zeta_i(s)^3 \cdot \Xi(h, s + it)^4 \cdot \Xi(2h, s + 2it)$$

also possesses a zero for $s = 1$, which would imply

$$\lim_{s \to 1} \log |\Theta(s)| = -\infty .$$

In particular for $s = \sigma > 1$, we have the product representation

$$\log |\Xi(h, \sigma + it)|$$

$$= \log 4 - \sum_{\omega} \log \left| 1 - \frac{e^{4ih \arg \omega}}{|\omega|^{2\sigma + 2it}} \right|$$

$$= \log 4 + \sum_{\omega} \sum_{n=1}^{\infty} \frac{\cos n(4h \arg \omega - 2t \log |\omega|)}{n|\omega|^{2n\sigma}}$$

which leads to the contradiction

$$\log |\Theta(s)| = 8 \log 4$$

$$+ \sum_{\omega} \sum_{n=1}^{\infty} \frac{3 + 4 \cos n(4h \arg \omega - 2t \log |\omega|) + \cos n(8h \arg \omega - 4t \log |\omega|)}{n|\omega|^{2n\sigma}}$$

$$\geq 0 .$$

Now we have reached our goal: $\psi_i(x)$ is a monotone non-decreasing function, defined for $x \geq 1$, with $\psi_i(x) = O(x)$. Hence the function defined by it

$$-\frac{\zeta_i'(s)}{\zeta_i(s)} = s \cdot \int_1^{\infty} \frac{1}{2} \psi_i(x) x^{-s-1} dx$$

is holomorphic in $\mathrm{Re}(s) > 1$, and

$$-\frac{\zeta_i'(s)}{\zeta_i(s)} - \frac{1}{s-1}$$

can be regarded as a holomorphic map, defined in some region containing the closed half-plane $\mathrm{Re}(s) \geq 1$. Therefore Corollary 1 implies

Proposition 2. *We have the asymptotic representation*

$$\psi_i(x) = \sum_{|\gamma|^2 \leq x} \Lambda(\gamma) \sim 2x \ .$$

Moreover, since for $h \neq 0$

$$e^{4ih \arg \gamma} \Lambda(\gamma) = O(\Lambda(\gamma)) \ ,$$

the Dirichlet series

$$-\frac{\Xi'(h,s)}{\Xi(h,s)} = \frac{1}{2} \sum_{\gamma \neq 0} \frac{1}{|\gamma|^{2s}} \cdot \Lambda(\gamma) e^{4ih \arg \gamma}$$

is holomorphic for $\mathrm{Re}(s) > 1$ and at $s = 1$ can be regarded as a holomorphic map without singularity in some region which contains the closed half-plane $\mathrm{Re}(s) \geq 1$, Corollary 3 implies

Proposition 3. *For integers $h \neq 0$*

$$\sum_{|\gamma|^2 \leq x} e^{4ih \arg \gamma} \Lambda(\gamma) = o(x)$$

holds.

These conclusions do not yet appear to be powerful. With their help it is nonetheless possible to calculate

$$\sum_{|\omega|^2 \leq x} e^{4ih \arg \omega}$$

and to show that, either for $h = 0$ the sum asymptotically agrees with $x/\log x$, or that for $h \neq 0$ it has order of magnitude equal to $o(x/\log x)$. Indeed the estimate

$$\sum_{k \geq 2} \sum_{|\omega|^{2k} \leq x} e^{4ih \arg \omega^k} \log |\omega|$$

$$= O\left(\log x \cdot \sum_{k \geq 2} \sum_{|\omega|^2 \leq x^{1/k}} 1 \right)$$

$$= O\left(\log x \cdot \sum_{2 \leq k \leq \log x / \log 2} \frac{\sqrt{x}}{\log \sqrt{x}} \right)$$

$$= O\left(\sqrt{x} \log x \right)$$

leads to

$$4 \cdot \sum_{|\omega|^2 \leq x} e^{4ih \arg \omega} \log |\omega|$$

$$= 4 \cdot \sum_{k=1}^{\infty} \sum_{|\omega|^{2k} \leq x} e^{4ih \arg \omega^k} \log |\omega| + O(\sqrt{x} \log x)$$

$$= \sum_{|\gamma|^2 \leq x} e^{4ih \arg \gamma} \Lambda(\gamma) + O(\sqrt{x} \log x)$$

$$= \begin{cases} 2x + o(x), & h = 0 \ , \\ o(x), & h \neq 0 \ . \end{cases}$$

By Abel transformation

$$\sum_{|\omega|^2 \leq x} e^{4ih \arg \omega}$$

$$= \sum_{2 \leq |\omega|^2 \leq x} e^{4ih \arg \omega} \log |\omega|^2 \cdot \frac{1}{\log |\omega|^2}$$

$$= \frac{1}{\log x} \cdot \sum_{2 \leq |\omega|^2 \leq x} e^{4ih \arg \omega} \log |\omega|^2$$

$$- \int_2^x \sum_{2 \leq |\omega|^2 \leq t} \log |\omega|^2 \cdot e^{4ih \arg \omega} \cdot \frac{-dt}{t \cdot \log^2 t}$$

$$= \frac{2}{\log x} \cdot \sum_{|\omega|^2 \leq x} e^{4ih \arg \omega} \log |\omega| + O\left(\int_2^x \frac{dt}{\log^2 t} \right)$$

$$= \begin{cases} \dfrac{x}{\log x} + o\left(\dfrac{x}{\log x}\right) + O\left(\dfrac{x}{\log^2 x}\right), & h = 0 \\[2ex] o\left(\dfrac{x}{\log x}\right) + O\left(\dfrac{x}{\log^2 x}\right), & h \neq 0 \end{cases}$$

$$\begin{cases} = \dfrac{x}{\log x} + o\left(\dfrac{x}{\log x}\right), & h = 0 \ , \\[2ex] = o\left(\dfrac{x}{\log x}\right), & h \neq 0 \end{cases}$$

giving the assertion. The consequence of all this is the generalisation of the prime number theorem to the ring $\mathbb{Z}(i)$:

Theorem 4: Hecke's Prime Number Theorem for $\mathbb{Z}(i)$. *If $\pi_i(x)$ counts the non-associated prime elements ω with $|\omega|^2 \leq x$, then*

$$\pi_i(x) \sim \frac{x}{\log x} \ .$$

If for $0 \le \alpha < \beta \le 2\pi$ $\pi_i(x; \alpha, \beta)$ counts the prime elements ψ of $\mathbb{Z}(i)$ lying in the corner region $\alpha \le \arg\gamma < \beta$ with $|\psi|^2 \le x$, then

$$\pi_i(x; \alpha, \beta) \sim \frac{2}{\pi}(\beta - \alpha) \cdot \frac{x}{\log x} \ .$$

Proof. On the one hand for $h = 0$ the formula above gives

$$\pi_i(x) = \sum_{|\omega|^2 \le x} 1 = \frac{x}{\log x} + o\left(\frac{x}{\log x}\right) \ .$$

On the other hand for $h \ne 0$, because

$$\lim_{x \to \infty} \frac{\log x}{x} \cdot \sum_{|\omega|^2 \le x} e^{4ih \arg \omega}$$

$$= \lim_{x \to \infty} \frac{1}{\pi_i(x)} \cdot \sum_{|\omega|^2 \le x} e^{2\pi i h(2/\pi \cdot \arg \omega)} = 0$$

satisfies Weyl's criterion from uniform distribution theory, for each Riemann integrable function $f(\varphi)$ it follows that

$$\lim_{x \to \infty} \frac{1}{4} \cdot \frac{\log x}{x} \cdot \sum_{|\psi|^2 \le x} f(\arg \psi)$$

$$= \frac{1}{2\pi} \int_0^{2\pi} f(\varphi) d\varphi \ .$$

The factor $\frac{1}{4}$ takes care of the fact that ψ does not only run over prime elements ω, but also over all their associates. By setting f equal to the characteristic function for $[\alpha, \beta[$, the second assertion of the theorem follows. $\square$

Exercises on Chapter 5

1. Let p_n be the nth prime number. Show that $\lim_{n \to \infty} \frac{p_n}{n \log n} = 1$.

2. Show that for $0 < s < 1$, $\zeta(s) < 0$.

* 3. Show that the estimate $\pi(x) = \frac{x}{\log x} + O\left(\frac{x}{\log^3 x}\right)$ is false.

* 4. (i) If $c = \lim_{x \to \infty} \log x \cdot \prod_{p \le x} \left(1 - \frac{1}{p}\right)$, then $\underline{\lim}_{n \to \infty} \frac{\varphi(n)}{n} \log \log n \le c$.
    (ii) $\overline{\lim}_{n \to \infty} \frac{\log \tau(n)}{\log n} \log \log n \ge \log 2$.

5. For $n \ge 1$ let $P(n) = \prod_{\substack{p \mid n \\ p > \log n}} \left(1 - \frac{1}{p}\right)$. Show that $\lim_{n \to \infty} P(n) = 1$.

6. Let $c = \lim_{x \to \infty} \log x \cdot \prod_{p \le x} \left(1 - \frac{1}{p}\right)$ (Mertens' theorem!). Show that

$$\underline{\lim}_{n \to \infty} \frac{\varphi(n)}{n} \log \log n = c \ .$$

[One can show that $c = e^C$ where C is Euler's constant.]

7. For $n \geq 2$, let $f(n) > 1$, $n = \prod_p p^{\alpha_p(n)}$ be the prime factor decomposition of n, and $P(n) = \prod_{p \geq f(n)}(1 + \alpha_p(n))$. Show that $\log P(n) \leq \log 2 \cdot \frac{\log n}{\log f(n)}$.

8. For $n \geq 2$, let $f(n) = \frac{\log n}{\log^2 \log n}$, $n = \prod_p p^{\alpha_p(n)}$ be the prime factor decomposition of n, and $Q(n) = \prod_{p < f(n)}(1 + \alpha_p(n))$. Show that $\log Q(n) = O(f(n))$ as $n \to \infty$.

9. Show that $\overline{\lim}_{n \to \infty} \frac{\log r(n)}{\log n} \cdot \log \log n = \log 2$.

* 10. Let $f : \mathbb{R} \to \mathbb{C}$ be differentiable, $\int_{\mathbb{R}} |f'(x)| dx < \infty$ and suppose that $\int_{\mathbb{R}} f(x) dx$ exists (i.e. $\lim_{\substack{K \to \infty \\ L \to -\infty}} \int_L^K f(x) dx$ exists). Show that

$$\sum_{n \in \mathbb{Z}} f(n) = \sum_{k \in \mathbb{Z}} \int_{\mathbb{R}} f(x) e^{2\pi i k x} dx \ .$$

(Poisson sum formula)

11. Let $z \in \mathbb{C}$. Show that $\int_{\mathbb{R}} e^{-\pi x^2 + 2\pi i x z} dx = e^{-\pi z^2}$.

12. Let $\Theta : (0, \infty) \to \mathbb{R}$, $\Theta(x) = \sum_{n \in \mathbb{Z}} e^{-\pi n^2 x}$. Θ is called a *Theta function*. Show that, for $x > 0$, $\Theta(x) = \frac{1}{\sqrt{x}} \Theta\left(\frac{1}{x}\right)$.

13. Let $\Psi : (0, \infty) \to \mathbb{R}$, $\Psi(x) = \frac{1}{2}(\Theta(x) - 1) = \sum_{n=1}^{\infty} e^{-\pi n^2 x}$. Prove:
   (i) $z \mapsto \int_1^{\infty} \Psi(x) x^z dx$ is an entire function.
   (ii)   For $\mathrm{Re}(s) > 1$,

$$\Gamma\left(\frac{s}{2}\right) \pi^{-s/2} \zeta(s) = \int_0^{\infty} \Psi(x) x^{s/2} \frac{dx}{x} \ .$$

14. Show that for $\mathrm{Re}(s) > 1$

$$\int_0^{\infty} \Psi(x) x^{s/2} \frac{dx}{x} = \int_1^{\infty} \Psi(x) \left(x^{s/2} + x^{(1-s)/2}\right) \frac{dx}{x} + \frac{1}{s(s-1)} \ .$$

15. For $s \in \mathbb{C}$ let $\xi(s) = \Gamma\left(\frac{s}{2}\right) \pi^{-s/2} \zeta(s)$. Show that $\xi(s) = \xi(1-s)$. (Functional equation of the Riemann zeta function)

16. (i)   For all $n \in \mathbb{N}$ $\zeta(-2n) = 0$.
    (ii) $\zeta(0) = -\frac{1}{2}$; if $\zeta(s) = 0$, then either $0 < \mathrm{Re}\, s < 1$ or $s = -2n$ for some $n \in \mathbb{N}$.
    (iii) If $\zeta(s) = 0$, then $\zeta(\bar{s}) = 0$ also.

17. For $n \geq 1$ let $r(n)$ be the number of solutions of $\varphi(m) = n$. Show that $r(n)$ is finite, and that for $\mathrm{Re}\, s > 1$,

$$\sum_{n=1}^{\infty} \frac{r(n)}{n^s} = \prod_p \left(1 + \frac{1}{(p^s - 1)(1 - p^{-1})^s}\right) = \sum_{n=1}^{\infty} \varphi(n)^{-s} \ .$$

* 18. For $\mathrm{Re}(s) > 1$ let $f(s) = \sum_{n=1}^{\infty} \frac{r(n)}{n^s}$. Show that f can be meromorphically continued on the region $\{s \in \mathbb{C} : \mathrm{Re}(s) > 0\}$, and that f has a simple pole at $s = 1$ with residue $\prod_p \left(1 + \frac{1}{p(p-1)}\right)$.

* 19. Show that $\lim_{N \to \infty} \frac{1}{N} \sum_{n=1}^{N} r(n) = \prod_p \left(1 + \frac{1}{p(p-1)}\right)$.

* 20. Show that $(\log p_n)_{n \geq 1}$ is not uniformly distributed mod 1.

Hints for the Exercises on Chapter 5

3. Find a contradiction to Mertens' theorem.

4. If p_k is the kth prime number, consider $n_k = p_1 \ldots p_k$.

10. Show first that $\lim_{|x| \to \infty} f(x) = 0$. Then apply the Euler sum formula together with

$$\{x\} - \frac{1}{2} = -\frac{1}{2\pi i} \sum_{k \neq 0} \frac{1}{k} e^{2\pi i k x} \ .$$

18. For $\mathrm{Re}(s) > 1$ consider $\frac{f(s)}{\zeta(s)}$.

19. Apply the Tauberian theorem of Wiener and Ikehara.

20. Show with the help of the prime number theorem that

$$\int_2^N \frac{1}{x} \left| \pi(x) - \frac{x}{\log x} \right| dx = o(\pi(N)) \ .$$

Apply partial integration in the integral $\int_2^N \frac{e^{2\pi i \log x}}{\log x} dx$, and finally show that

$$\lim_{N \to \infty} \frac{1}{\pi(N)} \left| \sum_{p \leq N} e^{2\pi i \log p} \right| = \frac{1}{\sqrt{1 + 4\pi^2}} \ .$$

6. Characters of Groups of Residues

The congruences introduced by Gauss in his "Disquitiones Arithmeticae" endowed the set $\mathbb{Z}$ of all integers with an infinite multiplicity of finite abelian groups, the prime groups of residues modulo a coprime integer. Whereas the investigations of the two previous chapters have been valid for asymptotic calculations for number theoretic functions over $\mathbb{Z}$, in this chapter of our book we will present asymptotic calculations, which hold for number theoretic functions over groups of residues. The *character* of a group of residues makes its appearance as the central concept for such investigations. What do we understand by this?

It is well-known that a set G of elements $g, g', \ldots$ is called a *group*, if a multiplication is defined, which uniquely associates to each pair g', g'' of elements from G an element $g = g'g''$. This satisfies the associativity law $g'(g''g''') = (g'g'')g'''$, possesses an identity element g_0 (usually denoted by $g_0 = 1$) lying in G with $g = g_0 g = g g_0$ for all g in the set, and for each g from G there exists an inverse element g^{-1} lying in G with $g^{-1}g = gg^{-1} = g_0$. The group G is called *abelian* or commutative, if the commutivity law $g'g'' = g''g'$ always holds, and it is called *finite*, if it contains only finitely many elements. The number of elements lying in G is called its *order* $|G|$. In what follows only finite abelian groups are of interest.

If g is an element of the group, the set of all powers g^h with h from $\mathbb{Z}$ (where g^0 is the identity element) also forms a group, which is either identical with G, or forms a proper subgroup of G. Because of the finiteness of G not all g^h can be distinct from each other; from $g^h = g^k$ with $h < k$ it follows that $g^{k-h} = g^0$, where $k - h$ is a natural number. The smallest natural number n with $g^n = g^0$ is called the *order* of the element g, and agrees with the order of the group of elements $g^0, g^1, g^2, \ldots, g^{n-1}$ generated by g. If all of G is obtained in this way, G is called a *cyclic* group, and g is a generating element of the cyclic group G.

In general a finite abelian group does not need to be cyclic. Nonetheless one can unravel its structure by means of successive powers of elements. If G' is a proper subgroup of G and if g denotes a group element not lying in G', then – since some power of g equals the identity element lying in G' – there must exist an exponent n, with the property that g^n lies in G'. The smallest exponent with this property is called the *indicator* of g; denote it by $h = g : G'$. One sees easily that the set of all $g'g^k$ with $g' \in G'$ and $0 \le k < h$

also forms a subgroup G'' of G with $G'' = h|G'|$. Firstly for each two elements $g_1' g^k$ and $g_2' g^j$ we have the relation

$$\left(g_1' g^k\right)\left(g_2' g^j\right) = \left(g_1' g_2'\right) g^{k+j} = g' g^r \ ,$$

where r is the smallest non-negative residue of $k+j$ modulo h, $k+j = hq+r$, $0 \le r < h$, and $g' = g_1' g_2' \left(g^h\right)^q$ is an element of G. Secondly for $k = 0$ the inverse element for $g' g^k$ equals $g'^{-1} g^k$ and for $0 < k < h$ $(g_1')^{-1} g^{h-k}$ with $g_1' = g' g^h$. This already shows that G'' is a group. If g' runs through all elements of G' and k through all integers from 0 to $h-1$, then for $g' g^k$ one does indeed obtain $h|G'|$ elements. They are pairwise distinct, since the equation $g_1' g^k = g_2' g^j$ implies for $k \le j$ that $g^{j-k} = g_1' \left(g_2'\right)^{-1}$, from which it follows that $j-k < h$, $j-k = 0$, $j = k$ and hence also that $g_1' = g_2'$. The group generated in this way will be denoted $G'' = G' \vee g$. If $\vee g_1$ denotes the cyclic subgroup of G generated by g_1, then if $\vee g_1 = G$ one has already described the structure of G. Otherwise there exists some group element g_2 not belonging to $\vee g_1$, so that one can form $\vee g_1 \vee g_2$. If this subgroup agrees with G the structure of G is already determined – otherwise one continues this method of construction – and obtains a sequence of subgroups

$$\vee g_1, \ \vee g_1 \vee g_2, \ \vee g_1 \vee g_2 \vee g_3, \ \ldots, \ \vee g_1 \vee g_2 \vee \ldots \vee g_\ell, \ \ldots$$

having ever greater orders. Because of the finiteness of G this sequence must terminate for some index $\ell = L$, so that

$$G = \vee g_1 \vee g_2 \vee \ldots \vee g_L$$

holds. Thus the general element of the finite abelian group G has the form

$$g = g_1^{k_1} g_2^{k_2} \ldots g_L^{k_L}$$

with $0 \le k_\ell < h_\ell$, where h_1 is the order of g_1 and for $\ell = 2, \ldots, L$

$$h_\ell = g_\ell : \vee g_1 \vee \ldots \vee g_{\ell-1}$$

denotes the indicator of g_ℓ with respect to $\vee g_1 \vee g_2 \ldots \vee g_{\ell-1}$.

A character $\mathbf{c}$ of a finite abelian group G is a complex-valued function defined on G with the *homomorphism property*

$$\mathbf{c}(g' g'') = \mathbf{c}(g') \cdot \mathbf{c}(g'') \ .$$

From now on we exclude the trivial case that $\mathbf{c}(g) = 0$ for all g from G. It follows that for some element g from G we must have $\mathbf{c}(g) \ne 0$, and for the identity element, because of $\mathbf{c}(g) = \mathbf{c}(g_0 g) = \mathbf{c}(g_0)\mathbf{c}(g)$, necessarily $\mathbf{c}(g_0) = 1$. If g denotes an arbitrary element from G of order n, it follows from $\mathbf{c}(g^n) = 1 = \mathbf{c}(g)^n$, that $\mathbf{c}(g)$ must be an nth root of unity $e^{2\pi i h/n}$ with h integral. In this way characters map group elements onto complex numbers of absolute value 1, and in particular associate elements of order n with nth roots of unity.

If $G = \vee g$ is a cyclic group of order n, n possible characters $\mathbf{c}^0, \mathbf{c}^1, \ldots, \mathbf{c}^{n-1}$ arise, if one defines $\mathbf{c}^j(g) = e^{2\pi i j/n}$, obtaining $\mathbf{c}^j(g^k) = e^{2\pi i k j/n}$ for the general group element g^k.

Assuming that the structure of the characters of the group G' is already known, let h denote the indicator of g with respect to G'. Then each character $\mathbf{c}'$ of G' can be extended in h different ways to a character $\mathbf{c}$ of $G' \vee g$. Since

$$\mathbf{c}(g'g^k) = \mathbf{c}'(g') \cdot \mathbf{c}(g)^k$$

must hold, and the fact that g^h lies in G' shows that $\mathbf{c}(g)^h = \mathbf{c}'(g^h)$, in order to ensure the homomorphism property the calculation

$$\begin{aligned}
\mathbf{c}\left(g_1'g^k \cdot g_2'g^j\right) &= \mathbf{c}\left(g_1'g_2' \cdot g^{k+j}\right) \\
&= \mathbf{c}'\left(g_1'g_2'\right) \cdot \mathbf{c}(g)^{k+j} \\
&= \mathbf{c}'\left(g_1'\right) \cdot \mathbf{c}'\left(g_2'\right) \cdot \mathbf{c}(g)^k \cdot \mathbf{c}(g)^j \\
&= \mathbf{c}\left(g_1'g^k\right) \cdot \mathbf{c}\left(g_2'g^j\right)
\end{aligned}$$

implies that for $\mathbf{c}(g)$ only the h distinct hth roots

$$\mathbf{c}'(g^h)^{1/h} \cdot e^{2\pi i j/h}, \quad j = 0, 1, \ldots, h-1 \ ,$$

of $\mathbf{c}'(g^h)$ come into question. It follows from all these considerations that

Proposition 1. *A finite abelian group has exactly as many characters as its order.*

If for two characters $\mathbf{c}', \mathbf{c}''$ of the group G one defines a multiplication $\mathbf{c}'\mathbf{c}'' = \mathbf{c}$ by

$$\mathbf{c}(g) = \mathbf{c}'\mathbf{c}''(g) = \mathbf{c}'(g) \cdot \mathbf{c}''(g) \ ,$$

the set $\mathbf{C}$ of all characters of G acquires the structure of a finite abelian group – the axioms are easily verified – in particular the so-called *principal character* $\mathbf{c}_0$ with $\mathbf{c}_0(g) = 1$ for all $g \in G$ plays the role of the identity element in $\mathbf{C}$, and the character inverse to $\mathbf{c}$ is given by $\mathbf{c}^{-1} = \bar{\mathbf{c}}$ with

$$\bar{\mathbf{c}}(g) = \overline{\mathbf{c}(g)} = \frac{1}{\mathbf{c}(g)} = \mathbf{c}(g^{-1}) \ .$$

$\mathbf{C}$ is therefore called the *character group* of G. It is related to G by a certain duality: on the one hand it has precisely as many elements as G, and on the other, for group elements as for characters the so-called orthogonality relations hold.

Proposition 2: Orthogonality Relations for Group Elements. *For the principal character* $\mathbf{c}_0$ *of the group G one has*

$$\sum_{g \in G} \mathbf{c}_0(g) = |G| \ ,$$

and for all the remaining characters c

$$\sum_{g \in G} \mathbf{c}(g) = 0 \ .$$

Proof. Because $\mathbf{c}_0(g) = 1$ the first formula is obvious. If $\mathbf{c}$ is not the principal character, then there exists some g_1 in G with $\mathbf{c}(g_1) \neq 1$, and as g runs through all the elements of G so does $g_1 g$. Hence from

$$\sum_{g \in G} \mathbf{c}(g) = \sum_{g \in G} \mathbf{c}(g_1 g) = \mathbf{c}(g_1) \cdot \sum_{g \in G} \mathbf{c}(g) \ ,$$

$$(1 - \mathbf{c}(g_1)) \cdot \sum_{g \in G} \mathbf{c}(g) = 0$$

the second formula follows by division by $1 - \mathbf{c}(g_1)$. $\square$

One can describe this result as follows. If $\Gamma = (\mathbf{c}(g))$ describes the matrix of all possible function values of characters on group elements, where g serves as row index and $\mathbf{c}$ as column index, and if Γ^+ denotes the transposed and complex conjugated matrix to Γ (i.e. Hermitian conjugated with $\Gamma^+ = (\bar{\mathbf{c}}(g))$ with $\mathbf{c}$ as row index and g as column index), consider the matrix product

$$\Gamma\Gamma^+ = \left(\sum_{g \in G} \mathbf{c}(g) \cdot \overline{\mathbf{c}'}(g) \right) = \left(\sum_{g \in G} (\mathbf{c}\overline{\mathbf{c}'})(g) \right)$$

$$= \left(\sum_{g \in G} \left(\mathbf{c}\mathbf{c}'^{-1} \right)(g) \right) \ .$$

The result is only non-zero for $\mathbf{c}\mathbf{c}'^{-1} = \mathbf{c}_0$, i.e. for $\mathbf{c} = \mathbf{c}'$, and here agrees with $|G|$. If I denotes the unit matrix it follows that

$$\Gamma^+\Gamma = |G| \cdot I \ ,$$

which shows that $\Gamma/\sqrt{|G|} = (\mathbf{c}(g))/\sqrt{|G|}$ is a unitary matrix. In particular the two matrices commute, $\Gamma^+\Gamma = |G| \cdot I$ – in detail

$$\sum_{\mathbf{c} \in C} \overline{\mathbf{c}(g)} \cdot \mathbf{c}(g') = \sum_{\mathbf{c} \in C} \mathbf{c}(g^{-1})\mathbf{c}(g')$$

$$= \sum_{\mathbf{c} \in C} \mathbf{c}(g^{-1}g') \ .$$

This means that for $g = g'$ we obtain the result $|G|$ and otherwise zero. Hence

Proposition 3: Orthogonality Relations for Group Characters. *For the identity element g_0 of G one has*

$$\sum_{c \in C} c(g_0) = |G| \ ,$$

and for all other group elements

$$\sum_{c \in C} c(g) = 0 \ .$$

The duality between the finite abelian group G and its character group $\mathbf{C}$ appears even more clearly in the following theorem:

Proposition 4. *The character group $\mathbf{C}$ of each finite abelian group G is isomorphic to G.*

Proof. We know that the general structure of G may be expressed in the form

$$G = \vee g_1 \vee g_2 \vee \ldots \vee g_L,$$

where h_1 denotes the order of g_1 and h_ℓ is the indicator of g_ℓ with respect to $\vee g_1 \vee \ldots \vee g_{\ell-1}$. To each element g from G which can be described as

$$g = g_1^{k_1} g_2^{k_2} \ldots g_L^{k_L}$$

with $0 \le k_\ell < h_\ell$, we associate a character c_g as follows: write

$$c_g \left(g_1^{j_1} g_2^{j_2} \ldots g_L^{j_L} \right)$$
$$= e^{2\pi i j_1 k_1 / h_1} e^{2\pi i j_2 k_2 / h_2} \cdot \ldots \cdot e^{2\pi i j_L k_L / h_L} \ .$$

It is clear that mapping g to c_g is a homomorphism, i.e. one has the relation $c_{g'} c_{g''} = c_{g'g''}$. If c_g agrees with the principal character c_0, $c_g = c_0$, it follows from

$$c_g(g_\ell) = e^{2\pi i k_\ell / h_\ell} = 1$$

that necessarily $k_\ell = 0$, where ℓ can be chosen arbitrarily between 1 and L. But this implies that g coincides with the identity element g_0, and in this way one shows that the map g to c_g is injective. Because G and $\mathbf{C}$ have the same number of elements, this map has to be an isomorphism, as asserted. $\square$

In this way the structure of finite abelian groups and their character groups is abstractly described.

The groups of interest in elementary number theory are the groups of residues coprime with an integer m, which consist of $\varphi(m)$ sets $a + m\mathbb{Z}$ with g.c.d. $(a, m) = 1$. More precisely: the group elements are the sets $a + m\mathbb{Z}$, where a runs through all integers coprime with m. Since for $a \equiv a' (\mathrm{mod}\ m)$ we have $a + m\mathbb{Z} = a' + m\mathbb{Z}$, we do indeed have only a group of order $\varphi(m)$. Multiplication $(a' + m\mathbb{Z})(a'' + m\mathbb{Z}) = a'a'' + m\mathbb{Z}$ defines the group operation. Dirichlet proposed not to allow the group characters to operate directly on the group elements, i.e. on the classes of residues, but to consider them as *number theoretic functions*. If c represents a character of the prime group of residues modulo m in the abstract, original sense, the associated Dirichlet character χ is defined as a number theoretic function by setting

$$\chi(a) = c(a + m\mathbb{Z}), \text{ for } a \in \mathbb{Z} \text{ with g.c.d. } (a, m) = 1, \text{ and}$$
$$\chi(n) = 0 \text{ for g.c.d. } (n, m) > 1 \ .$$

In this way he arrives at

Proposition 5. *There exist $\varphi(m)$ distinct Dirichlet characters modulo the natural number m; each of them is strongly multiplicative and periodic with period m. This means that for all k, n we have*

$$\chi(kn) = \chi(k) \cdot \chi(n)$$

and

$$\chi(n + m) = \chi(n) \ .$$

Conversely each strongly multiplicative and periodic (with period m) number theoretic function χ, which for g.c.d.$(n, m) > 1$ takes the value $\chi(n) = 0$, is one of the Dirichlet characters modulo m.

Proof. For g.c.d.$(k, m) = $ g.c.d.$(n, m) = 1$ strong multiplicativity is exactly the homomorphism property. Since for g.c.d.$(k, m) > 1$ it is also true that g.c.d.$(nk, m) > 1$, strong multiplicativity certainly holds. Periodicity follows directly from the definition. Since the number of Dirichlet characters agrees with the number of characters in the abstract sense, and therefore by Proposition 1 equals the order $\varphi(m)$ of the group of prime classes of residues, all assertions in the proposition are clear.   □

Since Dirichlet characters arise as number theoretic functions, the next step is to study them by means of the Dirichlet series which they form

$$L(\chi, s) = \sum_{n=1}^{\infty} \frac{\chi(n)}{n^s} \ ,$$

known under the name of *Dirichlet L-series*. In the region of absolute convergence, from the strong multiplicativity of the character it follows that we have the product representation

$$L(\chi, s) = \prod_p \frac{1}{1 - \dfrac{\chi(p)}{p^s}} \ .$$

In particular for the principal character $\chi = \chi_0$

$$L(\chi_0, s) = \prod_{p \nmid m} \frac{1}{1 - \dfrac{1}{p^s}}$$

$$= \prod_{p \mid m} \left(1 - \frac{1}{p^s}\right) \cdot \prod_{p} \frac{1}{1 - \dfrac{1}{p^s}} \ ,$$

and in this way one obtains the important formula

$$L(\chi_0, s) = \prod_{p \mid m} \left(1 - \frac{1}{p^s}\right) \cdot \zeta(s) \ .$$

In the same way that the Riemann ζ-function was used for the proof of the prime number theorem and $\Xi(h, s)$ for the proof of Hecke's prime number theorem for $\mathbb{Z}(i)$, Dirichlet L-series will be applied in the proof of a prime number theorem of Dirichlet.

Lemma 1. *If χ is not the principal character, $L(\chi, s)$ can be considered as a holomorphic function on the half-plane of all s with $\mathrm{Re}(s) > 0$. If $\chi = \chi_0$ is the principal character, then $L(\chi_0, s)$ is analytic in $\mathrm{Re}(s) > 1$, and the map*

$$L(\chi_0, s) - \frac{\varphi(m)}{m} \cdot \frac{1}{s - 1}$$

can be analytically continued on $\mathrm{Re}(s) > 0$, i.e. $L(\chi_0, s)$ is a holomorphic function in $\mathrm{Re}(s) > 0$ up to a simple pole at $s = 1$ with residue $\varphi(m)/m$.

Proof. For $\chi = \chi_0$ the conclusion of the lemma follows from the connection between $L(\chi_0, s)$ and the Riemann ζ-function just described, that is from the extendability of $\zeta(s) - 1/(s - 1)$ as an analytic function on $\mathrm{Re}(s) > 1$ to one on $\mathrm{Re}(s) > 0$, and from the known representation

$$\varphi(m) = m \cdot \prod_{p \mid m} \left(1 - \frac{1}{p}\right)$$

of the Euler function. For $\chi \neq \chi_0$ the conclusion follows, since on the one hand, given the orthogonality relations, one concludes that

$$\sum_{n=P+1}^{P+m} \chi(n) = 0 \ ,$$

and hence

$$\left| \sum_{n=P}^{Q} \chi(n) \right| \leq m \ ,$$

and on the other, using an Abel transformation for real $s = \mathrm{Re}(s) > 0$,

$$\left| \sum_{n=P}^{Q} \frac{\chi(n)}{n^s} \right|$$

$$= \left| \frac{1}{(Q+1)^s} \cdot \sum_{n=P}^{Q} \chi(n) \right.$$

$$\left. + \sum_{n=P}^{Q} \left(\frac{1}{n^s} - \frac{1}{(n+1)^s} \right) \cdot \sum_{k=P}^{n} \chi(k) \right|$$

$$\leq \frac{m}{(Q+1)^s} + m \cdot \sum_{n=P}^{Q} \left(\frac{1}{n^s} - \frac{1}{(n+1)^s} \right)$$

$$= \frac{m}{P^s} \ .$$

This shows that the residual sum of $L(\chi, s)$ tends to zero. $\qquad \square$

For monotone decreasing positive number theoretic functions $f(n)$ the last calculation has the generalisation

$$\left| \sum_{n=P}^{Q} \chi(n)f(n) \right|$$

$$= \left| f(Q+1) \cdot \sum_{n=P}^{Q} \chi(n) \right.$$

$$\left. + \sum_{n=P}^{Q} (f(n) - f(n+1)) \cdot \sum_{k=P}^{n} \chi(k) \right|$$

$$\leq m \cdot f(Q+1) + m \cdot \sum_{n=P}^{Q} (f(n) - f(n+1))$$

$$= m \cdot f(P) \ .$$

Thus the convergence of $f(n)$ to zero as $n \to \infty$ implies the convergence of the series

$$\sum_{n=1}^{\infty} \chi(n)f(n) \ .$$

Lemma 2. *On the line* $\mathrm{Re}(s) = 1$ $\qquad L(\chi, s) \neq 0$ *for all characters* χ.

Proof. For $\chi = \chi_0$ the conclusion of the lemma follows from the connection between $L(\chi_0, s)$ and the Riemann ζ-function. Now assume that $\chi \neq \chi_0$ and $L(\chi, 1 + it) = 0$. In the case $t \neq 0$ or $\chi^2 \neq \chi_0$ Mertens' familiar method of argument admits translation. Thus

$$\Theta(s) = L(\chi_0, s)^3 \cdot L(\chi, s + it)^4 \cdot L(\chi^2, s + 2it)$$

would also possess a zero for $s = 1$, implying that

$$\lim_{s \to 1} \log |\Theta(s)| = -\infty \ .$$

In the special case $s = \sigma > 1$, given the product representation of $L(\chi, s)$

$$\log |L(\chi, \sigma + it)| = -\sum_{p} \log \left|1 - \frac{\chi(p)}{p^{\sigma+it}}\right|$$

$$= \sum_{p \nmid m} \sum_{n=1}^{\infty} \frac{\cos n(\arg \chi(p) - t \log p)}{n p^{n\sigma}}$$

leads to the contradiction

$$\log |\Theta(s)|$$

$$= \sum_{p \nmid m} \sum_{n=1}^{\infty} \frac{3 + 4 \cos n\bigl(\arg \chi(p) - t \log p\bigr) + \cos n\bigl(2 \arg \chi(p) - 2t \log p\bigr)}{n p^{n\sigma}}$$

$$\geq 0 \ .$$

Only in the case of a *real character* χ, which is distinguished by $\chi^2 = \chi_0$, and for $t = 0$, does this method not work. This is because the factor $L(\chi^2, s+2it) = L(\chi_0, s)$ possesses a pole for $s = 1$ and therefore $\Theta(s)$ does not have to take the value zero at $s = 1$. Here the following technical device helps to our goal: because of the multiplicativity of χ the sum function

$$S_\chi(n) = \sum_{k|n} \chi(k)$$

of the real character χ is also multiplicative, and for the prime power p^ν we have

$$S_\chi(p^\nu) = \sum_{j=0}^{\nu} \chi(p^j) = 1 + \sum_{j=1}^{\nu} \chi(p)^j \ .$$

$\chi(p)$ can only take the values $0, 1$ or -1, and as a consequence one has that $S_\chi(p^\nu) = 1$, $S_\chi(p^\nu) = 1 + \nu$ or $S_\chi(p^\nu) = 0$ for odd ν, and $S_\chi(p^\nu) = 1$ for even ν. In any event, for all natural numbers n, $S_\chi(n) \geq 0$ and for the perfect squares $n = k^2$ $S_\chi(n) = S_\chi(k^2) \geq 1$. Since therefore

$$\sum_{n \leq x} \frac{S_\chi(n)}{\sqrt{n}} \geq \sum_{n \leq x, n = k^2} \frac{1}{\sqrt{n}} = \sum_{k \leq \sqrt{x}} \frac{1}{k}$$

diverges as $x \to \infty$, the convergence abscissa σ_0 of the Dirichlet series formed by S_χ

$$F(s) = \sum_{n=1}^{\infty} \frac{S_\chi(n)}{n^s}$$

must satisfy the inequality $\sigma_0 \geq 1/2$. By Landau's theorem (Proposition 18 of Chapter 4) $F(s)$ has a singularity in the region $\mathrm{Re}(s) \geq 1/2$. Because of

$$F(s) = \sum_{n=1}^{\infty} \frac{1}{n^s} \sum_{k|n} \chi(k) = L(\chi, s) \cdot \zeta(s)$$

and the analytic continuability of $L(\chi, s)$ to $\operatorname{Re}(s) > 0$, this is only possible at the simple pole $s = 1$ of $\zeta(s)$. Since the singularity is not removable, the value $L(\chi, 1) = 0$ is excluded.

Lemma 3. *For* $\operatorname{Re}(s) > 1$ *and all characters* χ *we have*

$$-\frac{L'(\chi, s)}{L(\chi, s)} = \sum_{n=1}^{\infty} \frac{\chi(n)\Lambda(n)}{n^s} \ .$$

Proof. This follows from the transformation

$$-\frac{L'(\chi, s)}{L(\chi, s)} = -\frac{d}{ds}\left(\log L(\chi, s)\right)$$

$$= \frac{d}{ds}\left(\sum_p \log\left(1 - \frac{\chi(p)}{p^s}\right)\right)$$

$$= \sum_p \frac{\chi(p)}{1 - \frac{\chi(p)}{p^s}} \cdot \frac{\log p}{p^s} = \sum_p \frac{\log p}{p^s} \cdot \sum_{\nu=0}^{\infty} \frac{\chi(p)^{\nu+1}}{p^{s\nu}}$$

$$= \sum_p \sum_{k=1}^{\infty} \frac{\chi(p^k)\log p}{p^{ks}} = \sum_{n=1}^{\infty} \frac{\chi(n)\Lambda(n)}{n^s} \ .$$

One can answer the point that here one has too freely manipulated the logarithm of a complex argument and its functional equation by reading the formal proof given above backwards. Because of

$$\log\left(1 - \frac{\chi(p)}{p^s}\right) = -\frac{\chi(p)}{p^s} + O\left(p^{-2\sigma}\right) , \quad \sigma = \operatorname{Re}(s) ,$$

the series

$$H(s) = -\sum_p \log\left(1 - \frac{\chi(p)}{p^s}\right)$$

converges absolutely and uniformly in each compact subset contained in $\operatorname{Re}(s) > 1$, and thus represents a holomorphic function in $\operatorname{Re}(s) > 1$. Now the conclusion follows unassailably from the relation

$$e^{H(s)} = L(\chi, s) \ .$$

$\square$

The assumptions needed for the application of Corollary 3 of the previous chapter are thus satisfied: $f(n) = \Lambda(n)$ and $g(n) = \chi(n)\Lambda(n)$ denote two number theoretic functions, where $\Lambda(n)$ takes only non-negative values and the formulae

$$\chi(n)\Lambda(n) = O\big(\Lambda(n)\big)$$

and

$$\psi(x) = \sum_{n \leq x} \Lambda(n) = O(x)$$

hold. The Dirichlet series

$$-\frac{\zeta'(s)}{\zeta(s)} = \sum_{n=1}^{\infty} \frac{\Lambda(n)}{n^s} \ ,$$

$$-\frac{L'(\chi,s)}{L(\chi,s)} = \sum_{n=1}^{\infty} \frac{\chi(n)\Lambda(n)}{n^s}$$

which they form are holomorphic for $\mathrm{Re}(s) > 1$, and the maps

$$-\frac{\zeta'(s)}{\zeta(s)} - \frac{1}{s-1} \ ,$$

$$-\frac{L'(\chi_0,s)}{L(\chi_0,s)} - \frac{1}{s-1} \ ,$$

$$-\frac{L'(\chi,s)}{L(\chi,s)} - \frac{0}{s-1} \quad \text{for} \quad \chi \neq \chi_0$$

can be understood as holomorphic functions, defined in a region containing the closed half-plane $\mathrm{Re}(s) \geq 1$. Therefore one has

Proposition 6. *If $\chi = \chi_0$ is the principal character, then we have the asymptotic approximation*

$$\sum_{n \leq x} \chi_0(n)\Lambda(n) \sim x \ .$$

For all other Dirichlet characters $\chi \neq \chi_0$ we have

$$\sum_{n \leq x} \chi(n)\Lambda(n) = o(x) \ .$$

In order to extract an interesting number theoretic statement from this, note first that for g.c.d.$(a,m) = 1$, because of the orthogonality relations, the function

$$\psi(x;a,m) = \sum_{n \leq x : n \equiv a (\mathrm{mod}\ m)} \Lambda(n)$$

can be rewritten as

$$\psi(x;a,m) = \frac{1}{\varphi(m)} \cdot \sum_{\chi} \bar{\chi}(a) \cdot \sum_{n \leq x} \chi(n)\Lambda(n) \ .$$

Since, given Proposition 6, the right-hand sum is asymptotically equal to x only in the case $\chi = \chi_0$, and all other right-hand summands are of order of magnitude equal to $o(x)$, we have

$$\psi(x; a, m) = \frac{1}{\varphi(m)} \cdot x + o(x) \sim \frac{1}{\varphi(m)} \cdot x \ .$$

The conclusion allows us to count all prime numbers $p \equiv a(\mathrm{mod}\ m)$ with $p \le x$. On the one hand

$$\sum_{n \ge 2} \sum_{p^n \le x, p^n \equiv a(\mathrm{mod}\ m)} \log p$$

$$\le \sum_{n \ge 2} \sum_{p^n \le x} \log p$$

$$= O\left(\log x \cdot \sum_{n \ge 2} \sum_{p \le x^{1/n}} 1 \right)$$

$$= O\left(\log x \cdot \sum_{2 \le n \le \log x / \log 2} \frac{\sqrt{x}}{\log \sqrt{x}} \right)$$

$$= O\left(\sqrt{x} \log x \right) \ ,$$

from which

$$\sum_{p \le x : p \equiv a(\mathrm{mod}\ m)} \log p \sim \frac{1}{\varphi(m)} \cdot x$$

follows. On the other hand the Abel transformation gives

$$\sum_{p \le x : p \equiv a(\mathrm{mod}\ m)} 1 = \sum_{p \le x : p \equiv a(\mathrm{mod}\ m)} \frac{\log p}{\log p}$$

$$= \frac{1}{\log x} \cdot \sum_{p \le x : p \equiv a(\mathrm{mod}\ m)} \log p$$

$$+ \int_2^x \sum_{p \le t : p \equiv a(\mathrm{mod}\ m)} \log p \cdot \frac{dt}{t \cdot \log^2 t}$$

$$= \frac{1}{\log x} \cdot \frac{x}{\varphi(m)} + o\left(\frac{x}{\varphi(m) \log x} \right)$$

$$+ O\left(\int_2^x t \cdot \frac{dt}{t \cdot \log^2 t} \right)$$

$$= \frac{1}{\varphi(m)} \frac{x}{\log x} + o\left(\frac{x}{\log x} \right) \ .$$

Theorem 1: Prime Number Theorem for Arithmetic Progressions.
If $\pi(x; a, m)$ counts the prime numbers $p \le x$ congruent to a modulo the

natural number m, and if the g.c.d. $(a, m) = 1$, then one has the asymptotic representation [19]

$$\pi(x; a, m) \sim \frac{1}{\varphi(m)} \cdot \frac{x}{\log x} \ .$$

Corollary 1: Dirichlet's Theorem on Prime Numbers in Arithmetic Progression. *If the natural numbers m and a are coprime, then the arithmetic progression $a, a+m, a+2m, a+3m, \ldots$ contains infinitely many prime numbers, i.e. there exist infinitely many prime numbers of the form $p = a + nm$, $n = 0, 1, 2, 3 \ldots$*

Until now the strong multiplicativity of Dirichlet characters has been used above all else as a method of proof. In what follows *their periodicity with period m* will mark the investigations. We begin by expressing number theoretic functions with period m in terms of *Fourier series* (which here reduce to Fourier sums)

$$f(n) = \sum_{k=1}^{m} \tilde{f}(k) e^{2\pi i k n/m} \ .$$

The *Fourier coefficients* $\tilde{f}(k)$ are determined by the formula

$$\tilde{f}(k) = \frac{1}{m} \sum_{n=1}^{m} f(n) e^{-2\pi i k n/m}$$

and in this way $\tilde{f}$ enters as the "Fourier transformed" number theoretic function with period m associated to f. One way to check these formulae is by direct calculation – thus

$$\sum_{k=1}^{m} \tilde{f}(k) e^{2\pi i k n/m}$$

$$= \sum_{k=1}^{m} \frac{1}{m} \sum_{r=1}^{m} f(r) e^{-2\pi i k r/m} e^{2\pi i k n/m}$$

$$= \frac{1}{m} \sum_{r=1}^{m} f(r) \cdot \sum_{k=1}^{m} e^{2\pi i k (n-r)/m} \ ,$$

where only in the case $n \equiv r \pmod m$ does the inner sum consist of m summands equal to 1, leading to the result $f(r) = f(n)$. For $n \not\equiv r \pmod m$, given the formula for the geometric sum

$$e^{2\pi i(n-r)/m} \cdot \frac{e^{2\pi i m(n-r)/m} - 1}{e^{2\pi i(n-r)/m} - 1} = 0$$

no further contribution is made. Alternatively one may convince oneself that the association of n to $e^{2\pi i k n/m}$ for $k = 1, \ldots, m$ labels the m characters of the *additive* group of *all* classes of residues modulo m, and that the relations above are immediate consequences of the orthogonality relations for characters. This second, abstract method of argument leads to a generalisation of the Fourier formulae. One writes

$$f(n) = \sum_{k(m)} \tilde{f}(k)e^{2\pi ikn/m}$$

and

$$\tilde{f}(k) = \sum_{n(m)} f(n)e^{-2\pi ikn/m} \ ,$$

where the summation over $k(m)$ (resp. $n(m)$) implies that k (resp. n) runs through a family of representatives for the complete system of residues modulo m. Because of the symmetry of the two formulae above it is ultimately irrelevant whether the Dirichlet characters are developed in Fourier series or play the role of Fourier coefficients. Gauss plumped for the latter – the Fourier sums formed in this way

$$G(n,\chi) = \sum_{k(m)} \chi(k)e^{2\pi ikn/m} \ ,$$

for all natural (even all integral) numbers n and all Dirichlet characters χ of the system of prime classes of residues, are called *Gauss sums*. [21]

With the aid of Gauss sums the following investigations persue two goals – the proof of an inequality of Pólya and Vinogradov, which sharpens the trivial estimate

$$\left| \sum_{n \leq N} \chi(n) \right| \leq m$$

for a character χ different from the principal character, and a simple analytic proof of the quadratic reciprocity law of Gauss.

First Application of Gauss Sums: The Theorem of Pólya and Vinogradov

A divisor m' of the modulus m of a Dirichlet character χ is called a defining modulus, if for all natural numbers a, b with g.c.d.$(a, m) =$ g.c.d.$(b, m) = 1$ it follows from $a \equiv b(\mathrm{mod}\ m')$ that $\chi(a) = \chi(b)$.

Proposition 7. *The divisor m' of the modulus m is a defining modulus for the Dirichlet character χ, if and only if for all natural numbers n with g.c.d.$(n, m) = 1$ it follows for $n \equiv 1(\mathrm{mod}\ m')$ that $\chi(n) = 1$.*

Proof. If m' is a defining modulus, in order to verify the conclusion one has only to put $a = n$ and $b = 1$ in the definition. If on the other hand the condition in the proposition holds, then starting from natural numbers a, b with g.c.d.$(a, m) =$ g.c.d.$(b, m) = 1$ and $a \equiv b(\mathrm{mod}\ m')$, one can find some natural number a' with $aa' \equiv 1(\mathrm{mod}\ m)$, hence also with $aa' \equiv 1(\mathrm{mod}\ m')$, and finally with $\chi(aa') = 1$. In particular $\chi(a')$ is certainly different from zero. $ba' \equiv aa' \equiv 1(\mathrm{mod}\ m')$ implies that $\chi(b)\chi(a') = \chi(ba') = 1 = \chi(aa') = \chi(a)\chi(a')$ which after division by $\chi(a')$ gives the defining formula for the defining modulus, viz $\chi(a) = \chi(b)$.

Proposition 8. *The divisor m' of the modulus m is a defining modulus for the Dirichlet character χ if and only if there exists some Dirichlet character χ' modulo m', which after multiplication by the principal character χ_0 modulo m gives χ. Thus*

$$\chi(n) = \chi_0(n) \cdot \chi'(n) \ .$$

Proof. Since it follows from $n \equiv 1(\mathrm{mod}\ m')$ that $\chi'(n) = 1$ and from g.c.d.$(n, m) = 1$ that $\chi_0(n) = 1$, from the formula $\chi(n) = \chi_0(n)\chi'(n)$ one obtains the condition $\chi(n) = 1$ stated in Proposition 7.

In the other direction if m' is a defining modulus of χ, one carries out the construction of χ' in the following way. Given n with g.c.d.$(n, m') > 1$, put $\chi'(n) = 0$. Otherwise given n with g.c.d.$(n, m') = 1$, one can, when for example one chooses some prime number from the arithmetic progression $n, m'+n, 2m'+n, 3m'+n, \ldots$, find some $n' \equiv n(\mathrm{mod}\ m')$ with g.c.d.$(n', m) = 1$. Now put $\chi'(n) = \chi(n')$. This definition makes sense, because m' as defining modulus for χ leads to the value $\chi(n'') = \chi(n')$ for all other $n'' \equiv n(\mathrm{mod}\ m')$ with g.c.d.$(n'', m) = 1$. By construction it is clear that χ' forms a Dirichlet character modulo m', and that for g.c.d.$(n, m) = 1$ the validity of the formula $\chi(n) = \chi_0(n)\chi'(n)$ follows from g.c.d.$(n, m') = 1$, $\chi'(n) = \chi(n')$ for suitable $n' \equiv n(\mathrm{mod}\ m')$, and thus from $\chi(n) = \chi(n') = \chi'(n) = \chi_0(n)\chi'(n)$. For g.c.d.$(m, n) > 1$ the validity of this formula follows, since both sides equal zero, given that $\chi(n) = \chi_0(n) = 0$. Hence the formula always holds.   □

The smallest defining modulus m_χ of a Dirichlet character χ modulo m is called the *conductor* of the Dirichlet character. If the conductor equals 1, it is clear that we are dealing with the principal character; on the other hand, if the conductor equals m, $m_\chi = m$, one calls the character *primitive*. The formula required to prove the inequality of Pólya and Vinogradov reads: $G(n, \chi) = \bar{\chi}(n)G(1, \chi)$. In complete generality it holds only for primitive characters.

Proposition 9. *For each natural number n coprime with the modulus m, the Gauss sum of an arbitrary Dirichlet character χ modulo m can be written as*

$$G(n, \chi) = \bar{\chi}(n) \cdot G(1, \chi) \ .$$

If the Dirichlet character is primitive, this formula always holds, hence also for n with g.c.d.$(m, n) > 1$ one has $G(n, \chi) = 0$.

Proof. For n and m coprime there exists some natural number n' with $nn' \equiv 1(\mathrm{mod}\ m)$. Because $\chi(n') = \bar{\chi}(n)$ and with k also $h = nk$ runs through a complete system of residues modulo m, one has

$$G(n, \chi) = \sum_{k(m)} \chi(k) e^{2\pi i n k/m}$$

$$= \sum_{k(m)} \chi(n') \chi(nk) e^{2\pi i n k/m}$$

$$= \bar{\chi}(n) \cdot \sum_{k(m)} \chi(h) e^{2\pi i h/m} = \bar{\chi}(n) \cdot G(1, \chi) \ ,$$

as claimed in the first part of the proposition.

Suppose now that g.c.d.$(n, m) = r > 1$, then the natural number $m' = m/r$ also divides the modulus m. If χ is primitive, there must exist some $n' \equiv 1(\mathrm{mod}\ m')$ with g.c.d.$(n', m) = 1$ and $\chi(n') \neq 1$, since otherwise m' would be a defining modulus. Furthermore the relation

$$\sum_{k(m):k\equiv h(\mathrm{mod}\ m')} \chi(k)$$

$$= \sum_{n'k(m):n'k\equiv h(\mathrm{mod}\ m')} \chi(n'k)$$

$$= \sum_{k(m):k\equiv h(\mathrm{mod}\ m')} \chi(n'k)$$

$$= \chi(n') \cdot \sum_{k(m):k\equiv h(\mathrm{mod}\ m')} \chi(k)$$

holds for each integer h, so that, given $\chi(n') \neq 1$, for each natural number h it follows that

$$\sum_{k(m):k\equiv h(\mathrm{mod}\ m')} \chi(k) = 0 \ .$$

As claimed this has the consequence

$$G(n, \chi) = \sum_{k(m)} \chi(k) e^{2\pi i n k/m}$$

$$= \sum_{k(m)} \chi(k) e^{2\pi i n k/m'r}$$

$$= \sum_{h(m')} \sum_{k(m):k\equiv h(\mathrm{mod}\ m')} \chi(k) e^{2\pi i n k/m'r}$$

$$= \sum_{h(m')} \sum_{k(m):k\equiv h(\mathrm{mod}\ m')} \chi(k) e^{2\pi i n h/m'r}$$

$$= 0 = \bar{\chi}(n) \cdot G(1, \chi) \ . \qquad \qquad \square$$

Proposition 10. *For each natural number n coprime with the modulus m and for each primitive Dirichlet character χ modulo m the equation*

$$|G(n,\chi)| = \sqrt{m}$$

holds.

Proof. Given Proposition 9 it suffices to prove that $|G(1,\chi)|^2 = m$. Also for r with g.c.d.$(r,m) = 1$ both k and kr run through a complete system of residues modulo m. Combining the calculation

$$|G(1,\chi)|^2 = \sum_{r(m)} \sum_{k(m)} \chi(r)\bar{\chi}(k)e^{2\pi i(r-k)/m}$$

$$= \sum_{r(m):\text{g.c.d.}(r,m)=1} \sum_{k(m)} \chi(r)\bar{\chi}(kr)e^{2\pi i(r-kr)/m}$$

$$= \sum_{r(m):\text{g.c.d.}(r,m)=1} \sum_{k(m)} \bar{\chi}(k)e^{-2\pi ikr/m}e^{2\pi ir/m}$$

$$= \sum_{r(m):\text{g.c.d.}(r,m)=1} G(-r,\bar{\chi})e^{2\pi ir/m}$$

and the fact that the primitivity of the Dirichlet character χ makes the requirement g.c.d.$(r,m) = 1$ superfluous, and since those r not coprime with m have $G(-r,\bar{\chi}) = 0$ and make no contribution,

$$= \sum_{r(m)} e^{2\pi ir/m} \sum_{k(m)} \bar{\chi}(k)e^{-2\pi irk/m}$$

$$= \sum_{r(m)} \sum_{k(m)} \bar{\chi}(k)e^{2\pi ir(1-k)/m}$$

$$= \sum_{k(m)} \sum_{r(m)} \bar{\chi}(k)e^{2\pi ir(1-k)/m}$$

$$= \sum_{k(m)} \bar{\chi}(k) \cdot \sum_{r=1}^{m} \left(e^{2\pi i(1-k)/m}\right)^r$$

$$= \bar{\chi}(1) \cdot m = m \ .$$

The final geometric sum only contributes non-trivially for $k \equiv 1(\bmod\ m)$.

Theorem 2: The Theorem of Pólya and Vinogradov. *Each primitive Dirichlet character χ_0 modulo m satisfies the inequality*

$$\left|\sum_{n \leq N} \chi(n)\right| < \sqrt{m} \cdot \log m \ .$$

More generally for each Dirichlet character χ modulo m different from the principal character χ_0 we have

$$\sum_{n \leq N} \chi(n) = O(\sqrt{m} \cdot \log m) \ .$$

Proof. First let χ denote a primitive Dirichlet character modulo m. For $m < 10$ one can check the inequality given above directly; the following considerations apply to larger values of m. Since in the Fourier representation

$$\chi(n) = \sum_{k(m)} \tilde{\chi}(k) e^{2\pi i k n / m}$$

Proposition 9 describes the Fourier coefficients as

$$\tilde{\chi}(k) = \frac{1}{m} \cdot \sum_{n(m)} \chi(n) e^{-2\pi i k n / m}$$

$$= \frac{1}{m} \cdot G(-k, \chi) = \frac{1}{m} \cdot \bar{\chi}(-k) \cdot G(1, \chi)$$

we have

$$\chi(n) = \frac{G(1, \chi)}{m} \cdot \sum_{k(m)} \bar{\chi}(-k) e^{2\pi i k n / m} \ .$$

For the sum in question this implies that

$$\sum_{n \leq N} \chi(n) = \frac{G(1, \chi)}{m} \cdot \sum_{k(m)} \bar{\chi}(-k) \cdot \sum_{n \leq N} e^{2\pi i k n / m}$$

$$= \frac{G(1, \chi)}{m} \cdot \sum_{k=1}^{m-1} \bar{\chi}(-k) \cdot \sum_{n \leq N} e^{2\pi i k n / m} \ .$$

For $k = m$ the summand vanishes, since $\bar{\chi}(m) = 0$. Taking account of Proposition 10 the estimate of the absolute value reads

$$\left| \sum_{n \leq N} \chi(n) \right| \leq \frac{\sqrt{m}}{m} \cdot \sum_{k=1}^{m-1} \left| \sum_{n \leq N} e^{2\pi i k n / m} \right| \ .$$

One estimates the inner trigonometric sum as follows: write

$$\sum_{n \leq N} e^{2\pi i k n / m} = f(k) \ .$$

Since

$$f(m - k) = \sum_{n \leq N} e^{2\pi i (m-k) n / m}$$

$$= \sum_{n \leq N} e^{-2\pi i k n / m} = \overline{f(k)}$$

$|f(m-k)| = |f(k)|$, leading to the simplification

$$\left| \sum_{n \leq N} \chi(n) \right| \leq \frac{\sqrt{m}}{m} \cdot 2 \cdot \sum_{k \leq m/2} |f(k)| \ .$$

If without loss of generality one restrict oneself to natural numbers N, the geometric sum $f(k)$ can be expressed as:

$$
\begin{aligned}
f(k) &= \sum_{n=1}^{N} \left(e^{2\pi i k/m} \right)^{n} \\
&= e^{2\pi i k/m} \cdot \frac{e^{2\pi i k N/m} - 1}{e^{2\pi i k/m} - 1} \\
&= \frac{e^{\pi i k N/m} - e^{-\pi i k N/m}}{e^{\pi i k/m} - e^{-\pi i k/m}} \cdot e^{\pi i k(N+1)/m} \\
&= \frac{\sin \dfrac{\pi k N}{m}}{\sin \dfrac{\pi k}{m}} \cdot e^{\pi i k(N+1)/m} \ .
\end{aligned}
$$

Its absolute value equals

$$
|f(n)| = \frac{\left| \sin \dfrac{\pi k N}{m} \right|}{\left| \sin \dfrac{\pi k}{m} \right|} \leq \frac{1}{\left| \sin \dfrac{\pi k}{m} \right|} \ .
$$

Since for $1 \leq k \leq m/2$, $0 < \pi k/m < \pi/2$, and for the corner in which φ increases from 0 to $\pi/2$, one can apply the trivial Jordan inequality $\sin \varphi \geq 2\varphi/\pi$, we have

$$
|f(k)| \leq \frac{1}{\dfrac{2}{\pi} \cdot \dfrac{\pi k}{m}} = \frac{m}{2k} \ .
$$

Since $m \geq 10$, and

$$
\begin{aligned}
\left| \sum_{n \leq N} \chi(n) \right| &\leq \frac{\sqrt{m}}{m} \cdot 2 \cdot \sum_{k \leq m/2} \frac{m}{2k} \\
&= \sqrt{m} \cdot \sum_{k \leq m/2} \frac{1}{k} < \sqrt{m} \cdot \log m
\end{aligned}
$$

this gives the first assertion in Theorem 2.

If χ denotes a Dirichlet character other than the principal character with $m' = m_\chi$ as conductor, then by Proposition 8 there exists a decomposition

$$\chi(n) = \chi_0(n) \cdot \chi'(n) \ ,$$

where χ' denotes a primitive character modulo m'. Given the relations

$$\sum_{n\leq N} \chi(n) = \sum_{n\leq N:\text{g.c.d.}(n,m)=1} \chi'(n)$$

$$= \sum_{n\leq N} \chi'(n) \cdot \sum_{k|\text{g.c.d.}(n,m)} \mu(k)$$

$$= \sum_{n\leq N} \sum_{k|m,k|n} \mu(k)\chi'(n)$$

$$= \sum_{k|m} \mu(k) \cdot \sum_{r\leq N/k} \chi'(rk)$$

$$= \sum_{k|m} \mu(k)\chi'(k) \cdot \sum_{r\leq N/k} \chi'(r)$$

and the fact that $\mu(k)\chi'(k) \neq 0$ presupposes that $k = p_1 p_2 \ldots p_\ell$ is a square-free product of ℓ distinct prime numbers $p_1,\ldots,p_\ell$, and because $\chi'(k) \neq 0$, none of the prime factors $p_1,\ldots,p_\ell$ can divide the conductor m', so that k has to be a divisor of m/m', it follows that

$$\left|\sum_{n\leq N} \chi(n)\right| \leq \sum_{k|m/m'} 1 \cdot \left|\sum_{r\leq N/k} \chi'(r)\right|$$

$$\leq \tau\left(\frac{m}{m'}\right) \cdot \sqrt{m'}\log m' \leq \tau\left(\frac{m}{m'}\right) \cdot \sqrt{m'}\log m \ .$$

Because of Proposition 4 in Chapter 4 starting from

$$\tau\left(\frac{m}{m'}\right) = O\left(\frac{\sqrt{m}}{\sqrt{m'}}\right)$$

one obtains, as asserted, that

$$\sum_{n\leq N} \chi(n) = O\left(\frac{\sqrt{m}}{\sqrt{m'}}\right) \cdot \sqrt{m'}\log m = O(\sqrt{m}\log m) \ . \qquad \square$$

In order to be able to understand the importance of Theorem 2 we include two richly significant examples of its application. In them we assume that the modulus $m = p$ is prime. This simplifies the situation, not only because with it all Dirichlet characters other than the principal character are primitive, but more importantly because the group of prime classes of residue modulo the prime number p is well-known to be cyclic. There certainly exists some primitive root g and an index function ind dependent upon g, defined for all numbers coprime to p, such that $n \equiv g^{\text{ind}(n)}(\text{mod } p)$ holds. The character group must also be cyclic; concretely one can describe the Dirichlet characters by

$$\chi_r(n) = e^{2\pi i r\cdot\text{ind}(n)/(p-1)}, \quad r = 0,1,\ldots,p-2 \ ,$$

for g.c.d.$(n,p) = 1$. Given the decomposition $p - 1 = u \cdot v$, the number n coprime with p is then called a vth power residue if and only if it is possible to solve the congruence

$$x^v \equiv n(\bmod\ p)\ .$$

One can also describe this by means of

$$\mathrm{ind}(n) \equiv 0(\bmod\ v)\ .$$

Elementary number theory teaches us that n is a vth power residue in the prime classes of residues modulo p, and that these form a subgroup. If we consider the cosets $(v;0),(v;1),\ldots,(v;v-1)$, where all n with $\mathrm{ind}(n) \equiv h(\bmod\ v)$ come to lie in $(v;h)$, we recognise the quotient group obtained by algebraically factoring out the group of prime classes of residues by the subgroup of vth powers. Since this is also cyclic with generating element $(v;1)$, it is clear that its characters are $c_0, c_1, \ldots c_{v-1}$ with

$$\mathbf{c}_k(v;h) = e^{2\pi i h k/v}, \quad k = 0, 1, \ldots, v - 1\ .$$

There exists a unique reversible morphism between these characters and the Dirichlet characters χ_{uk} with

$$\chi_{uk}(n) = e^{2\pi i u k \cdot \mathrm{ind}(n)/(p-1)}$$
$$= e^{2\pi i k \cdot \mathrm{ind}(n)/v}\ .$$

In particular the fact that n belongs to $(v;h)$ gives the relation $\mathbf{c}_k(v;h) = \chi_{uk}(n)$ and conversely. Therefore the χ_{uk}, $k = 0, 1, \ldots, v - 1$, embody the Dirichlet characters of the quotient group introduced above. We understand the orthogonality relations to read that n is a vth power residue modulo p iff

$$\frac{1}{v} \cdot \sum_{k(v)} \chi_{uk}(n) = 1\ ,$$

while for all other n

$$\frac{1}{v} \cdot \sum_{k(v)} \chi_{uk}(n) = 0$$

holds.

The number of vth power residues modulo the prime number p lying between 1 and N thus reads

$$\sum_{n \leq N} \frac{1}{v} \cdot \sum_{k(v)} \chi_{uk}(n) = \frac{1}{v} \cdot \sum_{k=0}^{v-1} \sum_{n \leq N} \chi_{uk}(n)$$
$$= \frac{1}{v} \cdot \sum_{n \leq N} \chi_0(n) + \frac{1}{v} \cdot \sum_{k=1}^{v-1} \sum_{n \leq N} \chi_{uk}(n)\ .$$

If one assumes $N < p$ the first summand becomes N/v, while by Theorem 2 the absolute value of the second summand can be estimated by

$$\frac{v-1}{v} \cdot \sqrt{p}\log p < \sqrt{p}\log p \ .$$

This implies

Corollary 2. *Between 1 and the natural number $N < p$ there lie at least*

$$\frac{N}{v} - \sqrt{p}\log p$$

and at most

$$\frac{N}{v} + \sqrt{p}\log p$$

vth power residues modulo the prime p.

Since n modulo p is a primitive root if and only if

$$\mathrm{g.c.d.}(\mathrm{ind}(n), p-1) = 1 \ ,$$

the sum

$$\sum_{n\leq N: \mathrm{g.c.d.}(\mathrm{ind}(n),p-1)=1} 1$$

prints out the number of prime roots between 1 and N. Vinogradov's lemma (Lemma 2 in Chapter 4) allows the transformation of the sum above to

$$\sum_{v=1}^{\infty} \mu(v) \cdot \sum_{n\leq N: v|\mathrm{g.c.d.}(\mathrm{ind}(n),p-1)} 1$$
$$= \sum_{v|p-1} \mu(v) \cdot \sum_{n\leq N: v|\mathrm{ind}(n)} 1 \ .$$

Here the inner sum accounts for the number of vth power residues to be found between 1 and the natural number N; by Corollary 2 this is bounded below (resp. above) by $N/v \pm \sqrt{p}\log p$. Hence the number of primitive roots to be found between 1 and the natural number $N < p$ is at least

$$\sum_{n\leq N: \mathrm{g.c.d.}(\mathrm{ind}(n),p-1)=1} 1$$
$$\geq \sum_{v|p-1} \mu(v) \cdot \frac{N}{v} - \sum_{v|p-1} \sqrt{p}\log p$$
$$= N \cdot \sum_{v|p-1} \frac{\mu(v)}{v} - \tau(p-1) \cdot \sqrt{p}\log p \ .$$

Since the Euler φ-function has the description

$$\varphi(n) = \sum_{m|n} \mu(m) \cdot \frac{n}{m} = n \cdot \sum_{m|n} \frac{\mu(m)}{m} \ ,$$

the estimate above can be simplified to

$$\frac{N}{p-1} \cdot \varphi(p-1) - \tau(p-1) \cdot \sqrt{p}\log p \ .$$

If $N = g$ itself denotes the smallest positive primitive root modulo p, then as a consequence one has the inequality

$$1 \geq \frac{g}{p-1} \cdot \varphi(p \overset{\cdot}{-} 1) - \tau(p-1) \cdot \sqrt{p}\log p \ ,$$

$$\begin{aligned}
g &\leq \frac{p-1}{\varphi(p-1)} + \tau(p-1) \cdot \frac{p-1}{\varphi(p-1)} \cdot \sqrt{p}\log p \\
&= O\left(\tau(p-1) \cdot \frac{p-1}{\varphi(p-1)} \cdot \sqrt{p}\log p\right) \\
&= O\left((p-1)^{\epsilon/3} \cdot \frac{p-1}{(p-1)^{1-\epsilon/3}} \cdot \sqrt{p} \cdot p^{\epsilon/3}\right) \\
&= O\left(p^{\epsilon/3} \cdot \frac{p}{p^{1-\epsilon/3}} \cdot \sqrt{p} \cdot p^{\epsilon/3}\right) \\
&= O\left(p^{1/2+\epsilon}\right)
\end{aligned}$$

in which we apply the estimate of Propositions 1 and 4 of Chapter 4.

Corollary 3: Vinogradov's Theorem. *If $g(p)$ denotes the smallest positive primitive root modulo the prime number p, then for it the estimate*

$$g(p) = O\left(p^{1/2+\epsilon}\right)$$

holds, where the positive number ε may be taken to be arbitrarily small. [22]

Second Application of Gauss Sums: The Quadratic Reciprocity Law

Proposition 11. *Apart from the principal character, modulo the odd prime number p there exists only one real Dirichlet character χ.*

Proof. For a number n coprime with p a real value for

$$\chi_r(n) = e^{2\pi i r \cdot \mathrm{ind}(n)/(p-1)}$$

is equivalent to the divisibility of $2r \cdot \mathrm{ind}(n)$ by $p-1$. This (in the case $\mathrm{ind}(n) = 1$ for a primitive root n) only leaves open the possibility $r = 0$ – corresponding to the principal character – and also $r = (p-1)/2$. The second real character obtained in this way clearly agrees with the Legendre symbol of elementary number theory,

$$\chi_{(p-1)/2}(n) = e^{\pi i \cdot \mathrm{ind}(n)} = (-1)^{\mathrm{ind}(n)} \ .$$

This means that with the values $+1$ (resp. -1) it shows whether n is a quadratic residue (resp. non-residue) modulo p.

In what follows let p always be assumed to be an odd prime number, and let the real character distinct from the principal character be denoted simply by χ. Then for all n coprime with p,

$$\chi(n) = \left(\frac{n}{p}\right)$$

holds. Since χ is primitive, its Gauss sum satisfies the relation

$$G(n,\chi) = \bar{\chi}(n) \cdot G(1,\chi) = \left(\frac{n}{p}\right) \cdot G(1,\chi) \ .$$

However for the proof of the reciprocity law the calculation of the powers of its Gauss sums is of importance. For the determination of the square $G(n,\chi)^2$ one can restrict oneself to $n = 1$ and proceed analogously to the proof of Proposition 10. Thus

$$
\begin{aligned}
G(1,\chi)^2 &= \sum_{r=1}^{p} \sum_{k=1}^{p} \left(\frac{r}{p}\right) \left(\frac{k}{p}\right) e^{2\pi i (r+k)/p} \\
&= \sum_{r=1}^{p} \sum_{k=1}^{p} \left(\frac{r}{p}\right) \left(\frac{kr}{p}\right) e^{2\pi i (r+kr)/p} \\
&- \sum_{r=1}^{p} \sum_{k=1}^{p} \left(\frac{k}{p}\right) e^{2\pi i r(1+k)/p} \\
&= \sum_{k=1}^{p} \left(\frac{k}{p}\right) \cdot \sum_{r=1}^{p} \left(e^{2\pi i (1+k)/p}\right)^{r} \\
&= \left(\frac{-1}{p}\right) \cdot p \ .
\end{aligned}
$$

The last equality holds, because the geometric sum only delivers the value p in the case $k \equiv -1 (\mathrm{mod}\ p)$ and otherwise takes the value zero. From this follows

Lemma 4. *If the Dirichlet character χ represents the Legendre symbol modulo the odd prime p, then for its Gauss sum*

$$G(n,\chi)^2 = \left(\frac{-1}{p}\right) \cdot p \ ,$$

where n is an arbitrary natural number coprime with p.

If in what follows q denotes an odd prime number distinct from p, we can calculate the power $G(n,\chi)^q$ as

$$G(n,\chi)^q$$

$$= \sum_{r_1(p)} \cdots \sum_{r_q(p)} \left(\frac{r_1}{p}\right) \cdots \left(\frac{r_q}{p}\right) e^{2\pi i n(r_1+\ldots+r_q)/p}$$

$$= \sum_{r_1(p)} \cdots \sum_{r_q(p)} \left(\frac{r_1 \ldots r_q}{p}\right) e^{2\pi i n(r_1+\ldots+r_q)/p} \ .$$

Since here also we are concerned with a periodic function with period p, there must be a Fourier expansion

$$G(n,\chi)^q = \sum_{k(p)} \hat{G}(q,k) e^{2\pi i k n/p}$$

with Fourier coefficients

$$\hat{G}(q,k) = \frac{1}{p} \cdot \sum_{n(p)} G(n,\chi)^q e^{-2\pi i k n/p}$$

$$= \frac{1}{p} \cdot \sum_{r_1(p)} \cdots \sum_{r_q(p)} \left(\frac{r_1 \ldots r_q}{p}\right) \cdot \sum_{n(p)} e^{2\pi i n(r_1+\ldots+r_q-k)/p} \ .$$

The right-hand geometric sum only differs from zero in the case $r_1+\ldots+r_q \equiv k(\mathrm{mod}\ p)$, and in this case takes the value p. Hence on one side

$$\hat{G}(q,k) = \sum_{r_1(p)} \cdots \sum_{r_q(p):r_1+\ldots+r_q \equiv k(\mathrm{mod}\ p)} \left(\frac{r_1 \ldots r_q}{p}\right)$$

remains. On the other side from

$$G(n,\chi)^q = \left(\frac{n}{p}\right)^q G(1,\chi)^q = \left(\frac{n}{p}\right) \cdot G(1,\chi)^q$$

we can see the representation

$$\hat{G}(q,k) = \frac{1}{p} \cdot G(1,\chi)^q \cdot \sum_{n(p)} \left(\frac{n}{p}\right) e^{-2\pi i k n/p}$$

$$= \frac{1}{p} \cdot G(1,\chi)^q \cdot \overline{G(k,\chi)} = \frac{1}{p} \cdot G(1,\chi)^q \cdot \left(\frac{k}{p}\right) \cdot \overline{G(1,\chi)}$$

$$= \frac{1}{p} \cdot G(1,\chi) \overline{G(1,\chi)} \cdot \left(\frac{k}{p}\right) \cdot G(1,\chi)^{q-1}$$

$$= \left(\frac{k}{p}\right) \cdot G(1,\chi)^{q-1}$$

where at the end we apply Proposition 10.
From

$$G(1,\chi)^{q-1} = \left(\frac{k}{p}\right) \cdot \hat{G}(q,k)$$

together with the previous sum formula for the Fourier coefficients we obtain
the following formula, when g.c.d.$(k, p) = 1$:

$$G(1, \chi)^{q-1} = \left(\frac{k}{p}\right) \cdot \sum_{r_1(p)} \cdots \sum_{r_q(p):r_1+\ldots+r_q \equiv k(\mathrm{mod}\ p)} \left(\frac{r_1 \ldots r_q}{p}\right) .$$

With a q-tuple $r_1, \ldots, r_q$ each switch in sequence order gives a q-tuple with
the same sum and the same product; if $q_1, \ldots, q_\ell$ of the numbers $r_1, \ldots, r_q$ are
congruent to each other modulo p $(q_1 + q_2 + \ldots + q_\ell = q)$, then there are

$$\frac{q!}{q_1! \ldots q_\ell!}$$

possible switches. Only for $\ell = 1$, i.e. $q_1 = q$, is the polynomial coefficient

$$\frac{q!}{q_1! \ldots q_\ell!}$$

not divisible by the prime number q, because in this one exceptional case
it equals 1. With this exception, because of the multiply appearing equal
summands, the sum

$$\sum_{r_1(p)} \cdots \sum_{r_q(p):r_1+\ldots+r_q \equiv k(\mathrm{mod}\ p)} \left(\frac{r_1 \ldots r_q}{p}\right)$$

gives a number divisible by q and therefore congruent to zero modulo q.
Therefore modulo q only the summand with the mutually congruent modulo p $r_1 \equiv r_2 \equiv \ldots r_q \equiv r(\mathrm{mod}\ p)$ remains:

$$G(1, \chi)^{q-1}$$
$$\equiv \left(\frac{k}{p}\right) \cdot \sum_{r(p):r+\ldots+r \equiv k(\mathrm{mod}\ p)} \left(\frac{r \ldots r}{p}\right) (\mathrm{mod}\ q) .$$

Since for this only those values of r with $qr \equiv k(\mathrm{mod}\ p)$ come into question,
and in this case

$$G(1, \chi)^{q-1} \equiv \left(\frac{k}{p}\right)\left(\frac{r^q}{p}\right) = \left(\frac{qr}{p}\right)\left(\frac{r^q}{p}\right)$$
$$= \left(\frac{q}{p}\right)\left(\frac{r^{q+1}}{p}\right) = \left(\frac{q}{p}\right) (\mathrm{mod}\ q)$$

holds, this argument leads to

Lemma 5. *If p and q are distinct odd prime numbers, and if the Dirichlet character χ represents the Legendre symbol modulo p, then we have the congruence*

$$G(1,\chi)^{q-1} \equiv \left(\frac{q}{p}\right) \pmod{q} \ .$$

For $n = 1$ from Lemma 4 we obtain

$$G(1,\chi)^{q-1} = \left(\frac{-1}{p}\right)^{\frac{q-1}{2}} \cdot p^{\frac{q-1}{2}} \ .$$

By Euler's criterion from elementary number theory

$$\left(\frac{-1}{p}\right) = (-1)^{\frac{p-1}{2}} \ ,$$

$$\left(\frac{p}{q}\right) \equiv p^{\frac{q-1}{2}} \pmod{q} \ ,$$

so that as a consequence

$$G(1,\chi)^{q-1} \equiv (-1)^{\frac{p-1}{2} \cdot \frac{q-1}{2}} \cdot \left(\frac{p}{q}\right) \pmod{q} \ .$$

Lemma 5 justifies the deduction

$$\left(\frac{q}{p}\right) \equiv (-1)^{\frac{p-1}{2} \cdot \frac{q-1}{2}} \cdot \left(\frac{p}{q}\right) \pmod{q} \ ,$$

and since both sides of the congruence can only take the values ± 1, we even have equality.

Theorem 3: Gauss' Quadratic Reciprocity Law. *Distinct odd prime numbers p and q obey the rule*

$$\left(\frac{p}{q}\right)\left(\frac{q}{p}\right) = (-1)^{\frac{p-1}{2} \cdot \frac{q-1}{2}} \ .$$

A second more analytic proof of the quadratic reciprocity law succeeds with the help of direct mediation by the sum $G(1,\chi)$ concentrated at the prime modulus p. Here as before χ denotes the Legendre symbol. The transformation

$$G(n,\chi) = \sum_{k(p)} \chi(k) e^{2\pi i k n/p}$$

$$= \sum_{k(p):\left(\frac{k}{p}\right)=1} e^{2\pi i k n/p} - \sum_{k(p):\left(\frac{k}{p}\right)=-1} e^{2\pi i k n/p}$$

$$= \sum_{k(p):\left(\frac{k}{p}\right)=1} e^{2\pi i k n/p}$$

$$- \left(\sum_{k(p)} e^{2\pi i k n/p} - \sum_{k(p):\left(\frac{k}{p}\right)=1} e^{2\pi i k n/p} - 1 \right)$$

$$= 1 + 2 \cdot \sum_{k(p):\left(\frac{k}{p}\right)=1} e^{2\pi i k n/p}$$

$$- \sum_{k=0}^{p-1} \left(e^{2\pi i n/p} \right)^{k}$$

$$= 1 + 2 \cdot \sum_{k(p):\left(\frac{k}{p}\right)=1} e^{2\pi i k n/p}$$

$$= 1 + \sum_{r(p):r^2 \equiv k \not\equiv 0 (\bmod\ p)} e^{2\pi i r^2 n/p}$$

$$= \sum_{r(p)} e^{2\pi i r^2 n/p}$$

suggest the definition of the *quadratic Gauss sums*

$$g(n,m) = \sum_{r=0}^{m-1} e^{2\pi i r^2 n/m} \ .$$

For natural numbers n not divisible by p

$$G(n,\chi) = \left(\frac{n}{p}\right) \cdot G(1,\chi) = \left(\frac{n}{p}\right) \cdot g(1,p)$$

holds. The calculation of the quadratic Gauss sum $g(1,m)$ succeeds with the help of the following technical results.

Lemma 6. *If $f(x)$ denotes a differentiable complex-valued function, which is defined on the interval $[P,Q]$ with integral P and Q, and if the derivative $f'(x)$ is integrable over this interval, then*

$$\sum_{n=P}^{Q} f(n) = \frac{f(P) + f(Q)}{2} + \sum_{k=-\infty}^{\infty} \int_{P}^{Q} f(x) e^{2\pi i k x} dx$$

holds.

Proof. First adapt the Euler sum formula to show that

$$\sum_{n=P}^{Q} f(n) = \int_{P}^{Q} f(x)dx + \int_{P}^{Q} \{x\}f'(x)dx + f(P)$$

$$= \int_{P}^{Q} f(x)dx + \int_{P}^{Q}\left(\{x\} - \frac{1}{2}\right)f'(x)dx$$

$$+ \frac{1}{2}\cdot\int_{P}^{Q} f'(x)dx + f(P)$$

$$= \frac{f(P)+f(Q)}{2} + \int_{P}^{Q} f(x)dx$$

$$+ \int_{P}^{Q}\left(\{x\} - \frac{1}{2}\right)f'(x)dx \ .$$

Then substitute the Fourier expansion of the saw-tooth curve

$$\{x\} - \frac{1}{2} = -\frac{1}{2\pi i}\cdot\sum_{k\neq 0}\frac{1}{k}\cdot e^{2\pi i k x}$$

noting that the partial sums

$$\sum_{0<|k|\leq K}\frac{1}{k}\cdot e^{2\pi i k x}$$

remain uniformly bounded in K and x. Since

$$\int_{P}^{Q}\left(\{x\} - \frac{1}{2}\right)f'(x)dx$$

$$= -\frac{1}{2\pi i}\cdot\sum_{k\neq 0}\frac{1}{k}\cdot\int_{P}^{Q} f'(x)e^{2\pi i k x}dx \ ,$$

partial integration gives

$$-\frac{1}{2\pi i}\cdot\sum_{k\neq 0}\frac{1}{k}\left(f(Q) - f(P) - 2\pi i k\cdot\int_{P}^{Q} f(x)e^{2\pi i k x}dx\right)$$

which equals

$$\sum_{k\neq 0}\int_{P}^{Q} f(x)e^{2\pi i k x}dx$$

(in the passage to the limit as $K \to \infty$ the first summand makes no contribution, since the individual members cancel in pairs). This concludes the lemma. $\square$

Lemma 7. *For the natural number m we have the relation*

$$\sum_{k=-\infty:2\nmid k}^{\infty} \int_{k/2}^{k/2+1} e^{2\pi i m x^2}\,dx$$

$$= \sum_{k=-\infty:2\mid k}^{\infty} \int_{k/2}^{k/2+1} e^{2\pi i m x^2}\,dx$$

$$= \int_{-\infty}^{\infty} e^{2\pi i m x^2}\,dx \ .$$

Proof. First by substitution and partial integration one has

$$\int_{a}^{b} e^{i x^2}\,dx = \frac{1}{2}\cdot\int_{a^2}^{b^2} \frac{e^{iu}}{\sqrt{u}}\,du$$

$$= \frac{1}{2ib}\cdot e^{ib^2} - \frac{1}{2ia}\cdot e^{ia^2} + \frac{1}{4i}\cdot\int_{a^2}^{b^2} \frac{e^{iu}}{u\sqrt{u}}\,du \ .$$

This shows that for $0 < a < b$ and $a \to \infty$ the Fresnel integral

$$\int_{-\infty}^{\infty} e^{i x^2}\,dx$$

exists given the Cauchy criterion. But now the assertion of the lemma is easy to obtain from the formulae

$$\sum_{|k|\leq K:2\mid k} \int_{k/2}^{k/2+1} e^{2\pi i m x^2}\,dx$$

$$= \sum_{|2k|\leq K} \int_{k}^{k+1} e^{2\pi i m x^2}\,dx$$

$$\sum_{|k|\leq K:2\nmid k} \int_{k/2}^{k/2+1} e^{2\pi i m x^2}\,dx$$

$$= \sum_{|2k-1|\leq K} \int_{k-1/2}^{k+1/2} e^{2\pi i m x^2}\,dx$$

by passing to the limit $K \to \infty$. $\qquad\qquad\Box$

From Lemma 6 it follows that

$$\sum_{r=0}^{m} e^{2\pi i r^2/m} = 1 + \sum_{k=-\infty}^{\infty} \int_{0}^{m} e^{2\pi i\left(\frac{x^2}{m}+kx\right)}\,dx \ .$$

Since the integral simplifies to

$$\int_0^m e^{2\pi i\left(\frac{x^2}{m}+kx\right)}\,dx = \int_0^m e^{2\pi im\left(\frac{x}{m}+\frac{k}{2}\right)^2}\,dx \cdot e^{-\pi i\frac{mk^2}{2}}$$

$$= m \cdot e^{-\pi imk^2/2} \cdot \int_{k/2}^{k/2+1} e^{2\pi imu^2}\,du$$

applying Lemma 7 we have

$$g(1,m) = \lim_{K\to\infty}\left(\sum_{|k|\le K:2|k} m \cdot \int_{k/2}^{k/2+1} e^{2\pi imu^2}\,du \right.$$

$$\left. + \sum_{|k|\le K:2\nmid k} m \cdot e^{-\pi im/2} \cdot \int_{k/2}^{k/2+1} e^{2\pi imu^2}\,du \right)$$

$$= m \cdot \left(1 + e^{-\pi im/2}\right) \cdot \int_{-\infty}^{\infty} e^{2\pi imu^2}\,du$$

$$= \sqrt{m} \cdot \left(1 + i^{-m}\right) \cdot \int_{-\infty}^{\infty} e^{2\pi ix^2}\,dx \ .$$

If in particular we put $m = 1$, the relation

$$1 = (1 - i) \cdot \int_{-\infty}^{\infty} e^{2\pi ix^2}\,dx$$

gives the value of the Fresnel integral

$$\int_{-\infty}^{\infty} e^{2\pi ix^2}\,dx = \frac{1}{1-i} = \frac{1+i}{2} \ .$$

In this way we have succeeded in calculating $g(1,m)$. $\square$

Proposition 12. *If*

$$g(n,m) = \sum_{r=1}^{m} e^{2\pi inr^2/m}$$

denotes the quadratic Gauss sum, then for $n = 1$ it takes the value

$$g(1,m) = \sqrt{m} \cdot \frac{1 + i^{-m}}{1 - i} \ .$$

The most important consequence of this is that one can work out the Gauss sums for odd prime moduli.

Theorem 4: Signs of Gauss Sums. *If p denotes an odd prime number and χ denotes the Legendre symbol modulo p then for the Gauss sum $G(n,\chi)$ one has:*

$$G(1,\chi) = \sqrt{p}$$

if $p \equiv 1(\mathrm{mod}\ 4)$, and

$$G(1,\chi) = i\sqrt{p}$$

if $p \equiv 3(\mathrm{mod}\ 4)$. With

$$G(n,\chi) = \left(\frac{n}{p}\right) \cdot G(1,\chi)$$

the value of the Gauss sum for arbitrary n is determined.

Since Proposition 10 already gave $|G(n,\chi)| = \sqrt{p}$, the important part of Theorem 4 consists in the determination of the argument of $G(n,\chi)$ – this is what is meant by the label "sign" of the Gauss sums.

An easily proved technical result now permits the second demonstration of Theorem 3.

Lemma 8. *If m' and m'' are coprime, quadratic Gauss sums satisfy the relation*

$$g(n, m'm'') = g(nm', m'') \cdot g(nm'', m') \ .$$

Proof. Since $r = m''r' + m'r''$ runs through a complete system of classes of residues modulo $m'm''$, as r' and r'' run independently through complete systems of classes of residues modulo m' and m'', the required relation follows directly from

$$\sum_{r(m'm'')} e^{2\pi i n r^2 / m'm''}$$

$$= \sum_{r'(m')} \sum_{r''(m'')} e^{2\pi i n (r'm'' + r''m')^2 / m'm''}$$

$$= \sum_{r'(m')} \sum_{r''(m'')} e^{2\pi i n m'' r'^2 / m'} \, e^{2\pi i n m' r''^2 / m''} \ ,$$

because the factor with mixed product as exponent

$$e^{2\pi i n 2 r' r''}$$

is always 1. $\qquad\qquad\square$

For odd values of m Proposition 12 gives

$$g(1,m) = \sqrt{m} \cdot i^{(m-1)^2/4} \ ,$$

so, in particular, for distinct odd prime numbers p and q, and the Legendre symbol χ modulo p, we have

$$g(q,p) = G(q,\chi) = \left(\frac{q}{p}\right) \cdot G(1,\chi)$$

$$= \left(\frac{q}{p}\right) \cdot \sqrt{p} \cdot i^{(p-1)^2/4} \ .$$

Analogously the relation

$$g(p, q) = \left(\frac{p}{q}\right) \cdot \sqrt{q} \cdot i^{(q-1)^2/4}$$

must also exist. Lemma 8 has the consequence

$$g(q, p) \cdot g(p, q) = g(1, pq) = \sqrt{pq} \cdot i^{(pq-1)^2/4} \ .$$

This in turn leads to the calculation

$$\left(\frac{p}{q}\right)\left(\frac{q}{p}\right) = i^{((pq-1)^2 - (p-1)^2 - (q-1)^2)/4}$$

$$= (-1)^{\frac{p-1}{2} \cdot \frac{q-1}{2}}$$

and once again proves the quadratic reciprocity theorem.

Exercises on Chapter 6

In what follows let Γ_n be the group of Dirichlet characters mod n. If $\chi \in \Gamma_n$ we use the abbreviation $\tau(\chi) := G(1, \chi)$.

  1. Let m and n be coprime natural numbers. Show that $\omega : \Gamma_m \times \Gamma_n \to \Gamma_{mn}$, $\omega(\chi, \chi') = \chi\chi'$ is an isomorphism.

* 2. Let m and n be coprime natural numbers, $\chi \in \Gamma_m$ and $\chi' \in \Gamma_n$. Show:
    (i) $m_{\chi\chi'} = m_\chi \cdot m_{\chi'}$.
    (ii) $\chi\chi'$ is primitive in Γ_{mn} (resp. real) if and only if χ and χ' are primitive (resp. real).

  3. For odd $n \in \mathbb{Z}$ let $\chi_1(n) = (-1)^{(n-1)/2}$, $\chi_2(n) = \left(\frac{2}{n}\right)$, and for even $n \in \mathbb{Z}$ let $\chi_1(n) = \chi_2(n) = 0$. Write $\chi_3 = \chi_1\chi_2$. Show that χ_1, χ_2 and χ_3 are the non-trivial characters mod 8.

* 4. Let p be a prime number, k a natural number and $\chi \in \Gamma_{p^k}$ primitive and real. Then:
    (i)   If $p > 2$, then $k = 1$ and for all $n \in \mathbb{Z}$, $\chi(n) = \left(\frac{n}{p}\right)$.
    (ii) If $p = 2$, then $k \in \{2, 3\}$. In the case $k = 2$, $\chi(n) = (-1)^{(n-1)/2}$ for all odd $n \in \mathbb{Z}$. In the case $n = 3$, either $\chi(n) = \left(\frac{2}{n}\right)$ or $\chi(n) = (-1)^{(n-1)/2}\left(\frac{2}{n}\right)$, again for all odd $n \in \mathbb{Z}$.
    Conversely these characters are real and primitive.

  5. Let m be an odd natural number and $\chi \in \Gamma_m$ primitive and real. Show that m is square-free and that for all $n \in \mathbb{Z}$, $\chi(n) = \left(\frac{n}{m}\right)$. Conversely this character is real and primitive.

* 6. Let m and n be coprime natural numbers, and let $\chi \in \Gamma_m$ and $\chi' \in \Gamma_n$ be primitive and real. Show that the following statements are equivalent:
    (i) $\chi(n)\chi'(m) = -1$.
    (ii) $\chi(-1) = \chi'(-1) = -1$.

  7. Let p be prime, $k \in \mathbb{N}$ and $\chi \in \Gamma_{p^k}$ be primitive and real. Show that $\tau(\chi) = (\chi(-1)p)^{1/2}$.

8. Let m and n be coprime natural numbers, $\chi \in \Gamma_m$, $\chi' \in \Gamma_n$. Show that:

$$\tau(\chi\chi') = \chi(n)\chi'(m)\tau(\chi)\tau(\chi') \ .$$

* 9. Let m be a natural number, $\chi \in \Gamma_m$ primitive and real. Show that

$$\tau(\chi) = (\chi(-1)m)^{1/2} \ .$$

(Sign of a Gauss sum)

10. Show that if $d \in \mathbb{Z}$, then there exists some $m \in \mathbb{N}$ with $\mathbb{Q}(\sqrt{d}) \subseteq \mathbb{Q}\left(e^{2\pi i/m}\right)$.

* 11. Let $m \geq 2$ be a natural number and $\chi \in \Gamma_m$ primitive. Then:
   (i) $L(\bar{\chi}, 1) = -\frac{1}{\tau(\chi)} \sum_{k=1}^{m-1} \chi(k)\left(\log\left(\sin\frac{\pi k}{m}\right) + i\pi\frac{k}{m}\right)$.
   (ii) If χ is real and $\chi(-1) = 1$, then

$$L(\chi, 1) = -\frac{1}{\sqrt{m}} \sum_{k=1}^{m-1} \chi(k)\log\left(\sin\frac{\pi k}{m}\right)$$

   and

$$\sum_{k=1}^{m-1} k\chi(k) = 0 \ .$$

   (iii) If χ is real and $\chi(-1) = -1$, then

$$L(\chi, 1) = -\frac{\pi}{m^{3/2}} \cdot \sum_{k=1}^{m} k\chi(k)$$

   and

$$\sum_{k=1}^{m-1} \chi(k)\log\left(\sin\frac{\pi k}{m}\right) = 0 \ .$$

12. A character $\chi \in \Gamma_m$ is said to be *induced* by a character $\chi' \in \Gamma_n$, if $n \mid m$ and for the principal character $\chi_0 \in \Gamma_m$ we have $\chi = \chi_0\chi'$. Show that if χ is induced by χ', then for $\mathrm{Re}(s) > 1$,

$$L(\chi, s) = L(\chi', s) \prod_{p \mid m} \left(1 - \chi'(p)p^{-s}\right) \ .$$

13. Let $\omega > 0$ and for $\mathrm{Re}(s) > 1$ let $\zeta(s, \omega) = \sum_{n=0}^{\infty} (n + \omega)^{-s}$ be the Hurwitz zeta function (see Chapter 4, Ex. 20). Show that for $m \in \mathbb{N}$, $\chi \in \Gamma_m$ and $\mathrm{Re}(s) > 1$,

$$L(\chi, s) = m^{-s} \sum_{k=1}^{m} \chi(k)\zeta\left(s, \frac{k}{m}\right) \ .$$

In particular $L(\chi, .)$ is an entire function for $\chi \neq \chi_0$, and $L(\chi_0, .)$ is meromorphic in $\mathbb{C}$ with a single pole at $s = 1$.

* 14. Let $n \in \mathbb{N}$, $\chi \in \Gamma_n$ and $\chi \neq \chi_0$. Show that as $x \to \infty$

$$\sum_{p \le x} \frac{\chi(p)}{p} \log p = O(1) \ .$$

* 15. Show (without applying a Tauberian theorem) that, for $m \in \mathbb{N}$ and $\chi \in \Gamma_m$, as $x \to \infty$ we have

$$\sum_{n \le x} \sum_{k \mid n} \mu(k)\chi(k) = O(x) \ .$$

16. Show with the help of the Tauberian theorem of Ingham and Newman that, for $m \in \mathbb{N}$ and $\chi \in \Gamma_m$ real, we have

$$\sum_{n=1}^{\infty} \frac{1}{n}\mu(n)\chi(n) = \frac{1}{L(\chi,1)} \ .$$

17. Let $m \in \mathbb{N}$ and $\chi \in \Gamma_m$. Show that as $x \to \infty$, $\sum_{n \le x} \mu(n)\chi(n) = o(x)$.

18. Let $m \in \mathbb{N}$ and $\chi \in \Gamma_m$. Show that, for $\mathrm{Re}(s) > 1$,

$$L(\chi^2, 2s) = L(\chi, s) \sum_{n=1}^{\infty} \frac{\lambda(n)\chi(n)}{n^s} \ ,$$

and deduce from this that, as $x \to \infty$,

$$\sum_{n \le x} \lambda(n)\chi(n) = o(x) \ .$$

19. Let m be a natural number and a an integer. Show that, as $x \to \infty$,

(i) $$\sum_{\substack{n \le x \\ n \equiv a \,(\mathrm{mod}\, m)}} \mu(n) = o(x).$$

(ii) $$\sum_{\substack{n \le x \\ n \equiv a \,(\mathrm{mod}\, m)}} \lambda(n) = o(x).$$

20. Let m and n be coprime natural numbers and k an integer. For the Ramanujan sums show that $c_m(k)c_n(k) = c_{mn}(k)$.

Hints for the Exercises on Chapter 6

2. Show that $m_{\chi\chi'} \ge m_\chi m_{\chi'}$ and $m_{\chi\chi'} \le m_\chi m_{\chi'}$, by applying Proposition 7.

4. For $p > 2$ choose a primitive root g mod p^k. Then $\chi(g) = -1$ for $\chi \ne \chi_0$, so that there only exists one real character $\chi \ne \chi_0$ mod p^k.

6. Apply the previous questions, the quadratic reciprocity law and the supplementary laws.

9. Induction on the number of prime factors n of m. Apply (6).

11. Apply Proposition 9.

14. $O(1) = \sum_{n \le x} \frac{\chi(n)}{n} \log n = \sum_{n \le x} \frac{\chi(n)}{n} \sum_{d \mid n} \Lambda(d)$
    $= \sum_{d \le x} \frac{1}{d}\Lambda(d)\chi(d)L(\chi,1) + O(1).$

15. $1 = \sum_{n \le x} \mu(n)\chi(n)\frac{1}{n} \sum_{k \le \frac{x}{n}} \frac{\chi(k)}{k} = \sum_{n \le x} \mu(n)\frac{\chi(n)}{n}L(\chi,1) + O(1),$
    for $\chi \ne \chi_0$.

7. The Algorithm of Lenstra, Lenstra and Lovász

In this section we present an algorithm, which has several applications in number theory.

(1) It makes possible the factorisation of a polynomial $f(x) \in \mathbb{Q}[x]$ into irreducible factors inside a realistic time span.

(2) Suppose given rational numbers $\alpha_1, \ldots, \alpha_n$ and a rational number ε. (We note that each calculation programme only involves rational numbers.) Again in realistic time the algorithm allows us to find integers $p_1, \ldots, p_n$ and q, which for $1 \leq i \leq n$ satisfy

$$|p_i - \alpha_i q| \leq \varepsilon \quad \text{and} \quad 1 \leq q \leq 2^{n(n+1)/4}\varepsilon^{-n} \ .$$

One should compare this result with the Dirichlet approximation theorem, which shows that $1 \leq q \leq \varepsilon^{-n}$, without however being able to give a method for the determination of the numbers $p_1, \ldots, p_n, q$ (except by unrealistically long trial and error).

(3) Let $\alpha_1, \ldots, \alpha_n$ be real numbers. We look for integers $m_1, \ldots, m_n$ not equal to 0, with smallest possible absolute value, so that

$$\left| \sum_{i=1}^{n} m_i \alpha_i \right|$$

is as small as possible. If it is possible to make this sum equal to zero, then $\alpha_1, \ldots, \alpha_n$ are linearly dependent over $\mathbb{Q}$.

(4) By applying (3) to $\alpha_i = \alpha^{i-1}$ for some given $\alpha \in \mathbb{R}$, one can decide whether α is algebraic.

(5) The n real numbers $\alpha_1, \ldots, \alpha_n$ are called algebraically independent, if $\left\{ \alpha_1^{k_1} \ldots \alpha_n^{k_n} \,|\, k_1, \ldots, k_n \geq 0 \quad \text{and integral} \right\}$ is linearly independent over $\mathbb{Q}$, i.e. for each polynomial $p(x_1, \ldots, x_n) \neq 0$ with integral coefficients it follows that

$$p(\alpha_1, \ldots, \alpha_n) \neq 0 \ .$$

If one restricts the exponents by $0 \leq k_i \leq m_i$ $(1 \leq i \leq n)$ and applies (3), it is possible to decide whether the numbers $\alpha_1, \ldots, \alpha_n$ are algebraically dependent (with degree $\leq m_1 + \ldots + m_n$).

For the first named application we refer to the article of Lenstra, Lenstra and Lovász in Math. Annalen 1982, pp. 515-534.

In what follows let $<, >$ denote the standard scalar product in $\mathbb{R}^n$.

Definition 1. Let $(b_1, \ldots, b_n)$ be some basis in $\mathbb{R}^n$. Set $b_1^* = b_1$, and suppose that the vectors $b_1^*, \ldots, b_{i-1}^*$ have already been defined. Then let

$$b_i^* = b_i - \sum_{j=1}^{i-1} < b_i, b_j^* > \left| b_j^* \right|^{-2} b_j^* \; ,$$

and for $1 \leq i, j \leq n$ write $\mu_{ij} = < b_i, b_j^* > \left| b_j^* \right|^{-2}$.

This process is identical with the Gram-Schmidt-orthogonalisation process, i.e. $(b_1^*, \ldots, b_n^*)$ is an orthogonal basis of $\mathbb{R}^n$. It follows directly from the definitions that, for $1 \leq i \leq n$, $\mu_{ii} = 1$, and for $1 \leq i < j \leq n$, $\mu_{ij} = 0$. Furthermore

$$b_i = \sum_{j=1}^{n} \mu_{ij} b_j^* \; ,$$

i.e. the matrix $(\mu_{ij})_{1 \leq i, j \leq n}$ is the matrix of the basis change

$$(b_1^*, \ldots, b_n^*) \mapsto (b_1, \ldots, b_n) \; .$$

It is a triangular matrix with determinant 1, so that $\det(b_1, \ldots, b_n) = \det(b_1^*, \ldots, b_n^*)$.

Definition 2. Let $(b_1, \ldots, b_n)$ be some basis of $\mathbb{R}^n$. It is called *reduced* if, for $1 \leq j < i \leq n$,

$$|\mu_{ij}| \leq \frac{1}{2}$$

and, for $1 < i \leq n$, $\frac{3}{4} \left| b_{i-1}^* \right|^2 \leq \left| b_i^* + \mu_{i,i-1} b_{i-1}^* \right|^2$.

We propose to show that each lattice $\mathfrak{G}$ in $\mathbb{R}^n$ has a reduced basis, and demonstrate how one can find one such. For this it is useful to introduce the following notation:

Definition 3. Let $\mathfrak{G}$ be a lattice in $\mathbb{R}^n$, $1 \leq k \leq n+1$ and $B = (b_1, \ldots, b_n)$ a basis of $\mathfrak{G}$. Then write

$$A_k(B) : \iff \left(1 \leq j < i < k \Rightarrow |\mu_{ij}| \leq \frac{1}{2} \right) \quad \text{and}$$

$$\left(1 < i < k \Rightarrow \frac{3}{4} \left| b_{i-1}^* \right|^2 \leq \left| b_i^* + \mu_{i,i-1} b_{i-1}^* \right|^2 \right) \; .$$

We note that B is reduced if and only if $A_{n+1}(B)$ holds. For each basis B of $\mathfrak{G}$ $A_1(B)$ and $A_2(B)$ are true.

If besides B, B' is also a basis for $\mathbb{R}^n$, we label the numbers arising in Definition 1 by μ'_{ij} and the associated orthogonal basis by $(b_1'^*, \ldots, b_n'^*)$. We prove some technical propositions:

Lemma 1. *Let $\mathfrak{G}$ be a lattice in $\mathbb{R}^n$, $2 \leq k \leq n$ and $B = (b_1, \ldots, b_n)$ a basis of $\mathfrak{G}$ satisfying A_k. Let r be the nearest integer to $\mu_{k,k-1}$ and*

$$B' = (b_1, \ldots, b_{k-1}, b_k - r b_{k-1}, b_{k+1}, \ldots, b_n) \ .$$

Then B' is a basis of $\mathfrak{G}$ which satisfies A_k and has $\left|\mu'_{k,k-1}\right| \leq \tfrac{1}{2}$.

Proof. It is clear that B' is a basis of $\mathfrak{G}$. For $1 \leq i < k$, $b_i' = b_i$, hence also $b_i'^* = b_i^*$ and $\mu'_{ij} = \mu_{ij}$ for $1 \leq j < i < k$. Therefore $A_k(B')$ holds. For $1 \leq j < k$

$$\mu'_{kj} = <b_k', b_j^*> \left|b_j^*\right|^{-2} = \mu_{kj} - r\mu_{k-1,j} \ ,$$

and hence in particular

$$\left|\mu'_{k,k-1}\right| = |\mu_{k,k-1} - r| \leq 1/2 \ .$$

$\square$

Remark 1. In order to shorten the calculation time it is useful to know how μ'_{ij} and $b_i'^*$ are formed from μ_{ij} and b_i^*. Without difficulty one convinces oneself that

$$b_i'^* = b_i^* \quad \text{for} \quad 1 \leq i \leq n \ ,$$
$$\mu'_{ij} = \mu_{ij} \quad \text{for} \quad 1 \leq i \neq k \leq n, \quad \text{and}$$
$$\mu'_{kj} = \mu_{kj} - r\mu_{k-1,j} \quad \text{for} \quad 1 \leq j \leq k \ .$$

Lemma 2. *Let $\mathfrak{G}$ be a lattice in $\mathbb{R}^n$, $2 \leq k \leq n$ and $B = (b_1, \ldots, b_n)$ a basis for $\mathfrak{G}$, which satisfies A_k and*

$$\frac{3}{4} \left|b_{k-1}^*\right|^2 > \left|b_k^* + \mu_{k,k-1} b_{k-1}^*\right|^2 \ .$$

Let $B' = (b_1, \ldots, b_{k-2}, b_k, b_{k-1}, b_{k+1}, \ldots, b_n)$. Then B' is a basis for $\mathfrak{G}$, which satisfies A_{k-1} and

$$\left|b_{k-1}'^*\right|^2 < \frac{3}{4} \left|b_{k-1}^*\right|^2 \ .$$

Proof. It is clear that B' is a basis for $\mathfrak{G}$. For $1 \leq i < k-1$, $b_i' = b_i$, hence also $b_i'^* = b_i^*$ and for $1 \leq j < i < k-1$, $\mu'_{ij} = \mu_{ij}$. Therefore B' satisfies A_{k-1}. We have

$$b'^*_{k-1} = b_k - \sum_{j=1}^{k-2} < b_k, b^*_j > |b^*_j|^{-2} b^*_j$$

$$= b^*_k + < b_k, b^*_{k-1} > |b^*_{k-1}|^{-2} b^*_{k-1}$$

$$= b^*_k + \mu_{k,k-1} b^*_{k-1} \ ,$$

from which the last assertion follows. $\qquad\qquad\Box$

Remark 2. In order to shorten the calculation time, it is again useful to know how to calculate the new numbers μ'_{ij} and the new vectors b'^*_i from the old. Thus

$$\mu'_{k,k-1} = \mu_{k,k-1} |b^*_{k-1}|^2 \left(|b^*_k|^2 + \mu^2_{k,k-1} |b^*_{k-1}|^2 \right)^{-1} \ .$$

For $i \notin \{k-1, k\}$, $b'^*_i = b^*_i$, and

$$b'^*_{k-1} = b^*_k + \mu_{k,k-1} b^*_{k-1}, \ b'^*_k = b^*_{k-1} - \mu'_{k,k-1} b'^*_{k-1} \ .$$

For $1 \leq j < k-1$,

$$\begin{pmatrix} \mu'_{k-1,j} \\ \mu'_{k,j} \end{pmatrix} = \begin{pmatrix} \mu_{k,j} \\ \mu_{k-1,j} \end{pmatrix}$$

and for $k < i \leq n$

$$\begin{pmatrix} \mu'_{i,k-1} \\ \mu'_{i,k} \end{pmatrix} = \begin{pmatrix} 1 & \mu'_{k,k-1} \\ 0 & 1 \end{pmatrix} \begin{pmatrix} 0 & 1 \\ 1 & -\mu_{k,k-1} \end{pmatrix} \begin{pmatrix} \mu_{i,k-1} \\ \mu_{i,k} \end{pmatrix} \ .$$

Otherwise $\mu'_{ij} = \mu_{ij}$.

Lemma 3. *Let $\mathfrak{G}$ be a lattice in $\mathbb{R}^n$, $1 < k \leq n$, and B a basis for $\mathfrak{G}$, which satisfies A_k, $|\mu_{k,k-1}| \leq \frac{1}{2}$, but not A_{k+1}. Let $\frac{3}{4} |b^*_{k-1}|^2 \leq |b^*_k + \mu_{k,k-1} b^*_{k-1}|^2$, and $\ell < k$ maximal with respect to the property $|\mu_{k\ell}| > \frac{1}{2}$. Let r be the integer closest to $\mu_{k\ell}$ and*

$$B' = (b_1, \ldots, b_{k-1}, b_k - r b_\ell, b_{k+1}, \ldots, b_n) \ .$$

Then B' is a basis of $\mathfrak{G}$, which satisfies A_k, $\frac{3}{4} |b'^*_{k-1}|^2 \leq |b'^*_k + \mu'_{k,k-1} b'^*_{k-1}|^2$, and $|\mu'_{kj}| \leq \frac{1}{2}$ for all $\ell \leq j < k$.

Proof. Since B does not satisfy A_{k+1}, there exists some such ℓ. B' is clearly a basis for $\mathfrak{G}$. Since b^*_i is the projection of b_i onto the orthogonal complement of

$$\mathbb{R}b_1 + \ldots + \mathbb{R}b_{i-1} = \mathbb{R}b'_1 + \ldots + \mathbb{R}b'_{i-1} \ ,$$

it follows that $b'^*_i = b^*_i$ for $1 \leq i \leq n$. From this it follows that for $1 \leq j < i \neq k$, $\mu'_{ij} = \mu_{ij}$ and for $1 \leq j < k$, $\mu'_{kj} = < b_k - r b_\ell, b^*_j > |b^*_j|^{-2} = \mu_{kj} - r \mu_{\ell j}$. This equals μ_{kj} for $\ell < j < k$. By assumption $\ell < k - 1$. Therefore $\mu'_{k,k-1} = \mu_{k,k-1}$ and hence

$$\frac{3}{4}\left|\mathbf{b}_{k-1}^{\prime *}\right|^2 \le \left|\mathbf{b}_k^{\prime *} + \mu_{k,k-1}^{\prime}\mathbf{b}_{k-1}^{\prime *}\right|^2 , \left|\mu_{kj}^{\prime}\right| \le \frac{1}{2}, \text{ for } \ell < j < k ,$$

and

$$\left|\mu_{k\ell}^{\prime}\right| = \left|\mu_{k\ell} - r\right| \le \frac{1}{2} .$$

Moreover B' satisfies A_k. $\qquad\qquad\square$

Remark 3. Collecting together,

$$\begin{aligned}
\mathbf{b}_i^{\prime *} &= \mathbf{b}_i^* , &&\text{for } 1 \le i \le n , \\
\mu_{ij}^{\prime} &= \mu_{ij} , &&\text{for } i \ne k \text{ or } \ell < j < k, \quad\text{and} \\
\mu_{kj}^{\prime} &= \mu_{kj} - r\mu_{\ell j} &&\text{for } 1 \le j \le \ell .
\end{aligned}$$

Corollary 1. *Let $\mathfrak{G}$ be a lattice in $\mathbb{R}^n$, $1 \le k \le n$, and B a basis for $\mathfrak{G}$ which satisfies A_k. Let $k = 1$ or $\frac{3}{4}\left|\mathbf{b}_{k-1}^*\right|^2 \le \left|\mathbf{b}_k^* + \mu_{k,k-1}\mathbf{b}_{k-1}^*\right|^2$. Then there exists a basis B' for $\mathfrak{G}$, which satisfies A_{k+1}.*

Proof. In the case $k = 1$ this is trivial. Let $1 < k \le n$. Without loss of generality by Lemma 1 we can assume that $|\mu_{k,k-1}| \le \frac{1}{2}$. If B satisfies A_{k+1} we are finished. Otherwise one applies Lemma 3 repeatedly, and sees that there exists a basis B' of $\mathfrak{G}$, which satisfies A_k, $\frac{3}{4}\left|\mathbf{b}_{k-1}^{\prime *}\right|^2 \le \left|\mathbf{b}_k^{\prime *} + \mu_{k,k-1}^{\prime}\mathbf{b}_{k-1}^{\prime *}\right|^2$ and $\left|\mu_{kj}^{\prime}\right| \le \frac{1}{2}$ for $1 \le j < k$. Hence this basis satisfies A_{k+1}. $\qquad\square$

Suppose now that we are given a basis $B = (\mathbf{b}_1, \ldots, \mathbf{b}_n)$ for $\mathfrak{G}$. The assumptions of Lemma 2 and Corollary 1 are alternatives. We begin the algorithm for $k = 2$ (we know that $A_2(B)$ holds) and apply Lemma 2 or Corollary 1, one after the other, depending on which assumptions are satisfied. At each step the index k either increases by 1 or decreases by 1 (Lemma 2). If it equals $n + 1$, the algorithm breaks off, and the basis is reduced.

If we can show that the assumption of Lemma 2 only enters finitely many times, then we are finished.

Theorem 1. *Let $\mathfrak{G}$ be a lattice in $\mathbb{R}^n$. Then $\mathfrak{G}$ has a reduced basis, which by the preliminaries above is explicitly constructible.*

Proof. First we introduce some notation. Let $B = (\mathbf{b}_1, \ldots, \mathbf{b}_n)$ be a basis of $\mathfrak{G}$, and for $1 \le i \le n$ let $\mathfrak{G}_i(B) = \mathbb{Z}\mathbf{b}_1 + \ldots + \mathbb{Z}\mathbf{b}_i$. $\mathfrak{G}_i(B)$ is a lattice in an i-dimensional vector space with lattice constant $d(\mathfrak{G}_i(B))$. This can be calculated as follows: Let $L_i : \mathbb{R}\mathbf{b}_1^* + \ldots + \mathbb{R}\mathbf{b}_i^* \to \mathbb{R}\mathbf{b}_1 + \ldots + \mathbb{R}\mathbf{b}_i$ be the linear transformation which, for $1 \le j \le i$ satisfies $L_i\left(\mathbf{b}_j^*\right) = \mathbf{b}_j$. Its determinant is

$$\det\left(\mu_{k\ell}\right)_{1 \le k,\ell \le i} = 1, \quad \text{and}$$
$$L_i\left(\mathfrak{G}_i(B^*)\right) = \mathfrak{G}_i(B) .$$

Therefore $d(\mathfrak{G}_i(B)) = d(\mathfrak{G}_i(B^*))$. Since $\{b_1^*, \ldots, b_n^*\}$ is a set of orthogonal vectors, $d(\mathfrak{G}_i(B^*))$ is the volume of the fundamental parallelepiped, equal to $\prod_{j=1}^{i} |b_j^*|$, so that $d(\mathfrak{G}_i(B)) = \prod_{j=1}^{i} |b_j^*|$. Let $d(B) = \prod_{i=1}^{n-1} d(\mathfrak{G}_i(B))$.

We show next that there is a constant $c > 0$, so that for each basis B of $\mathfrak{G}$, $d(B) \geq c$. Let $m = \min\{|\mathfrak{x}| : \mathfrak{x} \in \mathfrak{G}, \ \mathfrak{x} \neq 0\}$; then $m > 0$. By Chapter 3, Corollary 3, there exists some $\mathfrak{x}_i \in \mathfrak{G}_i(B) \subseteq \mathfrak{G}$, $\mathfrak{x}_i \neq 0$, so that the coordinates of $\mathfrak{x}_i$ are at most equal to $d(\mathfrak{G}_i(B))^{1/i}$, i.e.

$$|\mathfrak{x}_i| \leq \sqrt{i}\, d(\mathfrak{G}_i(B))^{1/i} .$$

Hence

$$m \leq \sqrt{i}\, d(\mathfrak{G}_i(B))^{1/i}, \quad \text{and furthermore}$$

$$d(B) \geq \prod_{i=1}^{n-1} m^i\, i^{-i/2} =: c$$

for each basis B for $\mathfrak{G}$. Now let $B = (b_1, \ldots, b_n)$ be a basis for $\mathfrak{G}$, $B' = (b_1', \ldots, b_n')$ the basis for $\mathfrak{G}$ formed according to Lemma 1 or Lemma 3 (which means according to Corollary 1). Then $b_i'^* = b_i^*$ for $1 \leq i \leq n$, i.e. $d(B) = d(B')$. If on the other hand B' is formed according to Lemma 2, then for $i < k-1$, $d(\mathfrak{G}_i(B)) = d(\mathfrak{G}_i(B'))$, because $b_j'^* = b_j^*$ for $1 \leq j \leq i$. But

$$d(\mathfrak{G}_{k-1}(B')) = |b_{k-1}'^*| \prod_{i=1}^{k-2} |b_i^*| < \sqrt{\frac{3}{4}} \prod_{i=1}^{k-1} |b_i^*| = \sqrt{\frac{3}{4}} d(\mathfrak{G}_{k-1}(B)) .$$

For $i \geq k$, $\mathfrak{G}_i(B) = \mathfrak{G}_i(B')$ and so again $d(\mathfrak{G}_i(B)) = d(\mathfrak{G}_i(B'))$. This implies that $d(B') < \sqrt{\frac{3}{4}} d(B)$. Therefore if Lemma 2 is applied t times, in all we have

$$c \leq d\left(B^{(t)}\right) < \left(\frac{3}{4}\right)^{t/2} d(B) ,$$

which for t sufficiently large is not possible.

We note that because of $d(\mathfrak{G}_k(B')) = d(\mathfrak{G}_k(B))$, in the case of Lemma 2 we can settle on

$$|b_{k-1}'^*| \cdot |b_k'^*| = |b_{k-1}^*|\, |b_k^*|$$

(which also sidesteps a wearisome calculation). In this way we can set out the following programme scheme (put $B_j = |b_j^*|^2$):

```
for i = 1,...,n do
    b* := b ;
     i    i
    for j = 1,...,i-1 do
        μ   =< b , b* > /B ;
         ij      i   j      j
        b* := b* - μ  b* ;
         i    i    ij j
    end;
    B  :=< b*, b* >;
     i      i   i
```

```
    end;
    k := 2;
```

(1) for $\ell = k - 1$ do (*)
    if $B_k < \left(\frac{3}{4} - \mu_{k,k-1}^2\right) B_{k-1}$ go to (2);
    for $\ell = k - 2, k - 3, \ldots, 1$ do (*)
    if $k = n + 1$, end;
    $k := k + 1$;
    go to (1);

(2) $\mu := \mu_{k,k-1}$;
    $B := B_k + \mu^2 B_{k-1}$;
    $\mu_{k,k-1} := \mu B_{k-1}/B$;
    $B_k := B_{k-1} B_k/B$;
    $B_{k-1} = B$;
    $(\mathbf{b}_{k-1}, \mathbf{b}_k) := (\mathbf{b}_k, \mathbf{b}_{k-1})$;
    for $j = 1, \ldots, k - 2$ do
      $(\mu_{k-1,j}, \mu_{k,j}) := (\mu_{k,j}, \mu_{k-1,j})$;
    end;
    for $i = k + 1, \ldots, n$ do
$$\begin{pmatrix} \mu_{i,k-1} \\ \mu_{i,k} \end{pmatrix} := \begin{pmatrix} 1 & \mu_{k,k-1} \\ 0 & 1 \end{pmatrix} \begin{pmatrix} 0 & 1 \\ 1 & -\mu \end{pmatrix} \begin{pmatrix} \mu_{i,k-1} \\ \mu_{i,k} \end{pmatrix};$$
    end;
    if $k > 2$ then $k := k - 1$;
    end;
    go to (1).

(*) if $|\mu_{k,\ell}| > 1/2$ then;
      $r :=$ nearest integer to $\mu_{k,\ell}$;
      $\mathbf{b}_k := \mathbf{b}_k - r\mathbf{b}_\ell$;
      for $j = 1, \ldots, \ell - 1$ do
        $\mu_{k,j} := \mu_{k,j} - r\mu_{\ell,j}$;
      end;
      $\mu_{k,\ell} := \mu_{k,\ell} - r$;
    end;

We can now pose the question of how many calculation steps are necessary in order to transform $(\mathbf{b}_1, \ldots, \mathbf{b}_n)$ into a reduced basis. For special lattices $\mathfrak{G}$, Lenstra, Lenstra and Lovász have shown:

Let $\mathfrak{G}$ be a lattice in $\mathbb{R}^n$ and $(\mathbf{b}_1, \ldots, \mathbf{b}_n)$ a basis for $\mathfrak{G}$, so that $< \mathbf{b}_i, \mathbf{b}_j > \in \mathbb{Z}$ for $1 \leq i, j \leq n$. Let $B = \max\left(2, |\mathbf{b}_1|^2, \ldots, |\mathbf{b}_n|^2\right)$. The number of calculation steps necessary to transform $(\mathbf{b}_1, \ldots, \mathbf{b}_n)$ into a reduced basis is $\leq cn^4 \log B$, where $c > 0$ denotes an absolute constant.

We note that the assumption for $\mathfrak{G}$ is satisfied if $\mathfrak{G} \subseteq \mathbb{Z}^n$. If for $1 \leq i, j \leq n$ it happens that $< \mathbf{b}_i, \mathbf{b}_j > \in \mathbb{Q}$, then we have to multiply by some suitable number $m \in \mathbb{N}$ in order to reach $< m\mathbf{b}_i, m\mathbf{b}_j > \in \mathbb{Z}$. From a reduced basis for $m\mathfrak{G}$ we immediately obtain one for $\mathfrak{G}$.

An algorithm in $\mathbb{R}^n$, for which there exist constants $c, k > 0$ so that it breaks off after at most cn^k calculation steps, is called an algorithm which works in polynomial time.

Now we come to the applications. For these we need

Proposition 1. *Let $(b_1, \ldots, b_n)$ be a reduced basis for $\mathbb{R}^n$. Then*

(1) for $1 \leq j \leq i \leq n$, $|b_j|^2 \leq 2^{i-1} |b_i^*|^2$

(2) $\prod_{i=1}^n |b_i^*| \leq \prod_{i=1}^n |b_i| \leq 2^{n(n-1)/4} \prod_{i=1}^n |b_i^*|$

(3) $|b_1| \leq 2^{(n-1)/4} \prod_{i=1}^n |b_i^*|^{1/n}$.

Proof. (1) Because $< b_i^*, b_{i-1}^* >= 0$, for $1 < i \leq n$

$$|b_i^*|^2 \geq \frac{3}{4} |b_{i-1}^*|^2 - \mu_{i,i-1}^2 |b_{i-1}^*|^2 \geq \frac{1}{2} |b_{i-1}^*|^2 \ ,$$

and it follows by induction on $i - j$, that for $1 \leq j \leq i \leq n$,

$$|b_j^*|^2 \leq 2^{i-j} |b_i^*|^2 \ .$$

Furthermore, for $1 \leq i \leq n$,

$$|b_i|^2 = |b_i^*|^2 + \sum_{j=1}^{i-1} \mu_{ij}^2 |b_j^*|^2 \leq |b_i^*|^2 + \frac{1}{4} \sum_{j=1}^{i-1} 2^{i-j} |b_i^*|^2$$

$$= \left(\frac{1}{2} + 2^{i-2} \right) |b_i^*|^2 \leq 2^{i-1} |b_i^*|^2 \ .$$

Hence, for $1 \leq j \leq i \leq n$,

$$|b_j|^2 \leq 2^{j-1} |b_j^*|^2 \leq 2^{i-1} |b_i^*|^2 \ .$$

(2) For $1 \leq i \leq n$,

$$|b_i|^2 = \sum_{j=1}^i \mu_{ij}^2 |b_j^*|^2 \geq |b_i^*|^2 \ ,$$

from which the first inequality follows. The second follows from (1), since $|b_i| \leq 2^{(i-1)/2} |b_i^*|$ and $\sum_{i=1}^{n-1} i = \frac{1}{2}(n-1)n$.

(3) By (1), for $1 \leq i \leq n$, we have

$$|b_1| \leq 2^{(i-1)/2} |b_i^*| \ , \quad \text{so that}$$

$$|b_1|^n \leq 2^{n(n-1)/4} \prod_{i=1}^n |b_i^*| \ ,$$

which is the assertion. $\qquad\qquad\square$

Corollary 2. *There exist an algorithm (working in polynomial time), which for given rational numbers $\alpha_1, \ldots, \alpha_n$ and ε, $0 < \varepsilon < 1$, determines integers $p_1, \ldots, p_n$ and q, such that $1 \le q \le 2^{n(n+1)/4}\varepsilon^{-n}$ and, for $1 \le i \le n$,*

$$|p_i - q\alpha_i| \le \varepsilon \ .$$

Proof. Let $\mathbf{e}_1, \ldots, \mathbf{e}_{n+1}$ be the standard unit vectors in $\mathbb{R}^{n+1}$ and $\mathbf{b} = (-\alpha_1, \ldots, -\alpha_n, 2^{-n(n+1)/4}\varepsilon^{n+1})$. Let $B = (\mathbf{e}_1, \ldots, \mathbf{e}_n, \mathbf{b})$ and $\mathfrak{G}$ the lattice spanned by B. Then $d(\mathfrak{G}) = 2^{-n(n+1)/4}\varepsilon^{n+1}$ and for $1 \le i, j \le n$, $< \mathbf{e}_i, \mathbf{b} > \in \mathbb{Q}$, $< \mathbf{e}_i, \mathbf{e}_j > \in \mathbb{Z}$. Let $(\mathbf{b}_1, \ldots, \mathbf{b}_{n+1})$ be a reduced basis for $\mathfrak{G}$. Because $\mathbf{b}_1 \in \mathfrak{G}$, there exist $p_1, \ldots, p_n, q \in \mathbb{Z}$ with

$$\mathbf{b}_1 = \left(p_1 - q\alpha_1, \ldots, p_n - \alpha_n q, q2^{-(n+1)n/4}\varepsilon^{n+1} \right) \ .$$

By Proposition 1(3) we have $|\mathbf{b}_1| \le 2^{n/4}d(\mathfrak{G})^{1/(n+1)} = \varepsilon$. Therefore, for $1 \le i \le n$, $|p_i - \alpha_{iq}| \le \varepsilon$ and $|q| \cdot 2^{-n(n+1)/4}\varepsilon^{n+1} \le \varepsilon$, i.e. $|q| \le 2^{n(n+1)/4}\varepsilon^{-n}$. If q were equal to 0, then because $\varepsilon < 1$, for $1 \le i \le n$, $p_i = 0$ and hence $\mathbf{b}_1 = 0$, which is impossible. Thus $q \ne 0$. If $q < 0$, replace p_i by $-p_i$ and q by $-q$. $\quad\square$

We come now to application (3) announced at the start. For this we first need

Corollary 3. *There exists an algorithm (working in polynomial time), which for given rational numbers $\alpha_1, \ldots, \alpha_n$ and given ε, $0 < \varepsilon < 1$ (so that $\varepsilon^{-1-1/n}2^{(n+1)/4} \in \mathbb{Q}$) determines integers $m_1, \ldots, m_n$ and q so that, for $1 \le i \le n$,*

$$|m_i| \le 2^{(n+1)/4}\varepsilon^{-1/n}, (m_1, \ldots, m_n) \ne (0, \ldots, 0) \text{ and } \left| \sum_{i=1}^{n} \alpha_i m_i + q \right| \le \varepsilon \ .$$

Proof. Let $c = \varepsilon^{-1-1/n}2^{(n+1)/4}$ and, for $1 \le i \le n$, let $\mathbf{b}_i = \mathbf{e}_i + c\alpha_i \mathbf{e}_{n+1}$ and $\mathbf{b}_{n+1} = (0, \ldots, 0, c)$. Then, for $1 \le i, j \le n+1$, $< \mathbf{b}_i, \mathbf{b}_j > \in \mathbb{Q}$. Let $\mathfrak{G} = \mathbb{Z}\mathbf{b}_1 + \ldots + \mathbb{Z}\mathbf{b}_{n+1}$. Then $d(\mathfrak{G}) = c$. Let $(\mathbf{b}_1', \ldots, \mathbf{b}_n')$ be a reduced basis for $\mathfrak{G}$. Then

$$0 < |\mathbf{b}_1'| \le 2^{n/4}c^{1/(n+1)} = 2^{(n+1)/4}\varepsilon^{-1/n} \ .$$

There exist integers $m_1, \ldots, m_n$ and q with

$$\mathbf{b}_1' = \left(m_1, \ldots, m_n, c\sum_{i=1}^{n} m_i\alpha_i + cq \right) \ .$$

It follows that

$$\left| c \sum_{i=1}^{n} m_i \alpha_i + cq \right| \leq 2^{n/4} c^{1/n+1} , \quad \text{i.e.}$$

$$\left| \sum_{i=1}^{n} m_i \alpha_i + q \right| \leq 2^{n/4} c^{-n/(n+1)} = \varepsilon .$$

If all $m_i = 0$, then because $\varepsilon < 1$, $q = 0$ and hence also $\mathfrak{b}_1' = 0$, which is impossible. $\qquad\square$

We note that Corollary 2 and Corollary 3 only make sense, when ε is not made too small, i.e. the m_i cannot be taken too large. If the m_i reach the order of magnitude of the denominators of the α_i, then the existence becomes trivial. This holds also for the following corollary.

Corollary 4. *There exists an algorithm (working in polynomial time), which for given rational numbers $\alpha_1, \ldots, \alpha_n$ and ε ($0 < \varepsilon < 1/4$ and $\varepsilon^{-1/n} 2^{(n+1)/4} \in \mathbb{Q}$) determines integers $m_1, \ldots, m_n$ so that, for $1 \leq i \leq n$*

$$|m_i| \leq 2^{(n+1)/4} \varepsilon^{-1/n}, (m_1, \ldots, m_n) \neq 0 \quad \text{and}$$

$$\left| \sum_{i=1}^{n} m_i \alpha_i \right| \leq 4\varepsilon^{1-1/n} n K 2^{(n+1)/4} .$$

Here $K = \max_{1 \leq i \leq n} |\alpha_i|$.

Proof. Let k ($\in \mathbb{Q}$) be so chosen that

$$\varepsilon + kKn2^{(n+1)/4} \varepsilon^{-1/n} = 1/2 .$$

Using Corollary 3 choose integers $m_1, \ldots, m_n, q$ so that, for $1 \leq i \leq n$,

$$|m_i| \leq 2^{(n+1)/4} \varepsilon^{-1/n}, (m_1, \ldots, m_n) \neq 0 \quad \text{and} \quad \left| \sum_{i=1}^{n} k\alpha_i m_i + q \right| \leq \varepsilon .$$

Then

$$|q| \leq \varepsilon + k \sum_{i=1}^{n} |\alpha_i| |m_i| \leq \varepsilon + Kk2^{(n+1)/4} \varepsilon^{-1/n} n = 1/2 ,$$

so that $q = 0$. Therefore

$$\left| \sum_{i=1}^{n} \alpha_i m_i \right| \leq \varepsilon/k = 2\varepsilon Kn2^{(n+1)/4} \varepsilon^{-1/n} /(1 - 2\varepsilon) \leq 4\varepsilon^{1-1/n} Kn2^{(n+1)/4} .$$

Applications (4) and (5) now follow without further comment.

Remark. With the results obtained it is possible to give an algorithm, which decomposes the polynomial $f(x) \in \mathbb{Z}[x]$ into irreducible factors, or, which amounts to the same thing, finds a single irreducible factor. First translate

Corollary 4 to numbers in $\mathbb{Q}(i)$. Then look for an (almost) zero $\alpha \in \mathbb{Z}(i)$ for $f(x)$. This is possible in polynomial time (see S. Smale, Proc. Int. Congress of Mathematicians, 1986, Vol. 1, pp. 172-195). Then using the pattern suggested by application (4), we look for the minimal polynomial $m(x)$ of α. This will then be an irreducible factor of $f(x)$.

In the work of Lenstra, Lenstra and Lovász referred to, one finds yet another algorithm.

Addenda

[1])   H. Davenport and W. Schmidt (Symposia Mathematica, Vol. IV, INDAM, Rome, 1968/69 (Academic Press, London 1970)) proved the following: if $\alpha = [a_0; a_1, \ldots]$ is the continued fraction expansion of the irrational number α, and if

$$\gamma(\alpha) = \overline{\lim}_{n \to \infty} [0; a_{n+1}, \ldots] \cdot [0; a_n, a_{n+1}, \ldots] \ ,$$

then

$$\left| \alpha - \frac{p}{n} \right| < \frac{c}{nN}$$

has infinitely many solutions for sufficiently large N if and only if

$$c > \frac{1}{1 + \gamma(\alpha)} \ .$$

[2])   The inequality

$$\left| \alpha - \frac{p}{n} \right| < \frac{1}{n^2}$$

can, as shown by A. Hurwitz, be sharpened to

$$\left| \alpha - \frac{p}{n} \right| < \frac{1}{\sqrt{5}n^2}$$

(see Exercise 20 in Chap. 1). A larger factor than $\sqrt{5}$ in the denominator is not possible (compare Niven-Zuckerman, p. 189ff, 221ff, loc.cit.).

In 1948 A.V. Prasad (J. London Math. Soc. *23*) proved the following: if p_n/q_n is the nth convergent fraction of $(\sqrt{5} - 1)/2$, and if for $n \geq 1$

$$c_n = \frac{1}{2}(\sqrt{5} + 1) + \frac{p_{2n-1}}{q_{2n-1}} \ ,$$

then

$$\left| \alpha - \frac{p}{q} \right| < \frac{1}{c_n q^2}$$

has at least n solutions in coprime numbers (p, q). There exist numbers α – for example $\alpha = (\sqrt{5} - 1)/2$ – so that this inequality has exactly n solutions of this kind.

[3])   The proof of Corollary 6 given by R. Apéry was published in 1981 (Mathématiques, Bull. Sect. Sci. III). The proof given here is due to F. Beukers, Bull. London Math. Soc. *11* (1979).

[4]   If in Proposition 2 one puts $\alpha_1 = \alpha_2 = \ldots = \alpha_L = 1/N$ one sees that the inequality $n \leq N^L$ cannot be sharpened. If on the other hand one replaces

$$\left| \alpha_\ell - \frac{p_\ell}{n} \right| < \frac{1}{Nn}$$

by

$$\left| \alpha_\ell - \frac{p_\ell}{n} \right| \leq \frac{1}{Nn} \; ,$$

then G. Kaindl (Sitzber. Österr. Akad. Wiss., Math. naturw. Kl. II, *185* (1976)) has shown that

$$n \leq N^L - \frac{N^L - 1}{N - 1}$$

is attainable. F. Langmayr asserts (Monatshefte Math. *20* (1980)) that this inequality also cannot be sharpened.

[5]   H. Davenport and W. Schmidt (Acta Arith. *16* (1969/70)) have shown the following: If $\lambda < 1$ and if in Proposition 2 (resp. Proposition 3) we replace the inequalities

$$\left| \alpha_\ell - \frac{p_\ell}{n} \right| < \frac{1}{Nn} \quad \text{resp.} \quad \left| \sum_{l=1}^{L} \alpha_\ell n_\ell - p \right| < \frac{1}{N}$$

by

$$\left| \alpha_\ell - \frac{p_\ell}{n} \right| < \frac{\lambda}{Nn} \quad \text{resp.} \quad \left| \sum_{l=1}^{L} \alpha_\ell n_\ell - p \right| < \frac{\lambda}{N}$$

then (in the sense of Lebesgue) the corresponding conclusions no longer hold for almost all $\alpha_1, \ldots, \alpha_L$.

[6]   As R. Tichy (Monatshefte Math. *88* (1979)) has proved, one can replace the bound

$$N^{M/L}$$

by

$$\left(N^M - \frac{N^M - 1}{N - 1} \right)^{1/L}$$

in Theorem 2, provided one is satisfied with the weaker inequality

$$\left| \sum_{\ell=1}^{L} \alpha_{m\ell} n_\ell - p_m \right| \leq \frac{1}{N} \; .$$

[7]   Hurwitz' inequality (see Addendum [2]) can be altered for a non-integral β. H. Minkowski proved that one can find infinitely many pairs of integers (p, n) with the property

$$|\alpha n - \beta - p| < \frac{1}{4|n|} \; .$$

J. H. Grace showed subsequently that the constant $1/4$ is the best possible (Proc. London Math. Soc. (2), *17* (1918)). J. H. Grace carried out his proof with continued fractions. E. Hlawka found a direct argument (Monatshefte für Math. *47* (1938)).

[8]) Theorem 2 was generalised by Kronecker himself in the following way. Suppose that real numbers $\alpha_{\ell m}$ $(\ell = 1, \ldots, L, m = 1, \ldots, M)$ and β_ℓ $(\ell = 1, \ldots, L)$ are arbitrarily preassigned. For each arbitrarily chosen small $\varepsilon > 0$ there exist integers n_m $(m = 1, \ldots, M)$ and p_ℓ $(\ell = 1, \ldots, L)$ with the property

$$\left| \sum_{m=1}^{M} \alpha_{\ell m} n_m - \beta_\ell - p_\ell \right| < \varepsilon, \quad \ell = 1, \ldots, L ,$$

if and only if for all integral h_ℓ $(\ell = 1, \ldots, L)$ the integrality of the M numbers

$$\sum_{\ell=1}^{L} h_\ell \alpha_{\ell m}, \quad m = 1, \ldots, M ,$$

implies the integrality of

$$\sum_{l=1}^{L} h_\ell \beta_\ell$$

(compare Exercise 9 of Chap. 3).

[9]) H. Weyl formulated Corollary 5 even more generally. If $p(x)$ is a polynomial, in which, apart from the constant summand, at least one irrational coefficient occurs, then already modulo 1 the sequence $p(1), p(2), \ldots, p(n), \ldots$ is uniformly distributed.

[10]) Behind van der Corput's inequality hides a very general principle. One finds it in its most general form in R. J. Taschner (Monatshefte Math. *91* (1981)).

[11]) Basic facts from the theory of quadratic number fields can for example be found in Hlawka-Schoissengeier, p. 139ff., loc.cit. or in Niven-Zuckerman, p. 273ff., loc.cit.

Using the notation in the book of Hlawka-Schoissengeier we append a discussion by Rabinowitsch on prime factor decomposition in a class of quadratic number fields as an extension of unique decomposition in $\mathbb{Z}(i)$.

Let $\mathcal{O}_d$ be Euclidean, then one can write each α from $\mathcal{O}_d$ with $|N(\alpha)| > 1$ as a product of prime elements:

$$\alpha = \pi_1 \pi_2 \ldots \pi_K .$$

This representation is uniquely determined up to units and the order of the factors. We show this in an analogous way to the integers $\mathbb{Z}$ and the Gaussian integers $\mathbb{Z}(i)$, by first choosing for each pair $\alpha, \beta \neq 0$ from $\mathcal{O}_d$ a greatest common divisor δ. This is possible by repeated use of the Euclidean algorithm, which shows the existence of integral elements ξ, η with $\delta = \alpha\xi + \beta\eta$, and it follows that a prime element π dividing the product $\alpha\beta$ must also divide one of the factors α or β. The property of prime elements (Fundamental Lemma of Number Theory (compare Hlawka-Schoissengeier, p. 9, loc.cit.)) contained in the last assertion forms the key to the proof of uniqueness in the decomposition into prime factors.

In a non-Euclidean $\mathcal{O}_d$ the uniqueness of prime factorisation is not guaranteed. A counterexample is found in $\mathcal{O}_{-5}$ with the decompositions

$$21 = 3 \cdot 7 = (1 + 2\sqrt{-5})(1 - 2\sqrt{-5})$$

(compare Hlakwa-Schoissengeier, p. 155, loc.cit., Niven-Zuckerman, p. 17f., loc. cit.). Nevertheless it is conceivable that for non-Euclidean $\mathcal{O}_d$ decomposition of elements into prime factors may also be unique up to units and order. For this the validity of the fundamental lemma may suffice, i.e. that from $\pi \mid \beta\alpha$ for each prime element π, $\pi \mid \alpha$ or $\pi \mid \beta$ follows. At this point Rabinowitsch starts from the following assertion:

If for each non-integral element ρ from $\mathbb{Q}(\sqrt{d})$ one can always construct integral elements ξ, η from $\mathcal{O}_d$ with

$$0 < |N(\rho\xi - \eta)| < 1$$

then the fundamental lemma holds in $\mathcal{O}_d$.

Indeed were the fundamental lemma to be false, there would have to exist some prime element π and two integral elements α, β with $\pi \mid \alpha\beta$ but with $\pi \nmid \alpha$ and $\pi \nmid \beta$. For such a prime element π let λ, μ be the pair of integral elements α, β with the property above, and for which the norm $|N(\lambda\mu)|$ takes the smallest possible value among all norms $|N(\alpha\beta)|$. Depending on whether $|N(\lambda)| \geq |N(\pi)|$ or $|N(\lambda)| < |N(\pi)|$, set $\rho = \lambda/\pi$ or $\rho = \pi/\lambda$. Since π does not divide λ, λ/π is not integral. If conversely λ were to divide π, then because π is a prime element, λ would have to be associated to π, because $\pi \mid \lambda\mu$ and $\pi \nmid \mu$ with λ a unit is not possible. However then π would have to divide λ, and this also is excluded. Therefore in each case ρ fails to be an integral element. Given the assumption there exist integral elements ξ, η with

$$0 < |N(\rho\xi - \eta)| < 1 \ .$$

In the case $\rho = \lambda/\pi$ from $0 < |N(\xi\lambda/\pi - \eta)| < 1$ one obtains $0 < |N(\lambda\xi - \pi\eta)| < |N(\pi)| \leq |N(\lambda)|$. For $\rho = \pi/\lambda$ from $0 < |N(\xi\pi/\lambda - \eta)| < 1$ one obtains $0 < |N(\pi\xi - \lambda\eta)| < |N(\lambda)| < |N(\pi)|$. In all cases one can find two integral elements φ, ψ (which in some sequence are compatible with ξ, η), so that the integral element $\nu = \lambda\varphi - \pi\psi$ satisfies the inequality

$$0 < |N(\nu)| < \min\bigl(|N(\lambda)|, |N(\pi)|\bigr) \ .$$

$|N(\nu)| < |N(\pi)|$ proves immediately that $\pi \nmid \nu$. Neither does π divide μ, however $\nu\mu = \lambda\mu\varphi - \pi\mu\psi$. Finally the formula

$$|N(\nu\mu)| = |N(\nu)||N(\mu)| < |N(\lambda)||N(\mu)| = |N(\lambda\mu)|$$

contradicts the original choices of λ and μ, so that the fundamental lemma cannot be false.

In order to prove the uniqueness of prime decomposition, following Rabinowitsch, we must show that in $\mathbb{Q}(\sqrt{d})$ there are no "bad elements", i.e. elements ρ, for which for arbitrarily chosen integral elements ξ, η it follows that $|N(\rho\xi - \eta)| \geq 1$ for $\rho\xi - \eta \neq 0$. The following five properties of bad elements are of importance for the subsequent considerations.

(i)   A bad element ρ always satisfies $|N(\rho)| \geq 1$.
(ii) A bad element ρ cannot be a rational number.
(iii) For all integral elements λ, if ρ is bad, then so is $\rho + \lambda$.
(iv) For all integral elements λ, ρ bad implies $\rho\lambda$ bad, except when $\rho\lambda$ is itself an element of $\mathcal{O}_d$.
(v)   If in $\mathcal{O}_d$ the uniqueness of prime decomposition fails, then there exists a bad element of the form

$$\rho = \frac{n - \vartheta}{p}$$

where p denotes a prime number, n is an integer with $0 \leq n < p$ and $\vartheta = \sqrt{d}$ (resp. $\vartheta = (1 + \sqrt{d})/2$) depending on whether $d \not\equiv 1$ or $d \equiv 1 \pmod 4$.

Property (i) follows from the fact that in the contrary case, for $0 < |N(\rho\xi - \eta)| < 1$ one could set $\eta = 0$ and $\xi = 1$. The second property follows from this, since for ρ from $\mathbb{Q}$, one could choose $\xi = 1$ and η the nearest integer to ρ. The third and fourth properties come from the formulae

$$0 < |N((\rho + \lambda)\xi - \eta)|$$
$$= |N(\rho\xi - (\eta - \lambda\xi))| < 1 \ ,$$
$$0 < |N((\rho\lambda)\xi - \eta)|$$
$$= |N(\rho(\lambda\xi) - \eta)| < 1 \ .$$

In order to check the fifth property we first prove that there must exist at least one bad element, which we can write in the form

$$\rho = \frac{a + b\vartheta}{c}$$

with a, b, c integral. The fraction is already so reduced, that no prime p dividing c also divides a and b. Therefore since

$$\rho_1 = \frac{c}{p} \cdot \rho = \frac{a + b\vartheta}{p}$$

also fails to be an integral element, because of (iv) ρ_1 must also be bad. If b/p were integral, because of (iii) we would obtain a bad rational number

$$\frac{a}{b} = \rho_1 - \frac{b}{p} \cdot \vartheta \ ,$$

contradicting property (ii). Therefore b and p are coprime, we can find integers x, y with $bx - py = -1$, and by (iii) and (iv) obtain the bad element

$$\rho_2 = \rho_1 x - y\vartheta = \frac{ax + (bx - py)\vartheta}{p}$$
$$= \frac{ax - \vartheta}{p} \ .$$

If we put $z = [ax/p]$, then from $ax = pz + n$ with $0 \le n < p$ we construct a bad element of the form

$$\rho_3 = \rho_2 - z = \frac{n - \vartheta}{p} \ ,$$

as required.

Up to this point our considerations are completely independent of d. From now on suppose that $d < 0$ and $d \equiv 1 \pmod 4$, i.e. that d has the form

$$d = 1 - 4m$$

for some natural number m. We have

$$\vartheta = \frac{1 + \sqrt{d}}{2} \ ,$$

and the norm of $x + y\vartheta = (2x + y + y\sqrt{d})/2$ can be calculated as

$$N(x + y\vartheta) = \frac{(2x + y)^2}{4} - d \cdot \frac{y^2}{4}$$
$$= \frac{4x^2 + 4xy + y^2}{4} + (4m - 1)\frac{y^2}{4}$$
$$= x^2 + xy + my^2 \ .$$

If

$$\rho = \frac{n - \vartheta}{p}$$

denotes a bad element in $\mathbb{Q}(\sqrt{d})$, because $N(\rho) \geq 1$ it must be that

$$N(n - \vartheta) = n^2 - n + m \geq p^2 = N(p) \ .$$

$n < p$ leads to

$$p^2 \leq n(n - 1) + m < p(p - 1) + m$$
$$= p^2 - p + m \ ,$$

i.e. to $p < m$. Since by our previous work at least one bad element of the form above exists, this implies:

Prime decomposition in $\mathcal{O}_d$, for $d = 1 - 4m$, can fail to be unique only, if among the finitely many elements $(n - \vartheta)/p$ with prime number $p < m$ and rational integer n with $0 \leq n < p$ at least one bad element occurs. As we will subsequently show, this is at worst the case, if one of the numbers $n^2 - n + m$ with $n = 0, 1, \ldots, m - 2$ is distinct from a prime number.

If indeed all numbers

$$N(n - \vartheta) = n^2 - n + m, \quad n = 0, 1, \ldots, m - 2 \ ,$$

are prime, there remain for a prime number $p < m$ only the alternatives: $p = N(n - \vartheta)$ or p is coprime with $N(n - \vartheta)$. In the case $p = N(n - \vartheta)$, because

$$0 < N\left(\frac{n - \vartheta}{p}\right) = \frac{N(n - \vartheta)}{p^2} = \frac{1}{p} < 1$$

and on account of property (i), $(n - \vartheta)/p$ is certainly not bad. In the second case one can find integers x, y with $N(n - \vartheta)x - py = 1$. Since $N(n - \vartheta) = (n - \vartheta)(n - \vartheta')$, after division by p it follows that

$$\frac{n - \vartheta}{p} \cdot (n - \vartheta')x - y = \frac{1}{p} \ .$$

For the integral elements $\xi = (n - \vartheta')x$, $\eta = y$, this leads to

$$0 < N\left(\frac{n - \vartheta}{p} \cdot \xi - \eta\right) = \frac{1}{p^2} < 1 \ ,$$

and in this case also $(n - \vartheta)/p$ cannot be bad. Even more actually holds:

If one of the numbers $N(n - \vartheta) = n^2 - n + m$, with $n = 0, 1, \ldots, m - 1$ is not a prime number, then prime decomposition in $\mathcal{O}_d$ cannot be unique.

Two considerations precede the proof. Firstly for $y \neq 0$ and $x + y\vartheta$ integral $N(x + y\vartheta) \geq m$. For on the one hand we have

$$N(x + y\vartheta) = x^2 + xy + my^2$$
$$= \left(x + \frac{y}{2}\right)^2 + \left(m - \frac{1}{4}\right) \cdot y^2 \geq m - \frac{1}{4}$$

and on the other $N(x + y\vartheta)$ is integral. Secondly for all $n = 0, 1, \ldots, m - 1$, $n - \vartheta$ shows itself to be a prime element, since an integer other than ± 1 can certainly not divide $n - \vartheta$. If a decomposition $n - \vartheta = \alpha\beta$ with integral elements α, β not belonging to $\mathbb{Z}$ were to exist, then for $N(\alpha) \geq m$, $N(\beta) \geq m$ we would have

$$m^2 \leq N(\alpha)N(\beta) = N(n - \vartheta)$$
$$= n^2 - n + m < m^2 - m + m = m^2$$

which would be absurd.

Furthermore, if for $n = 0, 1, \ldots, m - 1$ the natural number $N(n - \vartheta)$ were not prime, then in addition to the decomposition

$$N(n - \vartheta) = (n - \vartheta)(n - \vartheta')$$

into non-rational prime elements $n - \vartheta$, $n - \vartheta'$, there would have to exist a decomposition

$$N(n - \vartheta) = p_1 p_2 \ldots p_K$$

into prime numbers with $K \geq 2$. Even if this product could be split further into a decomposition in $\mathcal{O}_d$ consisting of prime elements, it would nonetheless remain essentially distinct from the decomposition above, because then more than two prime elements show up as factors.

Next one defines the polynomial

$$f(x) = x^2 + x + m$$

subject to the condition that $N(n - \vartheta) = f(n - 1)$. Since $N(-\vartheta) = N(\vartheta)$ and by the uniqueness of prime decomposition $N(m - 1 - \vartheta) = f(m - 2)$ is also a prime number, one can collect the earlier considerations together in the following way.

If given the natural number m, $d = 1 - 4m$, then the theorem on unique decomposition into prime factors holds in $\mathcal{O}_d$ if and only if

$$f(x) = x^2 + x + m$$

is a prime number for all integers $x = 0, 1, \ldots, m - 2$.

For $m = 2, 3$ the $f(x)$, $x = 0, 1, \ldots, m - 2$ are clearly prime numbers. In this way one obtains the Euclidean domains $\mathcal{O}_{-7}$ and $\mathcal{O}_{-11}$ (compare Niven-Zuckerman, p. 282f., loc.cit.). $m = 4$ does not satisfy the condition, since already $f(4) = 4$ is no prime number. Calculating $f(x)$ for $m = 5$ and $x = 0, 1, 2, 3$, gives only prime numbers; the same holds for $m = 11, 17, 41$ and the corresponding $f(x)$. Euler had already studied the polynomials $x^2 + x + 17$ and $x^2 + x + 41$. It is worthy of note that the quadratic polynomial $x^2 + x + 41$ gives only prime numbers for the 40 successive values of the argument $x = 0, 1, \ldots, 39$.

The theorem of unique decomposition into prime factors holds in the domain $\mathcal{O}_d$ with

$$d = -1, -2, -3, -7, -11, -19, -43, -67, -163 \ .$$

For $d = -1, -2, -3, -7, -11$ this is true because the domains are Euclidean (compare Niven-Zuckerman, p. 282f., loc.cit.), and for the remaining d one has only to substitute $m = 5, 11, 17, 41$ in $d = 1 - 4m$.

Gauss already conjectured that for only finitely many negative values d does $\mathcal{O}_d$ process the property of unique factorisation. Heilbronn succeeded in giving the first proof of this conjecture by applying analytic techniques. Lehmer calculated that apart from the values introduced above in discussing the unique prime decomposition of elements from $\mathcal{O}_d$ only values of $d < -6 \cdot 10^9$ were in question. After a flawed proof of Heegner, H. M. Stark and A. Baker (Michigan Math. J. *14* (1967)) showed that only the values of d written down above possess the property that prime decomposition in $\mathcal{O}_d$ is unique.

[12]) The estimate $O(\sqrt{N}/N)$ of the remainder is in no way sharp. The *Dirichlet divisor problem* poses the question of the infimum of all Θ, for which

$$\frac{1}{N} \cdot \sum_{n=1}^{N} \tau(n) = \log N + 2C - 1 + O\left(\frac{N^\Theta}{N}\right)$$

remains correct. G. H. Hardy and E. Landau proved that $\Theta \geq 1/4$ (Proc. London Math. Soc. (2), *15* (1916)). In the other direction H. Iwaniec and J. Mozzochi were able to prove $\Theta \leq 7/22$ (J. Number Theory *29* (1988) 60–93). For the best result until now we refer to W. Müller, W. G. Nowak (L.N. 1452, Springer-Verlag).

[13]) The estimate $O(\sqrt{N}/N)$ of the remainder is in no way sharp. The *Gauss circle problem* poses the question of the infimum of all Θ for which

$$\frac{1}{N} \cdot \sum_{n=1}^{N} r(n) = \pi + O\left(\frac{N^\Theta}{N}\right)$$

remains correct. G. H. Hardy and E. Landau proved that $\Theta \geq 1/4$ (see Landau, Vorlesungen über Zahlentheorie, 2. Band, p. 183ff., Hirzel, Leipzig, 1927). In the other direction H. Iwaniec and J. Mozzochi were able to prove $\Theta \leq 7/22$ (J. Number Theory *29* (1988) 60–93). For the best result until now we refer to W. Müller, W. G. Nowak (L.N. 1452, Springer-Verlag).

[14]) The Riemann ζ-function satisfies the functional equation (see Exercise 15, Chap. 5).

$$\pi^{-\frac{s}{2}} \Gamma\left(\frac{s}{2}\right) \cdot \zeta(s) = \pi^{-\frac{1-s}{2}} \Gamma\left(\frac{1-s}{2}\right) \cdot \zeta(1-s)$$

and can therefore be treated as a meromorphic function defined on all of $\mathbb{C}$ with a single pole at $s = 1$ (compare R. Taschner, Funktionentheorie, p. 135ff., Manzsche Verlags- und Universitätsbuchhandlung, Wien, 1982). From the functional equation one reads off that the only zeros of $\zeta(s)$ with $\mathrm{Re}(s) < 0$ lie at $s = -2, -4, -6, \ldots$ (see Exercise 16, Chap. 5). By a conjecture of Riemann all the zeros with $0 \leq \mathrm{Re}(s) \leq 1$ lie on the line $\mathrm{Re}(s) = 1/2$. Until now this Riemann conjecture has been neither proved nor disproved. The functional equation shows further that the zeros must be symmetrically positioned about the line $\mathrm{Re}(s) = 1/2$; because $\overline{\zeta(s)} = \zeta(\bar{s})$, they also lie symmetrically about the real axis (see Exercise 16, Chap. 5).

[15]) In spite of the fact that the polynomials $x^2 + x + 17$ and $x^2 + x + 41$ possess only prime values for the first 16 or 40 non-negative integral arguments (see Addendum [11]), there can exist no polynomial $p(x)$ with the property that it only represents prime numbers as x runs through all integral values, except for the trivial example of a constant polynomial. Indeed if $p(n)$ is a prime number, for each integer m $p(n)$ divides the expression $p(n + mp(n)) - p(n)$ and hence also $p(n + mp(n))$. Accepting the requirement above $p(n + mp(n)) = p(n)$ for all m, and this is impossible for a non-constant polynomial. Given this a result of V. Matijasevič (Sov. Math. Doklady *11* (1970)) is even more remarkable, by which there exists a polynomial in several variables, which on the substitution of natural numbers always delivers prime numbers, provided the value is positive, and in this way generates the set of all prime numbers. If one admits the degree 15905 for the polynomial, one can

limit the number of variables to 10 (Zapiski nautsch., Sem. Leningrad Otd. mat. Inst. Steklov *68* (1977)). J. P. Jones constructed a polynomial of this kind in the 26 variables $a, b, \ldots, y, z$ (Notices of the A.M.S. 1975) namely

$$
\begin{aligned}
(k+2)(1 &- (wz+h+j-q)^2 - ((gk+2g+k+1)(h+j)+h-z)^2 \\
&- (16(k+1)^3(k+2)(n+1)^2+1-f^2)^2 \\
&- (2n+p+q+z-e)^2 - (e^3(e+2)(a+1)^2+1-o^2)^2 \\
&- ((a^2-1)y^2+1-x^2)^2 - (16r^2y^4(a^2-1)+1-u^2)^2 \\
&- (((a+u^2(u^2-a^2))^2-1)(n+4dy)^2+1-(x+cu)^2)^2 \\
&- ((a^2-1)\ell^2+1-m^2)^2 - (ai+k+1-\ell-i)^2 - (n+\ell+v-y)^2 \\
&- (p+\ell(a-n-1)+b(2an+2a-n^2-2n-2)-m)^2 \\
&- (z-pm+p\ell(a-p)+t(2ap-p^2-1))^2 \\
&- (q-x+y(a-p-1)+s(2ap+2a-p^2-2p-2))^2) \ .
\end{aligned}
$$

[16]) De la Vallée Poussin proved

$$
\pi(x) = \operatorname{li} x + R(x)
$$

with

$$
R(x) = O\left(x e^{-\delta\sqrt{\log x}}\right)
$$

for some positive δ. After important preparatory work by Vinogradov, N. M. Korobov (Uspechi Mat. Nauk (1958)) showed with the help of estimates of trigonometric sums that

$$
R(x) = O\left(x e^{-\delta(\log x)^{3/5}(\log\log x)^{-1/5}}\right) \ .
$$

Gauss had accepted the validity of the inequality $\pi(x) < \operatorname{li} x$. To date no numerical counterexample is known; nevertheless J. E. Littlewood disproved the conjecture, and also proved

$$
\varliminf_{x\to\infty} \frac{(\pi(x)-\operatorname{li} x)\log x}{\sqrt{x}\log\log x} < 0 \ ,
$$

$$
\varlimsup_{x\to\infty} \frac{(\pi(x)-\operatorname{li} x)\log x}{\sqrt{x}\log\log x} > 0 \ .
$$

(compare A. E. Ingham, The Distribution of Prime Numbers, Ch. V, Cambridge University Press, 1932). By te Riele the smallest x with $\pi(x) > \operatorname{li} x$ is at most $6.69 \cdot 10^{370}$ in size (Math. Comp., *48* (1987)).

[17]) From the prime number theorem it follows that, for arbitrarily chosen $\varepsilon > 0$, there is always some prime number lying in the interval $[x, x+\varepsilon x]$, so long as x is sufficiently large. Otherwise one would have

$$
\begin{aligned}
0 &= \pi(x+\varepsilon x) - \pi(x) \\
&= \left(\frac{x+\varepsilon x}{\log(x+\varepsilon x)} - \frac{x}{\log x}\right)(1+o(1)) \ ,
\end{aligned}
$$

hence

$$
1 = \frac{1+\varepsilon}{\log(x+\varepsilon x)} \cdot \log x \cdot (1+o(1)) \ ,
$$

which cannot be true. It is even conjectured that for $x \geq 1$ there is always some prime number lying in the interval $[x, x + \sqrt{x}]$. If one fixes $x = p_n$, the nth prime number, it follows from this that $p_{n+1} - p_n \leq \sqrt{p_n}$, a conjecture already formulated by Legendre. Hoheisel showed (Berl. Sitzungsber. 1930) that there exists some $\Theta < 1$ with the property that, for sufficiently large x, there always exists some prime number in the interval $[x, x + x^\Theta]$. By Mozzochi (J. Number Theory *24* (1986)) one has

$$p_{n+1} - p_n = O\left(p_n^{1051/1920}\right) \ .$$

Another old conjecture says that for $x \geq 2$, $y \geq 2$,

$$\pi(x + y) \leq \pi(x) + \pi(y) \ ,$$

i.e. that in the interval $[x, x + y]$ there lie at most as many prime numbers as in the interval $[1, y]$. Although in the special case of large x and $y = x$ this conjecture has been proved, there are reasons for supposing that in general it is false (see J. Richards, Bull. A.M.S. *80* (1974)). Montgomery and Vaughan (Mathematika *20* (1973)) have proved

$$\pi(x + y) \leq \pi(x) + \frac{2y}{\log y}$$

and

$$\pi(x + y) \leq \pi(x) + 2\pi(y) \ .$$

[18]) The Riemann conjecture (compare Addendum [14]) is closely connected with the asymptotic behaviour of

$$\sum_{n \leq x} \mu(n)$$

for large x. It is correct if and only if

$$\sum_{n \leq x} \mu(n) = O\left(x^{1/2 + \varepsilon}\right)$$

holds for each positive ε. A conjecture of Mertens, by which one already always has

$$\left| \sum_{n \leq x} \mu(n) \right| \leq \sqrt{x}$$

has been disproved by Odlyzko and Riele (Report of the Department of Numerical Mathematics, Center for Math. and Comp. Science (1985)).

[19]) If the *great Riemann conjecture*, by which the non-trivial zeros of $L(\chi, s)$ for primitive χ lie only on the line $\mathrm{Re}(s) = 1/2$, were to be proved, one could show the existence of a constant $C > 0$ with the property

$$\left| \pi(x; a, m) - \frac{1}{\varphi(m)} \mathrm{li}\, x \right| \leq C \cdot \sqrt{x} \log x$$

for $1 \leq m \leq x$, for all m coprime with a and for all x. Montgomery (Topics in Multiplicative Number Theory, Springer LN, vol. 227) showed

$$\pi(x; a, m) < \frac{2x}{\varphi(m) \log \frac{x}{m}} \ .$$

C. L. Siegel showed (Acta Arithmetica *1* (1936)) that for each $\varepsilon > 0$ there exists some $\delta(\varepsilon) > 0$ with the property that for all moduli m, all real Dirichlet characters modulo m, and likewise for all $s \geq 1 - \delta(\varepsilon)m^{-\varepsilon}$, $L(\chi, s) \neq 0$. From this one deduces the prime number theorem of Page, Siegel and Walfisz. It says: for each positive A there exist positive constants c and C with the property that for all moduli

$$m \leq (\log x)^A$$

and all a coprime with m the inequality

$$\left| \pi(x; a, m) - \frac{1}{\varphi(m)} \mathrm{li}\, x \right| \leq C \cdot x \cdot e^{-c\sqrt{\log x}}$$

holds.

By Rodosskii (Izv. Akad. Nauk. SSSR *12* (1948)) there exist two positive constants c, C so that for all $x \geq 3$ there exists at most one natural number $\bar{m}$ with the following property. If the modulus m satisfies

$$m \leq x^{c/\log\log x}$$

and is no multiple of $\bar{m}$, then for all a coprime with m one has

$$\left| \pi(x; a, m) / \frac{x}{\varphi(m) \log x} - 1 \right| < \frac{C}{\log x} \ .$$

If for coprime a and m $p(a, m)$ denotes the smallest prime number with $p \equiv a$ (mod m), it follows from the prime number theorem of Page, Siegel and Walfisz that

$$p(a, m) = O\left(e^{m^\varepsilon}\right)$$

for each $\varepsilon > 0$. On the other hand from the great Riemann conjecture it follows that

$$p(a, m) = O\left(m^{2+\varepsilon}\right)$$

for each $\varepsilon > 0$. In any case Linnik (Mat. Sbornik *15* (1947)) has shown that there must exist some positive constant c with

$$p(a, m) \leq m^c \ .$$

[20])   The elementary proof of the prime number theorem is due to Erdös (Proc. Nat. Acad. Sci. USA *35* (1949)) and to Selberg (Ann. Math. (2) *50* (1949)); the original proof to Hadamard (Bull. Soc. Math. France *24* (1896)) and to de la Vallée Poussin (Ann. Soc. Sci. Bruxelles (2) *20* (1896)). The simplest proof until recently resting on the full Tauberian theorem of Wiener and Ikehara can be sketched as follows.

By the *Riemann-Lebesgue lemma*, for each complex-valued and integrable function $f(x)$ defined on the interval $[a, b]$ the equation

$$\lim_{|\lambda| \to \infty} \int_a^b f(x) e^{i\lambda x} dx = 0$$

holds. For this the proof goes as follows: first suppose that $f(x) = c_J(x)$ is the characteristic function of an interval J with end points a' and b'. Then

$$\int_a^b f(x)e^{i\lambda x}\,dx = \int_{a'}^{b'} e^{i\lambda x}\,dx$$

$$= \frac{1}{i\lambda}\left(e^{i\lambda b'} - e^{i\lambda a'}\right) = O\left(\frac{1}{|\lambda|}\right) \ .$$

With this it is clear that the conclusion of the Lebesgue-Riemann lemma is true for step functions. For an arbitrarily chosen integrable function f and for an arbitrarily small positive ε, one can then find a step function g with

$$\int_a^b |f(x) - g(x)|\,dx < \varepsilon \ .$$

Since for some sufficiently large λ_0 for all λ with $|\lambda| \geq \lambda_0$ the inequality

$$\left|\int_a^b g(x)e^{i\lambda x}\,dx\right| < \varepsilon \ ,$$

holds, from

$$\left|\int_a^b f(x)e^{i\lambda x}\,dx\right|$$

$$\leq \left|\int_a^b (f(x) - g(x))e^{i\lambda x}\,dx\right| + \left|\int_a^b g(x)e^{i\lambda x}\,dx\right|$$

$$\leq \int_a^b |f(x) - g(x)|\,dx + \varepsilon < 2\varepsilon$$

the Riemann-Lebesgue lemma follows in full generality.

A second preparatory step concerns the evaluation of the integral

$$\frac{1}{2}\cdot \int_{-2}^2 \left(1 - \frac{|t|}{2}\right) e^{itu}\,dt = \frac{\sin^2 u}{u^2} \ ,$$

which is achieved by direct calculation. Thus

$$\frac{1}{2}\int_{-2}^2 \left(1 - \frac{|t|}{2}\right) e^{itu}\,dt$$

$$= \frac{1}{2iu}\left(e^{2iu} - e^{-2iu}\right) - \frac{1}{4}\int_0^2 te^{itu}\,dt + \frac{1}{4}\int_{-2}^0 te^{itu}\,dt$$

$$= \frac{1}{u}\sin 2u - \frac{1}{2}\int_0^2 t\cos tu\,dt$$

$$= \frac{1}{2u^2}\left(1 - \cos 2u\right) = \frac{\sin^2 u}{u^2} \ .$$

For $u = 0$ the conclusion remains true, if one extends continuously.

The statement of the Wiener-Ikehara Tauberian theorem reads: *if $\varphi(u)$ denotes a real-valued, non-negative and monotone decreasing function, defined for non-negative u, for which in the region $\mathrm{Re}(s) > 1$ the integral*

$$J(s) = \int_0^\infty \varphi(u)e^{-su}\,du$$

exists, and if

$$J(s) - \frac{1}{s-1} = F(s)$$

can be considered as a function continuous in the region $\mathrm{Re}(s) \geq 1$, *then*

$$\lim_{u \to \infty} \varphi(u)e^{-u} = 1 \ .$$

In contrast to Newman's method of proof *only* the continuity and not the holomorphic property of the function $F(s)$ is required, and *no* growth condition of the form

$$\varphi(u) = O(e^u)$$

is supposed . As a result the proof of the stated Tauberian theorem is much more subtle. It rests on a sequence of assertions:

(i)   *For each positive* ε *we have*

$$\varphi(u)e^{-u} = O(e^{\varepsilon u}) \ .$$

This follows from

$$\frac{1}{\varepsilon} + F(1+\varepsilon) = \int_0^\infty \varphi(u)e^{-u}e^{-\varepsilon u}\,du$$

$$\geq \int_x^\infty \varphi(u)e^{-u}e^{-\varepsilon u}\,du \geq \varphi(x) \cdot \int_x^\infty e^{-u(1+\varepsilon)}\,du$$

$$= \frac{1}{1+\varepsilon} \cdot \varphi(x)e^{-x(1+\varepsilon)} \ .$$

(ii)   *If* p *denotes the integral*

$$p = \int_{-\infty}^\infty \frac{\sin^2 u}{u^2}\,du \ ,$$

then for each positive λ *we have*

$$\lim_{x \to \infty} \int_0^\infty \frac{\sin^2 \lambda(x-u)}{\lambda(x-u)^2}\,du = p \ .$$

Indeed if one substitutes $\lambda(x-u) = \nu$, one obtains

$$\int_0^\infty \frac{\sin^2 \lambda(x-u)}{\lambda(x-u)^2}\,du = \frac{-1}{\lambda} \cdot \int_{\lambda x}^{-\infty} \frac{\sin^2 \nu}{\lambda \nu^2} \cdot \lambda^2 d\nu \ ,$$

from which the assertion follows.

(iii)   *For each positive* λ *we have*

$$\lim_{x \to \infty} \lim_{\varepsilon \to 0+0} \int_0^\infty \left(\varphi(u)e^{-u} - 1\right) e^{-\varepsilon u} \cdot \frac{\sin^2 \lambda(x-u)}{(x-u)^2}\,du = 0 \ .$$

This can be justified as follows: In the region $\mathrm{Re}(s) > 1$ we have

$$F(s) = \int_0^\infty \left(\varphi(u)e^{-u} - 1\right) e^{-(s-1)u}\,du \ .$$

If in (i) one replaces ε by $\varepsilon/2$, then from $|t| \leq 2$ and $u \geq 0$ we deduce that

$$\left(1 - \frac{1}{2}|t|\right) e^{it\lambda x} \left(\varphi(u)e^{-u} - 1\right) e^{-\varepsilon u} e^{-itu\lambda}$$

$$= O\left(e^{\varepsilon u/2} e^{-\varepsilon u}\right) = O\left(e^{-\varepsilon u/2}\right) \ .$$

From this we have

$$\int_0^\infty \left(\varphi(u)e^{-u} - 1\right) e^{-\varepsilon u} \cdot \frac{\sin^2 \lambda(x - u)}{(x - u)^2} \, du$$

$$= \frac{1}{2} \cdot \lambda^2 \cdot \int_0^\infty \left(\varphi(u)e^{-u} - 1\right) e^{-\varepsilon u} \cdot \int_{-2}^2 \left(1 - \frac{|t|}{2}\right) e^{it\lambda(x-u)} dt \, du$$

$$= \frac{1}{2} \cdot \lambda^2 \cdot \int_{-2}^2 \left(1 - \frac{|t|}{2}\right) e^{it\lambda x} \cdot \int_0^\infty \left(\varphi(u)e^{-u} - 1\right) e^{-u(\varepsilon + it\lambda)} du \, dt$$

$$= \frac{1}{2} \cdot \lambda^2 \cdot \int_{-2}^2 \left(1 - \frac{|t|}{2}\right) e^{it\lambda x} F(1 + \varepsilon + it\lambda) dt \ .$$

Choosing (ε, t) to lie in $0 \leq \varepsilon \leq 1$, $-2 \leq t \leq 2$ one can bound

$$\left(1 - \frac{|t|}{2}\right) e^{it\lambda x} F(1 + \varepsilon + it) \ .$$

Since $F(1 + it\lambda)$ is continuous for $-2 \leq t \leq 2$ we have

$$\lim_{\varepsilon \to 0+0} \int_0^\infty \left(\varphi(u)e^{-u} - 1\right) e^{-\varepsilon u} \cdot \frac{\sin^2 \lambda(x - u)}{(x - u)^2} \, du$$

$$= \frac{1}{2} \cdot \lambda^2 \cdot \int_{-2}^2 \left(1 - \frac{|t|}{2}\right) e^{it\lambda x} F(1 + it\lambda) dt \ .$$

With the variable t lying in the interval $-2 \leq t \leq 2$ the function

$$\left(1 - \frac{|t|}{2}\right) F(1 + it\lambda)$$

is continuous and hence integrable, and (iii) follows from the Riemann-Lebesgue lemma.

(iv)   *If p is defined as in* (ii), *then for each positive λ we have*

$$\lim_{x \to \infty} \int_0^\infty \varphi(u)e^{-u} \cdot \frac{\sin^2 \lambda(x - u)}{(x - u)^2} \, du = \lambda p \ .$$

Indeed

$$0 \leq e^{-\varepsilon u} \cdot \frac{\sin^2 \lambda(x - u)}{(x - u)^2} \leq \frac{\sin^2 \lambda(x - u)}{(x - u)^2}$$

and choosing the variable u to be non-negative

$$\frac{\sin^2 \lambda(x - u)}{(x - u)^2}$$

is integrable. Therefore

$$\lim_{\epsilon \to 0} \int_0^\infty e^{-\epsilon u} \cdot \frac{\sin^2 \lambda(x-u)}{(x-u)^2}\, du$$

$$= \int_0^\infty \frac{\sin^2 \lambda(x-u)}{(x-u)^2}\, du \ ,$$

so that from (ii)

$$\lim_{x \to \infty} \lim_{\epsilon \to 0+0} \int_0^\infty e^{-\epsilon u} \cdot \frac{\sin^2 \lambda(x-u)}{(x-u)^2}\, du = \lambda p$$

follows. If this is added to the result obtained in (iii), one obtains the formula

$$\lim_{x \to \infty} \lim_{\epsilon \to 0+0} \int_0^\infty \varphi(u) e^{-u} e^{-\epsilon u} \cdot \frac{\sin^2 \lambda(x-u)}{(x-u)^2}\, du = \lambda p \ .$$

The association of ϵ to the integrand is monotone decreasing, so, because of the positivity of the integrand, allowing ϵ to go to the limit and interchanging this operation with integration, justifies (iv).

(v) *The inequality* $\overline{\lim}_{x \to \infty} \varphi(x) e^{-x} \leq 1$ *holds.*
From now on let $\lambda \geq 1$ and ϵ always be positive. There exists some number $x_0(\epsilon, \lambda)$, so that for $x' \geq x_0(\epsilon, \lambda)$, because of (iv), we have

$$\varphi\left(x' - \frac{1}{\sqrt{\lambda}}\right) e^{-x' - 1/\sqrt{\lambda}} \cdot \int_{x'-1/\sqrt{\lambda}}^{x'+1/\sqrt{\lambda}} \frac{\sin^2 \lambda(x'-u)}{(x'-u)^2}\, du$$

$$\leq \int_{x'-1/\sqrt{\lambda}}^{x'+1/\sqrt{\lambda}} \varphi(u) e^{-u} \cdot \frac{\sin^2 \lambda(x'-u)}{(x'-u)^2}\, du$$

$$\leq \int_0^\infty \varphi(u) e^{-u} \cdot \frac{\sin^2 \lambda(x'-u)}{(x'-u)^2}\, du < \lambda p + \epsilon \ .$$

The first integral satisfies

$$\int_{x'-1/\sqrt{\lambda}}^{x'+1/\sqrt{\lambda}} \frac{\sin^2 \lambda(x'-u)}{(x'-u)^2}\, du = \lambda \cdot \int_{-\sqrt{\lambda}}^{\sqrt{\lambda}} \frac{\sin^2 \nu}{\nu^2}\, d\nu$$

and is independent of x'. For $x \geq x_0(\epsilon, \lambda)$ and $x' = x + 1/\sqrt{\lambda}$, from

$$\varphi(x) e^{-x-2/\sqrt{\lambda}} \leq \left(p + \frac{\epsilon}{\lambda}\right) \left(\int_{-\sqrt{\lambda}}^{\sqrt{\lambda}} \frac{\sin^2 \nu}{\nu^2}\, d\nu \right)^{-1}$$

one deduces

$$\overline{\lim}_{x \to \infty} \varphi(x) e^{-x} \leq e^{2/\sqrt{\lambda}} \left(p + \frac{\epsilon}{\lambda}\right) \left(\int_{-\sqrt{\lambda}}^{\sqrt{\lambda}} \frac{\sin^2 \nu}{\nu^2}\, d\nu \right)^{-1} \ ,$$

so that (v) follows by taking the limit as $\epsilon \to 0$ and $\lambda \to \infty$.

(vi)   *The inequality* $\underline{\lim}_{x\to\infty}\varphi(x)e^{-x} \geq 1$ *holds.*

For $x' \geq 1 \geq 1/\sqrt{\lambda}$ the formulae

$$\int_0^{x'-1/\sqrt{\lambda}} \frac{\sin^2 \lambda(x'-u)}{\lambda(x'-u)^2}\, du = \int_{-\lambda x'}^{-\sqrt{\lambda}} \frac{\sin^2 \nu}{\nu^2}\, d\nu$$

and

$$\int_{x'+1/\sqrt{\lambda}}^{\infty} \frac{\sin^2 \lambda(x'-u)}{\lambda(x'-u)^2}\, du = \int_{\sqrt{\lambda}}^{\infty} \frac{\sin^2 \nu}{\nu^2}\, d\nu$$

are valid. Because of (v) there exists a constant K with $\varphi(x)e^{-x} \leq K$; because of (iv) there exists some $x_1(\varepsilon,\lambda)$ so that for $x' \geq x_1(\varepsilon,\lambda) - 1$ we deduce that

$$\lambda p - \varepsilon$$

$$< \left(\int_{x'-1/\sqrt{\lambda}}^{x'+1/\sqrt{\lambda}} + \int_0^{x'-1/\sqrt{\lambda}} + \int_{x'+1/\sqrt{\lambda}}^{\infty} \right) \varphi(u)e^{-u} \cdot \frac{\sin^2 \lambda(x'-u)}{(x'-u)^2}\, du$$

$$\leq \varphi\left(x' + \frac{1}{\sqrt{\lambda}}\right) e^{-x'+1/\sqrt{\lambda}} \cdot \lambda \cdot \int_{-\sqrt{\lambda}}^{\sqrt{\lambda}} \frac{\sin^2 \nu}{\nu^2}\, d\nu + 2K\lambda \cdot \int_{\sqrt{\lambda}}^{\infty} \frac{\sin^2 \nu}{\nu^2}\, d\nu \ .$$

Starting from $x \geq x_1(\varepsilon,\lambda)$ and $x' = x - 1/\sqrt{\lambda} \geq x_1(\varepsilon,\lambda) - 1$ it follows that

$$\lambda p - \varepsilon < \varphi(x)e^{-x}e^{2/\sqrt{\lambda}} \cdot \lambda \cdot \int_{-\sqrt{\lambda}}^{\sqrt{\lambda}} \frac{\sin^2 \nu}{\nu^2}\, d\nu + 2K\lambda \cdot \int_{\sqrt{\lambda}}^{\infty} \frac{\sin^2 \nu}{\nu^2}\, d\nu \ .$$

This implies that

$$\lambda p - \varepsilon \leq \underline{\lim}_{x\to\infty}\varphi(x)e^{-x}e^{2/\sqrt{\lambda}} \cdot \lambda \cdot \int_{-\sqrt{\lambda}}^{\sqrt{\lambda}} \frac{\sin^2 \nu}{\nu^2}\, d\nu$$

$$+ 2K\lambda \cdot \int_{\sqrt{\lambda}}^{\infty} \frac{\sin^2 \nu}{\nu^2}\, d\nu \ .$$

Passage to the limit as $\varepsilon \to 0$ leads to the estimate

$$\frac{\left(p - 2K \cdot \int_{\sqrt{\lambda}}^{\infty} \frac{\sin^2 \nu}{\nu^2}\, d\nu\right) e^{-2/\sqrt{\lambda}}}{\int_{-\sqrt{\lambda}}^{\sqrt{\lambda}} \frac{\sin^2 \nu}{\nu^2}\, d\nu} \leq \underline{\lim}_{x\to\infty}\varphi(x)e^{-x} \ .$$

Passage to the limit as $\lambda \to \infty$ finally gives (vi) and hence the conclusion of the Tauberian theorem of Wiener and Ikehara.

[21])   As Landau showed the convergence

$$\sum_{n \leq x} \mu(n) = o(x)$$

is in elementary terms equivalent to the statement of the prime number theorem, i.e. from the convergence the prime number theorem can be deduced by elementary considerations (without the application of complex variable or Tauberian arguments), and conversely. Related to this are the convergence statements

$$\sum_{n=1}^{\infty} \frac{\mu(n)}{n} = 0, \quad \sum_{n=0}^{\infty} \frac{(-1)^n \mu(2n+1)}{2n+1} = \frac{4}{\pi} \ .$$

The second was shown first of all by Möbius, who had however to assume the existence of the sum , i.e. the convergence of the series. This can be justified as follows:

If $f(n)$ denotes a bounded number theoretic function with

$$\sum_{n \leq x} f(n) = o(x) \ ,$$

then

$$\sum_{n \leq x} \frac{f(n)}{n} = \frac{1}{x} \cdot \sum_{n \leq x} \sum_{k \mid n} f(n) + o(1)$$

holds.

For some arbitrarily chosen $\varepsilon > 0$ let $x_0(\varepsilon)$ be chosen so large that for $x \geq x_0(\varepsilon)$ one can deduce that

$$\left| \sum_{n \leq x} f(n) \right| \leq \varepsilon x \ .$$

If one writes $\delta = \sqrt{\varepsilon}$, $x \geq x_0(\varepsilon)/\delta$, for $\delta x < k \leq x$ it follows that

$$\left| \sum_{\delta x < j \leq k} f(j) \right| \leq \left| \sum_{j \leq k} f(j) \right| + \left| \sum_{1 \leq j \leq \delta x} f(j) \right|$$
$$\leq \varepsilon k + \varepsilon \delta x \leq 2\varepsilon x$$

and

$$\sum_{\delta x < k \leq x-1} \left| \left\{ \frac{x}{k} \right\} - \left\{ \frac{x}{k+1} \right\} \right|$$
$$= \sum_{\delta x < k \leq x-1} \left| \frac{x}{k} - \frac{x}{k+1} - \left[\frac{x}{k} \right] + \left[\frac{x}{k+1} \right] \right|$$
$$\leq \sum_{\delta x < k \leq x-1} \left(\frac{x}{k} - \frac{x}{k+1} \right) + \sum_{\delta x < k \leq x-1} \left(\left[\frac{x}{k} \right] - \left[\frac{x}{k+1} \right] \right)$$
$$\leq \frac{1}{\delta} + \left[\frac{1}{\delta} \right] \leq \frac{2}{\delta} \ .$$

Abel transformation therefore gives

$$\left| \sum_{\delta x < k \leq x} f(k) \left\{ \frac{x}{k} \right\} \right|$$
$$= \left| \sum_{\delta x < k \leq x-1} \left(\left\{ \frac{x}{k} \right\} - \left\{ \frac{x}{k+1} \right\} \right) \cdot \sum_{\delta x < j \leq k} f(j) + \left\{ \frac{x}{[x]} \right\} \cdot \sum_{\delta x < j \leq x} f(j) \right|$$
$$\leq 2\varepsilon x \cdot \sum_{\delta x < k \leq x-1} \left| \left\{ \frac{x}{k} \right\} - \left\{ \frac{x}{k+1} \right\} \right| + 2\varepsilon x$$
$$\leq 2\varepsilon x \left(\frac{2}{\delta} + 1 \right) < 6 \cdot \frac{\varepsilon}{\delta} \cdot x = 6\sqrt{\varepsilon} \cdot x \ .$$

If one uses the boundedness of f – let M denote a bound for $|f(n)|$ – from

$$\left| \sum_{1 \leq k \leq x} f(k) \left\{ \frac{x}{k} \right\} \right| = \left| \sum_{1 \leq k \leq \delta x} f(k) \left\{ \frac{x}{k} \right\} + \sum_{\delta x < k \leq x} f(k) \left\{ \frac{x}{k} \right\} \right|$$

$$\leq M \cdot \delta x + 6\sqrt{\varepsilon} \cdot x = (M + 6)\sqrt{\varepsilon} \cdot x$$

one deduces the asymptotic representation

$$\sum_{k \leq x} f(k) \left\{ \frac{x}{k} \right\} = o(x) \ .$$

This shows, as claimed, that

$$\sum_{n \leq x} \frac{f(n)}{n} = \frac{1}{x} \cdot \sum_{n \leq x} f(n) \left[\frac{x}{n} \right] + \frac{1}{x} \cdot \sum_{n \leq x} f(n) \left\{ \frac{x}{n} \right\}$$

$$= \frac{1}{x} \cdot \sum_{k \leq x} f(k) \cdot \sum_{n : kn \leq x} 1 + o(1)$$

$$= \frac{1}{x} \cdot \sum_{n \leq x} \sum_{k | n} f(k) + o(1) \ .$$

For $f(n) = \mu(n)\chi(n)$, with a real Dirichlet character χ modulo m the right-hand side of this relation can be calculated. *One has*

$$\lim_{x \to \infty} \frac{1}{x} \cdot \sum_{n \leq x} \sum_{k | n} \mu(k)\chi(k) = \frac{1}{L(\chi, 1)} \ .$$

Indeed, since for each prime power

$$\sum_{k | p^\nu} \mu(k)\chi(k) = 1 - \chi(p)$$

holds, we have

$$\sum_{k | n} \mu(k)\chi(k) = \prod_{p | n} (1 - \chi(p)) \geq 0 \ .$$

If χ is not the principal character, then because of

$$\frac{\zeta(s)}{L(\chi, s)} = \sum_{n=1}^{\infty} \frac{1}{n^s} \cdot \sum_{k | n} \mu(k)\chi(k)$$

one can apply the Wiener-Ikehara Tauberian theorem (see Addendum [20]) to the function

$$\frac{L(\chi, 1)\zeta(s)}{L(\chi, s)} - \frac{1}{s - 1}$$

noting the positivity of

$$L(\chi, s) = \prod_p \frac{1}{1 - \frac{\chi(p)}{p^s}}$$

and therefore also of $L(\chi, 1)$. This leads to the stated limit relation. If $\chi = \chi_0$ denotes the principal character modulo m, and if $p_1, \ldots, p_K$ are the prime factors of the modulus m, then clearly one has

$$\sum_{n \le x} \sum_{k|n} \mu(k)\chi(k) = \sum_{n=p_1^{\nu_1}\cdots p_K^{\nu_K} \le x} 1 \ .$$

Given the condition under the sum sign the multiplicities ν_k are bounded by

$$\nu_k \le \left[\frac{\log x}{\log 2}\right]$$

so that again

$$\sum_{n \le x} \sum_{k|n} \mu(k)\chi(k) \le \left(\frac{\log x}{\log 2}\right)^K = o(x)$$

is valid.

In this way for real characters χ one reaches the conclusion

$$\sum_{n=1}^{\infty} \frac{\mu(n)\chi(n)}{n} = \frac{1}{L(\chi,1)} \ .$$

If one substitutes $m = 1$ (resp. $m = 4$ and for χ the Dirichlet character modulo 4 distinct from the principal character) one deduces the relations

$$\sum_{n=1}^{\infty} \frac{\mu(n)}{n} = 0, \quad \sum_{n=0}^{\infty} \frac{(-1)^n \mu(2n+1)}{2n+1} = \frac{4}{\pi} \ .$$

There is a further application in this direction for the so-called *Ramanujan sums*

$$c_n(k) = \sum_{j(n):\mathrm{g.c.d.}(j,n)=1} e^{2\pi ijk/n} \ ,$$

which agree with the Gauss sums calculated with the principal character modulo n. An application of Vinogradov's lemma gives the representation

$$c_n(k) = \sum_{h|n} \mu(h) \cdot \sum_{j:h|j|n} e^{2\pi ijk/n} = \sum_{h|n} \mu(h) \cdot \sum_{\ell \le n/h} e^{2\pi ih\ell k/n}$$

$$= \sum_{h|n} \mu(h) \cdot \sum_{\ell=1}^{n/h} \left(e^{2\pi ik/(n/h)}\right)^{\ell} \ .$$

The inner geometric sum is only non-zero in the case of divisibility of k by n/h, and here takes the value n/h, so that by putting $r = n/h$

$$c_n(k) = \sum_{r|\mathrm{g.c.d.}(n,k)} r \cdot \mu\left(\frac{n}{r}\right)$$

remains. For $\mathrm{Re}(s) > 1$ this gives the representation linked with a Dirichlet character modulo m, i.e.

$$\sum_{n=1}^{\infty} \frac{c_n(k)\chi(n)}{n^s} = \sum_{n=1}^{\infty} \sum_{r|n,r|k} r\mu\left(\frac{n}{r}\right)\chi(n)n^{-s}$$

$$= \sum_{r|k} r \cdot \sum_{n=1}^{\infty} \mu(n)\chi(nr)(nr)^{-s} = \sum_{r|k} \chi(r)r^{1-s} \cdot \sum_{n=1}^{\infty} \frac{\mu(n)\chi(n)}{n^s}$$

$$= \frac{1}{L(\chi,s)} \cdot \sum_{r|k} \chi(r)r^{1-s} \ .$$

This also holds for $s = 1$, if in the transformation

$$\sum_{n \le x} c_n(k)\chi(n)n^{-1} = \sum_{n \le x} \sum_{r|n,r|k} r\mu\left(\frac{n}{r}\right)\chi(n)n^{-1}$$

$$= \sum_{r|k} \sum_{h \le x/r} r\mu(h)\chi(hr)(hr)^{-1} = \sum_{r|k} \chi(r) \cdot \sum_{h \le x/r} \frac{\mu(h)\chi(h)}{h}$$

one passes to the limit as $x \to \infty$.

If one puts $m = 4$ and takes χ the Dirichlet character distinct from the principal character, one obtains

$$\frac{4}{\pi} \cdot \sum_{\ell|k} \chi(\ell) = \sum_{n=0}^{\infty} \frac{(-1)^n c_{2n+1}(k)}{2n+1} \ .$$

The sum on the left-hand side

$$4 \cdot \sum_{\ell|k} \chi(\ell)$$

here agrees with $r(k)$ the of representations of k as a sum of two squares.

If one puts $m = 1$, then for $s = 2$ one obtains the formula

$$\sum_{n=1}^{\infty} \frac{c_n(k)}{n^2} = \frac{6}{\pi^2} \cdot \sum_{\ell|k} \ell^{-1} = \frac{6}{\pi^2 k} \sum_{\ell|k} \ell$$

and in general for $\mathrm{Re}(s) > 1$

$$\sum_{n=1}^{\infty} \frac{c_n(k)}{n^s} = \frac{1}{\zeta(s)} \cdot \sum_{\ell|k} \ell^{1-s} \ .$$

As has just been shown this relation also holds for $s = 1$, and because of

$$-\sum_{n=1}^{\infty} \frac{c_n(k)\log n}{n^s}$$

$$= -\frac{\zeta'(s)}{\zeta(s)^2} \cdot \sum_{\ell|k} \ell^{1-s} - \frac{1}{\zeta(s)} \cdot \sum_{\ell|k} \ell^{1-s} \log \ell$$

and the limit relations

$$\lim_{s \to 1+0} \frac{1}{\zeta(s)} = 0, \qquad \lim_{s \to 1+0} \frac{\zeta'(s)}{\zeta(s)^2} = -1$$

differentiation leads to

$$\sum_{n=1}^{\infty} \frac{c_n(k)\log n}{n} = -\sum_{\ell|k} 1 \ .$$

If one puts all these calculations together, one recognises that the elementary number theoretic functions $\mu(n)$, $\varphi(n)$, $\tau(n)$, $\sigma(n)$, $r(n)$ can be represented by Ramanujan sums, using the formulae

$$\mu(n) = c_n(1) \ ,$$
$$\varphi(n) = c_n(n) \ ,$$
$$\tau(n) = -\sum_{m=1}^{\infty} \frac{c_m(n)\log m}{m} \ ,$$
$$\sigma(n) = \frac{\pi^2 n}{6} \cdot \sum_{m=1}^{\infty} \frac{c_m(n)}{m^2} \ ,$$
$$r(n) = \pi \cdot \sum_{m=0}^{\infty} \frac{(-1)^m c_{2m+1}(n)}{2m+1} \ .$$

Essentially these relations date back to Ramanujan himself (Trans. Cambridge Phil. Soc. *22* (1918)).

[22])   R. Gupta and M. Ram Murty have shown (Invent. math. *78* (1984)) that there exists some integer serving as a primitive root for infinitely many prime numbers. In this connection Artin conjectured, that *each* integer other than -1 or a perfect square is a primitive root for infinitely many prime numbers. Following a written communication from P. Turan to E. Hlawka, there are infinitely many prime numbers p, such that the smallest primitive root $g(p) \geq 1$ modulo p certainly satisfies $g(p) \geq c\log p$ for some constant c.

Bibliography

Introductory Works on Number Theory (preparation for reading this book)

Hlawka, E., Schoißengeier, J.: Zahlentheorie. Eine Einführung. Manz, Wien 1979
Niven, I., Zuckerman, H.S.: Einführung in die Zahlentheorie. Bibliogr. Inst., Mannheim 1975, 1976

Literature on the History of Number Theory

Dickson, L.E.: History of the Theory of Numbers. Carnegie Inst., Washington 1919, 1920, 1923
Ore, O.: Number Theory and Its History. McGraw-Hill, New York 1949

Classical Books on Geometric and Analytic Number Theory

Dirichlet, P.G. Lejeune: Vorlesungen über Zahlentheorie (R. Dedekind, Hrsg.). Vieweg, Braunschweig 1894
Euler, L.: Introductio in Analysis Infinitorum. Bousquet, Lausanne 1748
Gauß, C.F.: Disquisitiones Arithmeticæ. Fleischer, Leipzig 1801
Legendre, A.M.: Essai sur la Théorie des Nombres. Duprat, Paris 1798
Minkowski, H.: Geometrie der Zahlen. Teubner, Leipzig 1910
Minkowski, H.: Diophantische Approximationen. Teubner, Leipzig 1927

Further Introductory and More Advanced Works

Apostol, T.M.: Introduction to Analytic Number Theory. Springer, New York 1976
Apostol, T.M.: Modular Functions and Dirichlet Series in Number Theory. Springer, New York 1976
Baker, A.: Transcendental Number Theory. Cambridge University Press 1975
Blanchard, A.: Initiation à la théorie analytique des nombres premiers. Dunod, Paris 1969
Cassels, J.W.S.: An Introduction to Diophantine Approximation. Cambridge Tracts in Mathematics, vol. 45, 1957
Cassels, J.W.S.: An Introduction to the Geometry of Numbers. Springer, Berlin 1959
Chandrasekharan, K.: Arithmetical Functions. Springer, New York 1970
Chandrasekharan, K.: Introduction to Analytic Number Theory. Springer, New York 1968
Cohn, H.: A Second Course in Number Theory. Wiley, New York 1962
Davenport, H.: Multiplicative Number Theory. Markham, Chicago 1967
Edwards, H.M.: Riemann's Zeta Function. Academic Press, New York 1974
Ellison, W.: Les Nombres Premiers. Hermann, Paris 1975
Esterman, T.: Introduction to Modern Prime Number Theory. Cambridge Tracts in Mathematics, vol. 41, 1952
Galambos, J.: Representation of Real Numbers by Infinite Series. Springer, New York 1976

Gel'fond, A.O.: Transcendental and Algebraic Numbers. Dover, New York 1960

Hardy, G.H., Riesz, M.: The General Theory of Dirichlet Series. Cambridge Tracts in Mathematics, vol. 18, 1915

Hardy, G.H., Wright, E.M.: An Introduction to the Theory of Numbers. Clarendon, Oxford ^{5}1979

Hasse, H.: Vorlesungen über Zahlentheorie. Springer, Berlin 1964

Hlawka, E.: Theorie der Gleichverteilung. Bibliogr. Inst., Mannheim 1979

Hlawka, E., Firneis, F., Zinterhof, P.: Zahlentheoretische Methoden in der numerischen Mathematik. Oldenbourg, Wien München 1981

Hua, L.-K.: Introduction to Number Theory. Springer, Berlin 1982

Hua, L.-K., Wang, Y.: Applications of Number Theory to Numerical Analysis. Science Press, Oxford Peking 1981

Huxley, M.N.: The Distribution of Prime Numbers. Clarendon, Oxford 1972

Ingham, A.E.: The Distribution of Prime Numbers. Cambridge Tracts in Mathematics, vol. 30, 1932

Khintchin, A.: Continued Fractions. Chicago University Press, 1964

Knopfmacher. J.: Abstract Analytic Number Theory. North-Holland, Amsterdam 1975

Koksma, J.F.: Diophantische Approximationen. Springer, Berlin 1937

Kuipers, L., Niederreiter, H.: Uniform Distribution of Sequences. Wiley, New York 1974

Landau, E.: Handbuch der Lehre von der Verteilung der Primzahlen. Teubner, Leipzig 1909

Lang, S.: Diophantine Geometry. Interscience, New York 1962

Lang, S.: Introduction to Diophantine Approximations. Addison-Wesley, Reading 1966

Lang, S.: Introduction to Transcendental Numbers. Addison-Wesley, Reading 1966

Lekkerkerker, C.G.: Geometry of Numbers, 1st edn. North-Holland, Amsterdam, 1969. Second edition by Lekkerkerker, C.G., Gruber, P.M. (to appear)

LeVeque, W.J.: Reviews in Number Theory. AMS, Providence 1974

LeVeque, W.J.: Topics in Number Theory. Addison-Wesley, Reading 1956

Montgomery, H.L.: Topics in Multiplicative Number Theory. Lecture Notes in Mathematics, vol. 227. Springer, Berlin Heidelberg New York 1971

Narkiewicz, W.: Elementary and Analytic Theory of Algebraic Numbers. 2nd, substantially revised and extended edition. Springer, Berlin Heidelberg New York, 1990, and Panstwoww Wydawnictwo Naukowe, Warzawa

Niven, I.: Irrational Numbers. Carus Math. Monographs, 1956

Perron, O.: Die Lehre von den Kettenbrüchen. Teubner, Leipzig Berlin 1929

Perron, O.: Irrationalzahlen. de Gruyter, Berlin 1910

Prachar, K.: Primzahlverteilung. Springer, Berlin ^{2}1978

Rademacher, H.: Topics in Analytic Number Theory. Springer, New York 1973

Rieger, G.J.: Zahlentheorie. Vandenhoeck & Ruprecht, Göttingen 1976

Rogers, C.A.: Packing and Covering. Cambridge Tracts in Mathematics, vol. 54, 1964

Schmidt, W.M.: Diophantine Approximations. Lecture Notes in Mathematics, vol. 785. Springer, Berlin Heidelberg New York 1980

Schmidt, W.M.: Irregularities of Distribution. Tata Institute, Bombay 1977

Schneider, T.: Einführung in die transzendenten Zahlen. Springer, Berlin 1957

Schwarz, W.: Einführung in die Methoden und Ergebnisse der Primzahltheorie. Bibliogr. Inst., Mannheim 1969

Serre, J.-P.: A Course in Arithmetic. Springer, New York 1973

Siegel, C.L.: Lectures on Advanced Analytic Number Theory. Tata Institute, Bombay 1961

Siegel, C.L.: Transzendente Zahlen. Bibliogr. Inst., Mannheim 1967

Titchmarsh, E.C.: The Theory of the Riemann Zeta Function. Clarendon, Oxford 1951
Vinogradov, I.M.: The Method of Trigonometrical Sums in the Theory of Numbers. Interscience, New York 1954
Walfisz, A.: Weylsche Exponentialsummen in der neueren Zahlentheorie. VEB, Berlin 1963
Zagier, D.B.: Zetafunktionen und quadratische Körper. Springer, Berlin 1981
Zaremba, S.K.: Applications of Number Theory to Numerical Analysis. Academic Press, New York 1972

Solutions for the Exercises

Chapter 1

1. Put $q = 7, p = 22$. Then $|7\pi - 22| = 0.008 \ldots < 0.02$.

2. Put $\alpha = \frac{1}{N}$. From $|\frac{q}{N} - p| < \frac{1}{N}$ it follows that $q = pN \geq N$.

3. (i) Let $\alpha = \frac{a}{b}, b \in \mathbb{N}$. Then for all $n \in \mathbb{N}$ and $\gamma \in \mathbb{R}$ $(nb)^\gamma \langle nb\alpha \rangle = 0$, hence $\lim\limits_{q \to \infty} q^\gamma \langle \alpha q \rangle = 0$.

   (ii) By Lemma 1 for $n \in \mathbb{N}$ there exists $(p_n, q_n) \in \mathbb{Z} \times \mathbb{N}$ with $|\alpha q_n - p_n| < q_n^{-1}$ and $\lim\limits_{n \to \infty} q_n = \infty$. It follows that $\langle \alpha q_n \rangle < q_n^{-1}$. Therefore for each $\varepsilon > 0$ $q_n^{1-\varepsilon} \langle \alpha q_n \rangle < q_n^{-\varepsilon}$, hence $\lim\limits_{n \to \infty} q_n^{1-\varepsilon} \langle \alpha q_n \rangle = 0$. Therefore for each $\varepsilon > 0, \eta(\alpha) \geq 1 - \varepsilon$.

4. For $n \geq 0$ we have $0 \leq G_n \alpha - P_n = G_n \sum\limits_{k=n+1}^{\infty} \frac{z_k}{g_1 \cdots g_k} = \frac{1}{g_{n+1}}\left(z_{n+1} + \sum\limits_{k=n+2}^{\infty} \frac{z_k}{g_{n+2} \cdots g_k}\right)$, so that $0 \leq (G_n \alpha - P_n)g_{n+1} = z_{n+1} + \sum\limits_{k=n+2}^{\infty} \frac{z_k}{g_{n+2} \cdots g_k}$. Further $\sum\limits_{k=n+2}^{\infty} \frac{z_k}{g_{n+2} \cdots g_k} \leq \sum\limits_{k=n+2}^{\infty} \frac{1}{g_{n+2}^{k-n-1}} = \sum\limits_{k=1}^{\infty} g_{n+2}^{-k} = (g_{n+2} - 1)^{-1} < 1$, therefore $z_{n+1} = [G_{n+1}\alpha - P_n g_{n+1}]$.

5. For $n \geq 0$ let $G_n = g_1 \ldots g_n, P_n = G_n \sum\limits_{k=1}^{n} \frac{z_k}{g_1 \cdots g_k}, P_n' = G_n \sum\limits_{k=1}^{n} \frac{w_k}{g_1 \cdots g_k}$. By (4) $z_{n+1} = [G_{n+1}\alpha - P_n g_{n+1}], w_{n+1} = [G_{n+1}\alpha - P_n' g_{n+1}]$. Therefore $z_1 = [G_1 \alpha] = w_1$. If for $1 \leq i \leq n$ we have already proved that $z_i = w_i$, then it follows that $P_n = P_n'$, hence $z_{n+1} = w_{n+1}$.

6. By contradiction. Thus there exists an infinite set $N \subseteq \mathbb{N}$ and some $K > 0$, so that for all $n \in N, q_n \leq K$. Then $|p_n| \leq \varphi(q_n) + |\alpha| q_n \leq 1 + K|\alpha|$, so that $\{(p_n, q_n) : n \in N\}$ would have to be finite.

7. The map $(z_n)_{n \geq 1} \mapsto \sum\limits_{n=1}^{\infty} z_n 2^{-n}$ is a bijection between Z and $(0, 1]$, so that Z has the same cardinality as $\mathbb{R}$. Because $q_2 > 2$, (5) implies that $\psi : Z \to \mathbb{R}$ is injective. By the example in Chapter 1 we have $0 < |\psi((z_n)_{n \geq 1}) - \frac{P_n}{G_n}| \leq \frac{1}{G_N(q_{N+1}-1)} < \frac{\varphi(G_n)}{G_N}$, so that $\psi : Z \to A(\varphi)$. Hence $A(\varphi)$ has the same cardinality as Z and hence also as $\mathbb{R}$.

8. For $q \in \mathbb{N}$ let $\psi(q) = \min_{1 \leq b \leq q} \frac{\varphi(bq)}{b}$. By (7) $A(\psi) \neq \emptyset$. Let $\alpha \in A(\psi)$. We show that $\alpha + \mathbb{Q} \subseteq A(\varphi)$. Let $\frac{a}{b} \in \mathbb{Q}, b \in \mathbb{N}$. By (6) there exist infinitely many $(p_k, q_k) \in \mathbb{Z} \times \mathbb{N}$ with $|\alpha q_k - p_k| < \psi(q_k)$ and $q_k \geq b$. It follows that $|\alpha + \frac{a}{b} - \frac{p_k b + q_k a}{q_k b}| = |\alpha - \frac{p_k}{q_k}| \leq \frac{1}{q_k}\psi(q_k) < \frac{1}{bq_k}\varphi(bq_k)$, so that $\alpha + \mathbb{Q} \subseteq A(\varphi)$.

9. Let $N \in \mathbb{N}, \alpha \in A(\varphi) \cap [-g, g]$. By (6) there exists some $q \geq N$ and $p \in \mathbb{Z}$ with $|\alpha - \frac{p}{q}| < \frac{\varphi(q)}{q}$, i.e. with $\alpha \in (\frac{p}{q} - \frac{\varphi(q)}{q}, \frac{p}{q} + \frac{\varphi(q)}{q})$. We have $|p| < |\alpha| q + \varphi(q) \leq 1 + gq$, from which it follows that $A(\varphi) \cap [-g, g] \subseteq \bigcap_{N=1}^{\infty} \bigcup_{q=N}^{\infty} \bigcup_{|p| \leq gq} (\frac{p}{q} - \frac{\varphi(q)}{q}, \frac{p}{q} + \frac{\varphi(q)}{q})$.

If λ denotes the Lebesgue measure of a set of real numbers, then for each $N \in \mathbb{N}$ it follows that $\lambda(A(\varphi) \cap [-g, g]) \leq \sum_{q=N}^{\infty} \sum_{|p| \leq gq} 2\frac{\varphi(q)}{q} = \sum_{q=N}^{\infty} 2(1 + 2gq)\frac{1}{q}\varphi(q) \leq$

$8g \sum_{q=N}^{\infty} \varphi(q)$. Hence $\lambda(A(\varphi) \cap [-g, g]) = 0$.

10. (i) $\Rightarrow$ (ii)   Let $k \in \mathbb{N}$. For all $t \geq k$ there exists $(p_t, q_t) \in \mathbb{Z} \times \mathbb{N}$ with $|\alpha - \frac{p_t}{q_t}| < q_t^{-t-1} \leq q_t^{-k-1}$ and $q_t > 1$. Hence $|\alpha q_t - p_t| < \varphi_k(q_t)$. Because $q_t > 1$ we have $\lim_{t \to \infty} \frac{p_t}{q_t} = \alpha \notin \mathbb{Q}$, so that $\{(p_t, q_t) : t \geq k\}$ is infinite. Therefore $\alpha \in A(\varphi_k)$.

(ii) $\Rightarrow$ (iii)   Let $k \in \mathbb{N}$. Then by (6) there exists $(p_k, q_k) \in \mathbb{Z} \times \mathbb{N}$ with $q_k > k$ and $|\alpha - \frac{p_k}{q_k}| < q_k^{-k-2}$, so that $\langle \alpha q_k \rangle \leq |\alpha q_k - p_k| \leq q_k^{-k-1}$. It follows that $\lim_{k \to \infty} q_k^k \langle \alpha q_k \rangle = 0$ and hence, for $\gamma > 0$, that $\lim_{k \to \infty} q_k^\gamma \langle \alpha q_k \rangle = 0$. Therefore for each $\gamma \geq 0, \eta(\alpha) \geq \gamma$.

(iii) $\Rightarrow$ (i)   Let $k \in \mathbb{N}$. Because $\varliminf_{q \to \infty} q^{k-1} \langle q\alpha \rangle = 0$ there exists some $q > 1$ with $\langle q\alpha \rangle < q^{1-k}$. There exists some $p \in \mathbb{Z}$ with $\langle \alpha q \rangle = |q\alpha - p|$, from which $\alpha \in \mathcal{L}$ follows.

11. Let $q(X) \in \mathbb{Z}[X], q(\alpha) = 0, q(X) \neq 0$, so that $n = \deg q(X) \geq \deg m(X)$, and it suffices to prove the assertion for $m(X)$. Let $\frac{p}{q} \in \mathbb{Q}$. If $|\alpha - \frac{p}{q}| > 1$, then $|\alpha - \frac{p}{q}| > cq^{-n}$ and we are finished. Otherwise $m(\frac{p}{q}) \neq 0$, because $m(X)$ is irreducible, and $q^n m(\frac{p}{q}) \in \mathbb{Z}$, so that $q^{-n} \leq |m(\frac{p}{q})| = |m(\alpha) - m(\frac{p}{q})| = |\alpha - \frac{p}{q}||m'(x_0)|$ for some x_0 between α and $\frac{p}{q}$. It follows that $|x_0 - \alpha| \leq |\alpha - \frac{p}{q}| \leq 1$, hence $q^{-n} \leq |\alpha - \frac{p}{q}|c_1 \leq |\alpha - \frac{p}{q}|c^{-1}$.

12. (i)   By contradiction. Let $\alpha \in \mathcal{L}$ be algebraic, $f(x) \in \mathbb{Z}[X]$ with $f(\alpha) = 0$. Let $n = \deg f(X)$. For $k \in \mathbb{N}$ there exists $\frac{p_k}{q_k} \in \mathbb{Q}, q_k > 1$, with $|\alpha - \frac{p_k}{q_k}| < q_k^{-k}$. By (11) there exists some $c > 0$, so that $cq_k^{-n} \leq |\alpha - \frac{p_k}{q_k}| \leq q_k^{-k}$, which gives $c \leq q_k^{n-k}$. Allowing $k \to \infty$ we obtain a contradiction.

(ii) and (iii): Let $\varphi : \mathbb{N} \to (0, 1], \varphi(q) = e^{-q}$ and for $k \in \mathbb{N}$ write $\varphi_k(q) = q^{-k}$. We show that $A(\varphi) \subseteq A(\varphi_k)$. Suppose namely that $\alpha \in A(\varphi)$. Then there exist infinitely many $(p_t, q_t) \in \mathbb{Z} \times \mathbb{N}$ with $|\alpha q_t - p_t| < e^{-q_t}$. For sufficiently large t we have $e^{-q_t} < q_t^{-k}$, so that $\alpha \in A(\varphi_k)$. It follows from (10) that $A(\varphi) \subseteq \mathcal{L}$. Because of (7) and (8) $A(\varphi)$ is dense and has the same cardinality as $\mathbb{R}$, hence also $\mathcal{L}$.

(iv) By (10) $\mathcal{L} \subseteq A(\varphi_2)$. Since $\sum_{q=1}^{\infty} \varphi_2(q) < \infty$, by (9) $A(\varphi_2)$ is a set of measure zero, hence also $\mathcal{L}$.

13. Let $\beta = \frac{a}{b} \in \mathbb{Q}, b \in \mathbb{N}$. Let $(p_k, q_k) \in \mathbb{Z} \times \mathbb{N}$ be such that $q_k > 1, |\alpha - \frac{p_k}{q_k}| < q_k^{-k}$, that is, $\alpha \in \mathcal{L}$.

(i) We show that $\alpha\beta \in \mathcal{L}$. Because $a \neq 0$ we have $|a| + b - 2 \leq |a|\,b - 1$, hence $|a|\,b^{k-1} \leq (|a|b)^k \leq (2^{|a|-1}2^{b-1})^k \leq (2^{|a|b-1})^k \leq (q_{k|a|b}^{|a|b-1})^k$. Therefore $|\alpha\beta - \frac{a p_{|a|bk}}{b q_{|a|bk}}| < \frac{|a|}{b} q_{|a|bk}^{-|a|bk} \leq \frac{|a|}{b} q_{|a|bk}^{-k}(|a|b^{k-1})^{-1} = (bq_{|a|bk})^{-k}$, and hence $\alpha\beta \in \mathcal{L}$.

(ii) We have $b \leq 2^{b-1} \leq q_{bk}^{b-1}$, so that $|\alpha + \beta - \frac{bp_{bk}+aq_{bk}}{bq_{bk}}| = |\alpha - \frac{p_{bk}}{q_{bk}}| < q_{bk}^{-bk} = (q_{bk}^{1-b})^k q_{bk}^{-k} \leq (bq_{bk})^{-k}$, so that $\alpha + \beta \in \mathcal{L}$.

14. Let $\alpha \in \mathbb{R}$. We must show that there exist $\beta, \gamma \in \mathcal{L}$ with $\alpha = \beta + \gamma$. We distinguish between 4 cases.

(i) $\alpha \in \mathbb{Q}$. Choose $\beta \in \mathcal{L}$ arbitrarily, then $\gamma = \alpha - \beta \in \mathcal{L}$ (by (13)).

(ii) $\alpha \in \mathcal{L}$. Choose $\beta = \gamma = \frac{\alpha}{2} \in \mathcal{L}$ (by (13)).

(iii) $\alpha \in (0,1), \alpha \notin \mathbb{Q} \cup \mathcal{L}$. We apply the hint: we have $p_k \in \mathbb{Z}$ and
$$|\beta - \tfrac{p_k}{q_k}| = \sum_{j=(2k)!}^{\infty} b_j 2^{-j} = \sum_{j=(2k+1)!}^{\infty} b_j 2^{-j} \leq \sum_{j=(2k+1)!}^{\infty} 2^{-j} = 2^{1-(2k+1)!} =$$
$2 \cdot (2^{-(2k)!})^{2k+1} = 2 \cdot (2q_k)^{-2k-1} < q_k^{-k}$, so that $\beta \in \mathbb{Q} \cup \mathcal{L}$. Analogously $\gamma \in \mathbb{Q} \cup \mathcal{L}$. If β were in $\mathbb{Q}$, then because $\alpha = \beta + \gamma$, α would have to belong to $\mathbb{Q} \cup \mathcal{L}$. Therefore $\beta \in \mathcal{L}$ and analogously $\gamma \in \mathcal{L}$.

(iv) If $\alpha \in \mathbb{R}$, let $g \in \mathbb{Z}$ be such that $\alpha - g \in [0,1)$. Let $\beta, \gamma \in \mathcal{L}$ be such that $\alpha - g = \beta + \gamma$. Then by (13) $\beta + g \in \mathcal{L}$ and $\alpha = (\beta + g) + \gamma$.

15. The first assertion follows from the fact that for $n \geq 1$ $\mathbb{Z} \subseteq \mathcal{F}_n$. Let $\frac{a}{b}, \frac{c}{d}$ be distinct elements of $\mathcal{F}_n$, then it follows that $|\frac{a}{b} - \frac{c}{d}| = \frac{|ad-bc|}{bd} \geq \frac{1}{bd} \geq \frac{1}{n^2}$, implying the second assertion.

16. W.l.o.g. let $\frac{a}{b} < \frac{c}{d}$. Because $cb - ad = 1$ Cramer's rule gives $x = - \begin{vmatrix} e & c \\ f & d \end{vmatrix} = cf - ed, y = - \begin{vmatrix} a & e \\ b & f \end{vmatrix} = eb - af$. From this follows: $\frac{a}{b} < \frac{e}{f} < \frac{c}{d} \Leftrightarrow eb - af > 0$ and $cf - ed > 0 \Leftrightarrow x > 0$ and $y > 0$.

17. (i) By contradiction. Let $\frac{a}{b} < \frac{e}{f} < \frac{c}{d}$ and $\frac{e}{f} \in \mathcal{F}_n$. Because $|cb-ad| = 1$ by (16) there exist $x, y \in \mathbb{N}$ with $ax + cy = e, bx + dy = f$. Hence $f \geq b \cdot 1 + d \cdot 1 > n$, a contradiction.

(ii) With $x = 1 = y$ it follows from (16) that $\frac{a}{b} < \frac{a+c}{b+d} < \frac{c}{d}$, if w.l.o.g. we assume that $\frac{a}{b} < \frac{c}{d}$. From $(a + c)b - a(b + d) = 1$ it follows that g.c.d.$(a + c, b + d) = 1$, so that $\frac{a}{b}$ and $\frac{c}{d}$ are not neighbours in $\mathcal{F}_{d+b} \subseteq \mathcal{F}_n$.

18. (i) $\Rightarrow$ (ii) We have $d < y_{t+1} = y_t + d$, hence $y_t \in \mathbb{N}$, and also $\frac{x_t}{y_t} - \frac{c}{d} = \frac{\operatorname{sgn}(ad-bc)}{dy_t}$, so that $\frac{x_t}{y_t} > \frac{c}{d} \Leftrightarrow ad - bc > 0 \Leftrightarrow \frac{a}{b} > \frac{c}{d}$.
By (17) $\frac{x_t}{y_t}$ and $\frac{c}{d}$ are neighbours in $\mathcal{F}_d$, and indeed $\frac{x_t}{y_t}$ is the upper neighbour of $\frac{c}{d}$, if this is so for $\frac{a}{b}$. Since this is uniquely determined, and $\frac{a}{b}, \frac{c}{d}$ are

neighbours in $\mathcal{F}_d$, it follows that $\frac{x_t}{y_t} = \frac{a}{b}$, hence $x_t = a, y_t = b$. Therefore $ad - bc = \mathrm{sgn}(ad - bc)$.

(ii) $\Rightarrow$ (i) follows from (17).

19. Let $\frac{a}{b} < \frac{a+c}{b+d} < \frac{c}{d}$. W.l.o.g. let $\frac{a}{b} \leq \alpha < \frac{a+c}{b+d} < \frac{c}{d}$. We argue by contradiction: we have $\alpha - \frac{a}{b} \geq \frac{1}{\sqrt{5}b^2}, \frac{a+c}{b+d} - \alpha \geq \frac{1}{\sqrt{5}(b+d)^2}, \frac{c}{d} - \alpha \geq \frac{1}{\sqrt{5}d^2}$. By (18) $cb - ad = 1$, so that by addition we obtain $\frac{1}{b(b+d)} = \frac{a+c}{b+d} - \frac{a}{b} \geq \frac{1}{\sqrt{5}}(\frac{1}{b^2} + \frac{1}{(b+d)^2})$ and $\frac{1}{bd} = \frac{c}{d} - \frac{a}{b} \geq \frac{1}{\sqrt{5}}(\frac{1}{b^2} + \frac{1}{d^2})$. If we put $x = \frac{d}{b}$, then it follows that $\sqrt{5} \geq x + \frac{1}{x}$ and $\sqrt{5} \geq x + 1 + \frac{1}{x+1}$. Therefore $x^2 - \sqrt{5}x + 1 \leq 0$, i.e. $(x - \frac{\sqrt{5}}{2})^2 \leq \frac{1}{4}$, so that $x \in [\frac{\sqrt{5}-1}{2}, \frac{\sqrt{5}+1}{2}]$ and $x + 1 \in [\frac{\sqrt{5}-1}{2}, \frac{\sqrt{5}+1}{2}]$. This gives $x = \frac{\sqrt{5}-1}{2} \notin \mathbb{Q}$, a contradiction.

20. For $n \in \mathbb{N}$ let $\frac{a_n}{b_n}$ and $\frac{c_n}{d_n}$ be neighbours in $\mathcal{F}_n$ with $\frac{a_n}{b_n} \leq \alpha < \frac{c_n}{d_n}$, and let $\frac{p_n}{q_n} \in \left\{ \frac{a_n}{b_n}, \frac{a_n+c_n}{b_n+d_n}, \frac{c_n}{d_n} \right\}$ be such that $|\alpha - \frac{p_n}{q_n}| < \frac{1}{\sqrt{5}q_n^2}$. We need to show that $\varepsilon := \inf_n |\alpha - \frac{p_n}{q_n}| = 0$. Let $\varepsilon > 0$. Since $\bigcup_{k=1}^{\infty} \mathcal{F}_k = \mathbb{Q}$ is dense in $\mathbb{R}$, there exist $k, l \in \mathbb{N}$ with $(\alpha - \varepsilon, \alpha) \cap \mathcal{F}_k \neq \emptyset, (\alpha, \alpha + \varepsilon) \cap \mathcal{F}_l \neq \emptyset$. Let $m = \max(k, l)$. Then $(\alpha - \varepsilon, \alpha) \cap \mathcal{F}_m \neq \emptyset, (\alpha, \alpha + \varepsilon) \cap \mathcal{F}_m \neq \emptyset$, so that $\alpha - \varepsilon < \frac{a_m}{b_m} < \alpha < \frac{c_m}{d_m} < \alpha + \varepsilon$, and therefore $|\alpha - \frac{p_m}{q_m}| < \varepsilon$, contradicting the definition of ε.

Chapter 2

1. Let $\alpha = (\alpha_1, \ldots, \alpha_n) \in \mathbb{R}^n$ be linearly independent over $\mathbb{Z}, \mathfrak{G} = \mathbb{Z}^n$ and $\mathfrak{H} = \mathbb{Z}\alpha$. Let $\beta \in \mathbb{R}^n$ and $\varepsilon > 0$. By Theorem 2 there exists some $\mathfrak{p} \in \mathfrak{G}$ and $q \in \mathbb{Z}$, so that for $1 \leq i \leq n$ $|\alpha_i q - \beta_i - p_i| < \frac{\varepsilon}{\sqrt{n}}$. Therefore $|\alpha q - \mathfrak{p} - \beta| < \varepsilon$. If we put $\mathfrak{x} := \alpha q - \mathfrak{p} \in \mathfrak{G} + \mathfrak{H}$, then $|\mathfrak{x} - \beta| < \varepsilon$. Hence $\mathfrak{G} + \mathfrak{H}$ is dense in $\mathbb{R}^n$.

2. By Theorem 1 there exist infinitely many $(p, q) \in \mathbb{Z} \times \mathbb{N}$ with $|\alpha q - p - \beta - \varepsilon/2| < \varepsilon/2$, i.e. $-\varepsilon/2 < \alpha q - p - \beta - \varepsilon/2 < \varepsilon/2$. Therefore $0 < \alpha q - p - \beta < \varepsilon$.

3. In the following log denotes the logarithm to base 10. Then $\log 2 \notin \mathbb{Q}$, because $2 = 10^{p/q}$, hence $2^q = 10^p$ is not possible. Let $\beta := z_1 + z_2 10^{-1} + \ldots + z_N 10^{-N+1}$. Then $\beta \geq 1$ and $\varepsilon := \log(1 + 10^{-N+1}\beta^{-1}) > 0$. By (2) there exists $(K, n) \in \mathbb{Z} \times \mathbb{N}$ with $n \log 2 - \log \beta - \varepsilon > N$ and $0 \leq n \log 2 - K - \log \beta < \varepsilon$. Because $K > N, 10^K \beta \in \mathbb{Z}$, and it follows that $10^K \beta \leq 2^n < \beta 10^K + 10^{K-N+1}$, i.e. $z_1 10^K + \ldots + z_N 10^{K-N+1} + 0 \cdot 10^{K-N} + \ldots + 0 \cdot 1 \leq 2^n \leq z_1 10^K + \ldots + z_N 10^{K-N+1} + 9 \cdot 10^{K-N} + \ldots + 9 \cdot 1$.

4. Let b' be an integer with $|b' - \beta q| \leq 1/2$. Because g.c.d. $(p, q) = 1$, there exist $x_0, y_0 \in \mathbb{Z}$ with $b' = px_0 - qy_0$. Let t be an integer with $|t + \frac{x_0}{q}| \leq 1/2, x_t = x_0 + qt, y_t = y_0 + pt$. Then $b' = px_t - qy_t, |x_t| \leq q/2$ and $q|\alpha x_t - y_t - \beta| = q|(\alpha - \frac{p}{q})x_t - y_t - \beta + \frac{px_t}{q}| \leq \frac{|x_t|}{\sqrt{5}q} + |px_t - qy_t - \beta q| \leq \frac{1}{2\sqrt{5}} + |b' - \beta q| \leq \frac{1}{2}(1 + \frac{1}{\sqrt{5}})$. Here the equality sign cannot hold, if one distinguishes between the cases $x_t = 0$ and $x_t \neq 0$.

5. Let $(p_n, q_n) \in \mathbb{Z} \times \mathbb{N}$ be so chosen for $n \geq 1$, that $|\alpha q_n - p_n| < \frac{1}{\sqrt{5} q_n}$ and $b_n' \in \mathbb{Z}$ be such that $|b_n' - \beta q_n| \leq 1/2$. Let $(x_n, y_n) \in \mathbb{Z} \times \mathbb{Z}$ be so chosen that $p_n x_n - y_n q_n = b'$, and let $x_n(t) = x_n + q_n t, y_n(t) = y_n + p_n t$. Let $t_n \in \mathbb{Z}$ be so chosen, that $1 \leq x_n(t_n) \leq q_n, u_n = x_n(t_n), v_n = y_n(t_n)$. It follows that $q_n |\alpha u_n - v_n - \beta| = q_n |(\alpha - \frac{p_n}{q_n}) u_n - v_n - \beta + u_n \frac{p_n}{q_n}| < \frac{u_n}{\sqrt{5} q_n} + |b_n' - \beta q_n| \leq \frac{1}{\sqrt{5}} + \frac{1}{2}$, so that $|\alpha u_n - v_n - \beta| < \left(\frac{2\sqrt{5}+5}{10} \right) \frac{1}{q_n} \leq \frac{2\sqrt{5}+5}{10} \cdot \frac{1}{u_n}$. Now let $A = \{(u_n, v_n) : n \geq 1\}$ be finite. Then there exists some n_0, so that for infinitely many n, $(u_n, v_n) = (u_{n_0}, v_{n_0})$. Because $\lim\limits_{n \to \infty} q_n = \infty$ it follows that $\beta = \alpha u_{n_0} - v_{n_0}$. However then we have $|\alpha(q_n + u_{n_0}) - (v_{n_0} + p_n) - \beta| = |\alpha q_n - p_n| < \frac{1}{\sqrt{5} q_n} < \frac{2\sqrt{5}+5}{10}(q_n + u_{n_0})^{-1}$ for sufficiently large n.

6. Let $p_1, \ldots, p_n$ be distinct prime numbers and $a_0, a_1, \ldots, a_n$ integers with $a_0 + a_1 \log p_1 + \ldots + a_n \log p_n = 0$. Then $p_1^{a_1} \ldots p_n^{a_n} = e^{-a_0}$, which is transcendental for $a_0 \neq 0$. Hence $a_0 = 0$ and $p_1^{a_1} \ldots p_n^{a_n} = 1$. From the uniqueness of the prime decomposition it follows that $a_1 = \ldots = a_n = 0$.

7. Let $\{g_1^{1/2}, \ldots, g_k^{1/2}\}$ be linearly dependent over $\mathbb{Z}$ and k chosen as small as possible, $k \geq 2$. Let $K = \mathbb{Q}(\sqrt{g_1}, \ldots, \sqrt{g_{k-1}})$ and choose $a_1, \ldots, a_k \in \mathbb{Q}$ in such a way that $\sqrt{g_k} = \sum\limits_{i=1}^{k-1} a_i \sqrt{g_i} + a_k$. Since, for $1 \leq i < k, \{\sqrt{g_j} : 1 \leq j \leq k, j \neq i\}$ is linearly independent over $\mathbb{Z}$, we must have $a_i \neq 0$ for $1 \leq i < k$. Let $\sigma : K \to \mathbb{C}$ be a field homomorphism. Then, for $1 \leq i \leq k, \sigma(\sqrt{g_i})^2 = \sigma(g_i) = g_i$, so that there exists $\varepsilon_i(\sigma) \in \{1, -1\}$ with $\sigma(\sqrt{g_i}) = \varepsilon_i(\sigma)\sqrt{g_i}$. In particular $\sigma : K \to K$, so that K is normal, and hence Galois over $\mathbb{Q}$. Let G be the Galois group of K over $\mathbb{Q}$, so that $[K : \mathbb{Q}] = |G|$. For $\sigma \in G$ we have $\sum\limits_{i=1}^{k-1} a_i \varepsilon_k(\sigma)\sqrt{g_i} + \varepsilon_k(\sigma)a_k = \varepsilon_k(\sigma)\sqrt{g_k} = \sigma(\sqrt{g_k}) = \sum\limits_{i=1}^{k-1} a_i \sigma(\sqrt{g_i}) + a_k = \sum\limits_{i=1}^{k-1} a_i \varepsilon_i(\sigma)\sqrt{g_i} + a_k$. Since $\{g_1^{1/2}, \ldots, g_{k-1}^{1/2}\}$ is linearly independent over $\mathbb{Z}$, it follows that $a_i \varepsilon_k(\sigma) = a_i \varepsilon_i(\sigma)$ for $1 \leq i \leq k$, so that $\varepsilon_k(\sigma) = \varepsilon_i(\sigma)$ for $1 \leq i \leq k$. For $\sigma, \mu \in G$ $\varepsilon_k(\sigma \cdot \mu)\sqrt{g_k} = \sigma(\mu(\sqrt{g_k})) = \sigma(\varepsilon_k(\mu)\sqrt{g_k}) = \varepsilon_k(\mu)\varepsilon_k(\sigma)\sqrt{g_k}$, i.e. $\varepsilon_k(\sigma \circ \mu) = \varepsilon_k(\sigma) \cdot \varepsilon_k(\mu)$. Therefore $\varepsilon_k : G \to \{1, -1\}$ is a homomorphism. If $\varepsilon_k(\sigma) = 1$, then for $1 \leq i \leq k, \varepsilon_i(\sigma) = 1$, hence $\sigma(\sqrt{g_i}) = \sqrt{g_i}$, and $\sigma = \mathrm{Id}_K$. Therefore ε_k is injective, which implies that $|G| \leq 2$, i.e. $[K : \mathbb{Q}] \leq 2$. However then $k \leq 2$, and this is impossible.

8. Let $0 \leq x < y \leq 1$ and choose $p \in \mathbb{N}$ so that $e^p(e^y - e^x) > 1$. Then there exists some $n \in \mathbb{N}$ with $e^{p+x} < n < e^{p+y}$, i.e. $p + x < \log n < p + y$. From this it follows that $p = [\log n]$, so that $x < \{\log n\} < y$.

9. For $n \geq 1$ put $I_n = (x_n - \varepsilon 2^{-n-1}, x_n + \varepsilon 2^{-n-1})$ and $M = (0, 1) \cap \bigcup\limits_{n=1}^{\infty} I_n$. Then M is open and $\{x_n : n \geq 1\} \subseteq M$. Therefore for all $N \in \mathbb{N}$, $\frac{1}{N} \sum\limits_{n=1}^{N} c_M(x_n) = 1$.

We have $\lambda(M) \leq \sum\limits_{n=1}^{\infty} \lambda(I_n) = \varepsilon \sum\limits_{n=1}^{\infty} 2^{-n} = \varepsilon$. If $(x_n)_{n \geq 1}$ is uniformly distributed, and $\varepsilon < 1$, then c_M is not Riemann integrable, hence M not Jordan measurable.

10. Let $f : \mathbb{R} \to \mathbb{R}$, $f(x) = 0$ for $x \notin \mathbb{Q}$; $f(\frac{p}{q}) = \frac{1}{q}$ for $(p,q) \in \mathbb{Z} \times \mathbb{N}$, g.c.d.$(p,q) = 1$. Then f is Riemann integrable on $[0,1]$, $\int_0^1 f(x)dx = 0$ and f has period 1.

Therefore $\lim\limits_{N\to\infty} \frac{1}{N} \sum\limits_{n=1}^{N} \frac{1}{qn} = \lim\limits_{N\to\infty} \frac{1}{N} \sum\limits_{n=1}^{N} f(\frac{pn}{qn}) = \int_0^1 f(x)dx = 0$.

11. Let $f : \mathbb{R} \to \mathbb{R}$, $f(t) = c_{[0,x)}(\{\sin 2\pi t\})$. Then f has period 1 and is Riemann integrable on $[0,1]$. From this it follows that:

$$\lim\limits_{N\to\infty} \frac{1}{N} \sum\limits_{n=1}^{N} c_{[0,x)}(\{\sin 2\pi n\alpha\}) = \lim\limits_{N\to\infty} \frac{1}{N} \sum\limits_{n=1}^{N} f(n\alpha) = \int_0^1 f(t)dt =$$

$$\int_0^{1/2} c_{[0,x)}(\sin 2\pi t)dt + \int_{1/2}^{1} c_{[-1,x-1)}(\sin 2\pi t)dt =$$

$$\int_0^{1/2} c_{[0,x)}(\sin 2\pi t)dt + \int_{-1/2}^{0} c_{[-1,x-1)}(\sin 2\pi t)dt =$$

$$2\int_0^{1/4} c_{[0,x)}(\sin 2\pi t)dt + 2\int_{-1/4}^{0} c_{[-1,-1+x)}(\sin 2\pi t)dt =$$

$$2\left(\frac{1}{2\pi}\arcsin x + \frac{1}{2\pi}\arcsin(x-1) - \frac{1}{2\pi}\arcsin(-1)\right) =$$

$$\frac{1}{2} + \frac{1}{\pi}\arcsin x - \frac{1}{\pi}\arcsin(1-x).$$

This is not identically equal to x, because then we would have

$$1 = \frac{1}{\pi}\left(\frac{1}{\sqrt{1-x^2}} + \frac{1}{\sqrt{2x-x^2}}\right) \text{ for } x \in (0,1).$$

12. If $\alpha = \frac{p}{q}$ is rational, then $\{\{\sin 2\pi n\alpha\} : n \geq 1\} = \{\{\sin 2\pi \frac{np}{q}\} : 1 \leq n \leq q\}$ is finite, hence not dense in $[0,1]$.

Now let $\alpha \notin \mathbb{Q}$ and $g : [0,1] \to \mathbb{R}$, $g(x) = \frac{1}{2} + \frac{1}{\pi}\arcsin x - \frac{1}{\pi}\arcsin(1-x)$. For $0 < x < 1$, $g'(x) = \frac{1}{\pi}(1-x^2)^{-1/2} + \frac{1}{\pi}(2x-x^2)^{-1/2} > 0$, so that g is strictly increasing. Therefore, if $0 \leq x < y \leq 1$, then $\lim\limits_{N\to\infty} \frac{1}{N} \sum\limits_{n=1}^{N} c_{[x,y)}(\{\sin 2\pi n\alpha\}) = g(y) - g(x) > 0$, so that there must exist some $n \in \mathbb{N}$ with $\{\sin 2\pi n\alpha\} \in [x,y]$.

13. We observe that for $x,y \in \mathbb{R}$ $|e^{ix} - e^{iy}| = \left|\int_x^y e^{it}dt\right| \leq \left|\int_x^y dt\right| = |y-x|$, and that for $x > 0$ $\log(1+x) \leq x$.

Therefore for all $h \in \mathbb{Z}, h \neq 0$, we have:

$$\frac{1}{N}\left|\sum_{n=1}^{N} e^{2\pi ih(x_n + \log n)}\right| \leq \frac{1}{N}\left|\sum_{n=1}^{N-1}(e^{2\pi ih\log n} - e^{2\pi ih\log(n+1)})\sum_{j=1}^{n} e^{2\pi ihx_j}\right|$$

$$+ \frac{1}{N}\left|\sum_{n=1}^{N} e^{2\pi ihx_n}\right| \leq \frac{1}{N}\sum_{n=1}^{N-1} 2\pi|h|\log(1+\frac{1}{n})\left|\sum_{j=1}^{n} e^{2\pi ihx_j}\right| + \frac{1}{N}\left|\sum_{n=1}^{N} e^{2\pi ihx_n}\right|$$

$$\leq \frac{2\pi|h|}{N}\sum_{n=1}^{N-1} \frac{1}{n}\left|\sum_{j=1}^{n} e^{2\pi ihx_j}\right| + \frac{1}{N}\left|\sum_{n=1}^{N} e^{2\pi ihx_n}\right|.$$

Since $\lim\limits_{n\to\infty} \frac{1}{n}\left|\sum\limits_{j=1}^{n} e^{2\pi ihx_j}\right| = 0$, we also obtain $\lim\limits_{N\to\infty} \frac{1}{N}\left|\sum\limits_{n=1}^{N} e^{2\pi ih(x_n + \log n)}\right| = 0$.

14. From the Weyl criterion it follows that if $(x_n)_{n\geq 1}$ is uniformly distributed mod 1, then so ist $(-x_n)_{n\geq 1}$. If therefore $(\log n)_{n\geq 1}$ were uniformly distributed mod 1, then so would be $(-\log n)_{n\geq 1}$ and by (13) $(0)_{n\geq 1}$. This is absurd.

15. For each integer $h \neq 0$ we have $\dfrac{1}{N}\Big| \sum\limits_{n=1}^{N} (e^{2\pi i h(x_n+y_n)} - e^{2\pi i h x_n})\Big|$

$\leq \dfrac{1}{N}\sum\limits_{n=1}^{N} |e^{2\pi i h y_n} - 1| \leq 2\pi|h|\dfrac{1}{N}\sum\limits_{n=1}^{N} |y_n|$. The Weyl criterion gives the assertion.

16. Let $\alpha = \frac{1}{2}(1+\sqrt{5}), \alpha' = \frac{1}{2}(1-\sqrt{5})$. Then $\alpha^2 = \alpha + 1$ and $\alpha'^2 = \alpha' + 1$, so that, for $n \geq 1, \alpha^{n+2} = \alpha^{n+1} + \alpha^n$ and $\alpha'^{n+2} = \alpha'^{n+1} + \alpha'^n$. If one writes $T_n = \frac{1}{\sqrt{5}}(\alpha^{n+1} - \alpha'^{n+1})$, then for $n \geq 1$ it follows that $T_{n+1} = T_n + T_{n-1}$. We have $T_0 = \frac{1}{\sqrt{5}}(\alpha - \alpha') = 1$, and $T_1 = \frac{1}{\sqrt{5}}(\alpha^2 - \alpha'^2) = \frac{1}{\sqrt{5}}(\alpha - \alpha') = 1$, so that for $n \geq 0, F_n = \frac{1}{\sqrt{5}}(\alpha^{n+1} - \alpha'^{n+1})$. From the Weyl criterion it follows that, if $(x_n)_{n\geq 1}$ is uniformly distributed mod 1 and $c \in \mathbb{R}$, then $(x_n + c)_{n\geq 1}$ is also uniformly distributed mod 1. Since e is transcendental, $\log\alpha \notin \mathbb{Q}$, and therefore $(n\log\alpha)_{n\geq 1}$ is uniformly distributed mod 1. Hence the same holds for $((n+1)\log\alpha - \log\sqrt{5})_{n\geq 1}$. From $\lim\limits_{n\to\infty}\log(1 - (\frac{\alpha'}{\alpha})^{n+1}) = 0$, it follows that

$\lim\limits_{N\to\infty}\dfrac{1}{N}\sum\limits_{n=1}^{N}\Big|\log\big(1 - (\tfrac{\alpha'}{\alpha})^{n+1}\big)\Big| = 0$. By (15) we conclude that $(\log F_n)_{n\geq 1} =$

$\Big((n+1)\log\alpha + \log(1 - (\tfrac{\alpha'}{\alpha})^{n+1}) - \log\sqrt{5}\Big)_{n\geq 1}$ is uniformly distributed mod 1.

17. Let $h \in \mathbb{Z}, h \neq 0$ and $\{u,v\} = \{\sin,\cos\}$, where we do not wish to be more specific about which of the functions u, v is sin, resp. cos. From the Euler sum formula and the second mean-value theorem of integral calculus it follows that, for $N \in \mathbb{N}$ there exists some $\xi_N \in [0,N]$ with $\Big| \sum\limits_{n=1}^{N} u(2\pi h\, f(n))\Big|$

$= \Big| \frac{1}{2}(u(2\pi h\, f(1)) + u(2\pi h\, f(N))) + \int\limits_{1}^{N} u(2\pi h\, f(x))dx$

$+ \int\limits_{1}^{N}(\{x\} - \tfrac{1}{2})2\pi h\, f'(x)u'(2\pi h\, f(x))dx \Big|$

$\leq 1 + \frac{1}{2\pi|h|}\Big| \int\limits_{1}^{N}\frac{1}{f'(x)}(2\pi h\, f'(x)u(2\pi h\, f(x))dx \Big| + \int\limits_{1}^{N} 2\pi|h||f'(x)|dx$

$\leq 1 + 2\pi|h|(f(N) - f(1)) + \frac{1}{2\pi|h|}\Big| \int\limits_{1}^{N}\frac{1}{f'(x)}v(2\pi h\, f(x))'dx \Big| = 1 + 2\pi|h|(f(N) - f(1))$

$+ \frac{1}{2\pi|h|}\Big| \frac{1}{f'(1)}\int\limits_{1}^{\xi_N} v(2\pi h\, f(x))'dx + \frac{1}{f'(N)}\int\limits_{\xi_N}^{N} v(2\pi h\, f(x))'dx \Big|$

$\leq 1 + 2\pi|h|(f(N)) - f(1)) + \frac{1}{\pi|h|f'(1)} + \frac{1}{\pi|h|f'(N)}$.

The assertion follows after division by N together with the Weyl criterion.

18. Induction on k. $k = 1$: since f' is decreasing and $\lim\limits_{x\to\infty} f'(x) = 0$, we have $f' \geq 0$. If for some x_0 we had $f'(x_0) = 0$, then for $x \geq x_0, f'(x) = 0$ and $\lim\limits_{x\to\infty} x f'(x) = 0$, a contradiction. Therefore $f' > 0$ and the assertion follows from (17). We argue from $k-1$ to k. Let $h \in \mathbb{N}, g : [1,\infty) \to \mathbb{R}, g(x) = f(x+h) - f(x)$. By the mean-value theorem there exists some $\xi_x \in (x, x+h)$ with

$g^{(k-1)}(x) = hf^{(k)}(\xi_x)$. Then $\lim\limits_{x\to\infty} \xi_x = \infty$, $\lim\limits_{x\to\infty} \frac{x}{\xi_x} = 1$, $g^{(k)}(x) = f^{(k)}(x+h) - f^{(k)}(x) \leq 0$, so that $g^{(k-1)}$ is descreasing, $\lim\limits_{x\to\infty} g^{(k-1)}(x) = \lim\limits_{x\to\infty} h f^{(k)}(\xi_x) = 0$, and $\lim\limits_{x\to\infty} x\, g^{(k-1)}(x) = \lim\limits_{x\to\infty} x f^{(k)}(\xi_x) = \lim\limits_{x\to\infty} \frac{x}{\xi_x} \lim\limits_{x\to\infty} \xi_x f^{(k)}(\xi_x) = \infty$. By the inductive assumption $(g(n))_{n\geq 1}$ is uniformly distributed mod 1. By Theorem 4 $(f(n))_{n\geq 1}$ is uniformly distributed mod 1.

19. Since if $(x_n)_{n\geq 1}$ is uniformly distributed mod 1, so is $(-x_n)_{n\geq 1}$, it suffices to prove the assertion for $\alpha > 0$. Let $k = [\sigma]+1$ and $f : [1,\infty) \to \mathbb{R}$, $f(x) = \alpha x^{\sigma}$. Then $f^{(k)}(x) = \alpha\sigma(\sigma-1)\ldots(\sigma-k+1)x^{\sigma-k} = \alpha\sigma(\sigma-1)\ldots(\sigma-[\sigma])x^{\sigma-[\sigma]-1}$. From this it follows that $f^{(k)}$ is monotone decreasing and that $\lim\limits_{x\to\infty} f^{(k)}(x) = 0$. Because $\sigma \notin \mathbb{N}$, $\lim\limits_{x\to\infty} x f^{(k)}(x) = \alpha\sigma(\sigma-1)\ldots(\sigma-[\sigma]) \lim\limits_{x\to\infty} x^{\sigma-[\sigma]} = \infty$, so that the assertion follows from (18).

20. We observe that for $x,y \in [0,1)$ $\{x-y\} = x - y + c_{[0,y)}(x)$. Let c_1 and c_2 absolute constants. Then for $N \in \mathbb{N}$ we have

$$\left| \sum_{n=1}^{N} c_{[0,x)}(\{\log n\}) - N g_x(\{\log N\}) \right| = \left| \sum_{p \in \mathbb{Z}} \sum_{n=1}^{N} c_{[p,p+x)}(\log n) - N g_x(\{\log N\}) \right|$$

$$= \left| \sum_{0 \leq p \leq \log N} \sum_{n=1}^{N} c_{[e^p, e^{p+x})}(n) - N g_x(\{\log N\}) \right|$$

$$= \left| \sum_{0 \leq p \leq \log N - x} ([e^{p+x}] - [e^p] - c_{\mathbb{Z}}(e^{p+x}) + c_{\mathbb{Z}}(e^p)) \right.$$

$$+ \sum_{\log N - x < p \leq \log N} \sum_{n=1}^{N} c_{[e^p, e^{p+x})}(n) - N g_x(\{\log N\}) \Big|$$

$$\leq \left| \sum_{0 \leq p \leq [\log N - x]} (e^{p+x} - e^p) + c_{[0,x)}(\{\log N\})(N - e^{[\log N]}) - N g_x(\{\log N\}) \right|$$

$$+ c_1 \log N$$

$$= \left| \frac{e^{[\log N - x]+1} - 1}{e - 1}(e^x - 1) + N c_{[0,x)}(\{\log N\})(1 - e^{-\{\log N\}}) - N g_x(\{\log N\}) \right|$$

$$+ c_1 \log N$$

$$\leq N \left| e^{1-x} \cdot \frac{e^x - 1}{e - 1} \cdot e^{-\{\log N - x\}} + c_{[0,x)}(\{\log N\})(1 - e^{-\{\log N\}}) - g_x(\{\log N\}) \right|$$

$$+ c_2 \log N$$

$$= N \Big| e^{1-x} \cdot \frac{e^x - 1}{e - 1} \cdot c_{[0,x)}(\{\log N\}) e^{-\{\log N\}+x-1}$$

$$+ e^{1-x} \cdot \frac{e^x - 1}{e - 1} \cdot e^{-\{\log N\}+x} c_{[x,1]}(\{\log N\})$$

$$+ c_{[0,x)}(\{\log N\})(1 - e^{-\{\log N\}}) - g_x(\{\log N\}) \Big| + c_2 \log N$$

$$= N \Big| c_{[0,x)}(\{\log N\})(1 - e^{-\{\log N\}}(1 - \frac{e^x - 1}{e - 1}))$$

$$+ c_{[x,1]}(\{\log N\}) e^{-\{\log N\}+1} \cdot \frac{e^x - 1}{e - 1} - g_x(\{\log N\}) \Big| + c_2 \log N = c_2 \log N.$$

The function g_x is increasing in $[0,x)$ and decreasing in $[x,1]$, therefore $\max\limits_{t \in [0,1]} g_x(t) = \frac{e}{e-1}(1 - e^{-x})$. We have $\lim\limits_{t \to x-0} g_x(t) = 1 - \frac{e^{1-x}-1}{e-1} = g_x(x)$, so that g_x is continuous. Furthermore $g_x(0) = \frac{e^x - 1}{e - 1} = g_x(1)$, hence $\min\limits_{t \in [0,1]} g_x(t) = \frac{e^x - 1}{e - 1}$. By (8) $(\{\log N\})_{N\geq 1}$ is dense in $[0,1]$, hence $(g_x(\{\log N\}))_{N\geq 1}$ dense in $\left[\frac{e^x - 1}{e - 1}, \frac{e}{e-1}(1 - e^{-x}) \right]$. This gives the assertion, since the sequences $\left(\frac{1}{N} \sum\limits_{n=1}^{N} c_{[0,x)}(\{\log n\}) \right)_{N\geq 1}$ and $(g_x(\{\log N\}))_{N\geq 1}$ have the same accumulation points.

Chapter 3

1. Because $\mathfrak{v}_k \in \mathfrak{S}, \mathfrak{S} \neq \{0\}$. Since $\mathfrak{S}$ is discrete, $\mathfrak{S}$ is finite and $\alpha_k > 0$. Because $\mathfrak{v}_k \in \mathfrak{S}, \alpha_k \leq 1$. Now let $\mathfrak{x} \in \mathfrak{G}$. Since $\mathfrak{G}$ has rank k, and $\{\mathfrak{a}_1, \ldots, \mathfrak{a}_{k-1}, \mathfrak{v}_k\} \subseteq \mathfrak{G}$ is linearly independent, for each $1 \leq i \leq k$ there exists $\beta_i \in \mathbb{R}$, with $\mathfrak{x} = \sum_{i=1}^{k-1} \beta_i \mathfrak{a}_i + \beta_k \mathfrak{v}_k$. Let $\gamma_i = \beta_i - \alpha_i \left[\frac{\beta_k}{\alpha_k}\right]$ for $1 \leq i \leq k$, and $\mathfrak{w} = \sum_{i=1}^{k-1} [\gamma_i]\mathfrak{a}_i \in \mathfrak{G}$.

   Because $\mathfrak{a}_k \in \mathfrak{S} \subseteq \mathfrak{G}, \mathfrak{x} - \mathfrak{w} - [\frac{\beta_k}{\alpha_k}]\mathfrak{a}_k = \sum_{i=1}^{k-1} \left(\beta_i - [\gamma_i] - \alpha_i[\frac{\beta_k}{\alpha_k}]\right)\mathfrak{a}_i + (\beta_k - \alpha_k\left[\frac{\beta_k}{\alpha_k}\right])\mathfrak{v}_k = \sum_{i=1}^{k-1} \{\gamma_i\}\mathfrak{a}_i + \alpha_k\{\frac{\beta_k}{\alpha_k}\}\mathfrak{v}_k \in ([0,1)\mathfrak{a}_1 + \ldots + [0,1)\mathfrak{a}_{k-1} + [0,1]\mathfrak{v}_k) \cap \mathfrak{G} = \mathfrak{S}$. Since $\alpha_k\{\frac{\beta_k}{\alpha_k}\} < \alpha_k$, the definition of α_k implies that $\{\frac{\beta_k}{\alpha_k}\} = 0$, i.e. $\frac{\beta_k}{\alpha_k} \in \mathbb{Z}$.

   Therefore $\sum_{i=1}^{k-1} \gamma_i \mathfrak{a}_i = \sum_{i=1}^{k-1} (\{\gamma_i\} + [\gamma_i])\mathfrak{a}_i = \mathfrak{x} - \frac{\beta_k}{\alpha_k}\mathfrak{a}_k \in \mathfrak{G} \cap (\mathbb{R}\mathfrak{a}_1 + \ldots + \mathbb{R}\mathfrak{a}_{k-1}) = \mathbb{Z}\mathfrak{a}_1 + \ldots + \mathbb{Z}\mathfrak{a}_{k-1}$. Since $\{\mathfrak{a}_1, \ldots, \mathfrak{a}_{k-1}\}$ is linearly independent, it follows that $\gamma_i \in \mathbb{Z}$ for $1 \leq i \leq k$. Therefore $\mathfrak{x} = \mathfrak{w} + \frac{\beta_k}{\alpha_k}\mathfrak{a}_k \in \mathbb{Z}\mathfrak{a}_1 + \ldots + \mathbb{Z}\mathfrak{a}_{k-1} + \mathbb{Z}\mathfrak{a}_k$, i.e. $\mathfrak{G} \subseteq \mathbb{Z}\mathfrak{a}_1 + \ldots + \mathbb{Z}\mathfrak{a}_k$. The converse is trivial.

2. Induction on k. The case $k = 0$ is trivial. We argue from $k - 1$ to k. Let $\{\mathfrak{v}_1, \ldots, \mathfrak{v}_k\} \subseteq \mathfrak{G}$ be linearly independent, $\mathfrak{G}_0 = \mathfrak{G} \cap (\mathbb{R}\mathfrak{v}_1 + \ldots + \mathbb{R}\mathfrak{v}_{k-1})$. Then $\mathfrak{G}_0 \subseteq \mathfrak{G}$ is a discrete subgroup of $\mathbb{R}^n$ of rank $k - 1$, so that, by the inductive assumption there exist linearly independent vectors $\mathfrak{a}_1, \ldots, \mathfrak{a}_{k-1}$ with $\mathfrak{G} \cap (\mathbb{R}\mathfrak{v}_1 + \ldots + \mathbb{R}\mathfrak{v}_{k-1}) = \mathbb{Z}\mathfrak{a}_1 + \ldots + \mathbb{Z}\mathfrak{a}_{k-1}$. Because $\mathfrak{a}_i \in \mathbb{R}\mathfrak{v}_1 + \ldots + \mathbb{R}\mathfrak{v}_{k-1}$, for $1 \leq i < k, \mathbb{R}\mathfrak{a}_1 + \ldots + \mathbb{R}\mathfrak{a}_{k-1} = \mathbb{R}\mathfrak{v}_1 + \ldots + \mathbb{R}\mathfrak{v}_{k-1}$, so that $\mathfrak{G} \cap (\mathbb{R}\mathfrak{a}_1 + \ldots + \mathbb{R}\mathfrak{a}_{k-1}) = \mathbb{Z}\mathfrak{a}_1 + \ldots + \mathbb{Z}\mathfrak{a}_{k-1}$. By (1) there exists some $\mathfrak{a}_k \in \mathbb{R}^n$ with $\mathfrak{G} = \mathbb{Z}\mathfrak{a}_1 + \ldots + \mathbb{Z}\mathfrak{a}_k$. Since $\mathfrak{G}$ has rank $k, \mathfrak{a}_1, \ldots, \mathfrak{a}_k$ are linearly independent.

3. We show first that such a sequence $(\mathfrak{x}_p)_{p \geq 1}$ exists. Suppose namely that $\mathfrak{x}_0$ is an accumulation point of $\mathfrak{G}$. Then $\mathfrak{x}_0 \in \mathfrak{G}$ and there exists a sequence $(\mathfrak{y}_p)_{p \geq 1}$ in $\mathfrak{G}$ with $\mathfrak{y}_p \neq \mathfrak{x}_0$ (for $p \geq 1$), $|\mathfrak{y}_p - \mathfrak{x}_0| \leq 1$ and $\lim_{p \to \infty} \mathfrak{y}_p = \mathfrak{x}_p$. For $p \geq 1$ write $\mathfrak{x}_p = \mathfrak{y}_p - \mathfrak{x}_0$. Then $(k_p\mathfrak{x}_p)_{p \geq 1}$ is a sequence in the unit ball, which must have an accumulation point $\mathfrak{a}$. Let $\mathfrak{a} = \lim_{i \to \infty} k_{p_i}\mathfrak{x}_{p_i}$. Then $|(k_{p_i} + 1)\mathfrak{x}_{p_i}| > 1$ and $\mathfrak{a} = \lim_{i \to \infty} (k_{p_i} + 1)\mathfrak{x}_{p_i}$, so that $|\mathfrak{a}| \geq 1$. Therefore $\mathfrak{a} \neq 0$. Finally suppose that $t \in \mathbb{R}$. Because $|\{tk_{p_i}\}\mathfrak{x}_{p_i}| \leq |\mathfrak{x}_{p_i}|, t\mathfrak{a} = \lim_{i \to \infty} tk_{p_i}\mathfrak{x}_{p_i} = \lim_{i \to \infty} (\{tk_{p_i}\} + [tk_{p_i}])\mathfrak{x}_{p_i} = \lim_{i \to \infty} [tk_{p_i}]\mathfrak{x}_{p_i} \in \mathfrak{G}$, since $\mathfrak{G}$ is closed, and $[tk_{p_i}]\mathfrak{x}_{p_i} \in \mathfrak{G}$. Therefore $\mathbb{R}\mathfrak{a} \subseteq \mathfrak{G}$.

4. Let $\mathfrak{V} = \bigcup_{\mathbb{R}\mathfrak{a} \subseteq \mathfrak{G}} \mathbb{R}\mathfrak{a}$. Clearly for $\lambda \in \mathbb{R}, \lambda\mathfrak{V} \subseteq \mathfrak{V}$. Let $\mathfrak{x}, \mathfrak{y} \in \mathfrak{V}$. We show $\mathfrak{x} + \mathfrak{y} \in \mathfrak{V}$. There exist $\mathfrak{a}, \mathfrak{b} \in \mathbb{R}^n$ with $\mathbb{R}\mathfrak{a} \subseteq \mathfrak{G}, \mathbb{R}\mathfrak{b} \subseteq \mathfrak{G}, \mathfrak{x} \in \mathbb{R}\mathfrak{a}, \mathfrak{y} \in \mathbb{R}\mathfrak{b}$. Then $\mathbb{R}(\mathfrak{x} + \mathfrak{y}) \subseteq \mathbb{R}\mathfrak{x} + \mathbb{R}\mathfrak{y} \subseteq \mathbb{R}\mathfrak{a} + \mathbb{R}\mathfrak{b} \subseteq \mathfrak{G}$, hence $\mathfrak{x} + \mathfrak{y} \in \mathfrak{V}$. Therefore $\mathfrak{V}$ is a vector space, and clearly the largest with $\mathfrak{V} \subseteq \mathfrak{G}$. From $\mathbb{R}^n = \mathfrak{V} + \mathfrak{W}$ it follows that $\mathfrak{G} = (\mathfrak{V} \cap \mathfrak{G}) + (\mathfrak{W} \cap \mathfrak{G}) = \mathfrak{V} + (\mathfrak{W} \cap \mathfrak{G})$. $\mathfrak{W} \cap \mathfrak{G}$ is closed. If $\mathfrak{W} \cap \mathfrak{G}$ were not discrete, then by (3) there would exist some $\mathfrak{a} \in \mathfrak{W} \cap \mathfrak{G}, \mathfrak{a} \neq 0$, with $\mathbb{R}\mathfrak{a} \subseteq \mathfrak{W} \cap \mathfrak{G}$. It follows that $\mathfrak{a} \in \mathfrak{V} \cap \mathfrak{W}$, a contradiction.

5. Let $\mathfrak{V}$ be the largest subspace in $\mathbb{R}^n$ with $\mathfrak{V} \subseteq \mathfrak{G}$ and $p = \dim_\mathbb{R} \mathfrak{V}$. Let $\{\mathfrak{a}_1, \ldots, \mathfrak{a}_p\}$ be a basis of $\mathfrak{V}$, $\mathfrak{W}$ a subspace of $\mathbb{R}^n$ with $\mathfrak{V} + \mathfrak{W} = \mathbb{R}^n, \mathfrak{V} \cap \mathfrak{W} = \{0\}$. By (4) $\mathfrak{W} \cap \mathfrak{G}$ is discrete, so that by (2) there exist linearly independent

vectors $a_{p+1}, \ldots, a_{p+q} \in \mathbb{R}^n$ with $\mathfrak{W} \cap \mathfrak{G} = \mathbb{Z}a_{p+1} + \ldots + \mathbb{Z}a_{p+q}$. Because $\mathfrak{V} \cap \mathfrak{W} = \{0\}, \{a_1, \ldots, a_{p+q}\}$ is linearly independent and there exist vectors $a_{p+q+1}, \ldots, a_n$, so that $\{a_1, \ldots, a_n\}$ is a basis of $\mathbb{R}^n$. By (4) $\mathfrak{G} = \mathfrak{V} + (\mathfrak{W} \cap \mathfrak{G}) = \mathbb{R}a_1 + \ldots + \mathbb{R}a_p + \mathbb{Z}a_{p+1} + \ldots + \mathbb{Z}a_{p+q}$. Therefore $p + q$ is the rank of $\mathfrak{G}$.

6. For $\mathfrak{x} \in \mathbb{R}^n$, let $f_{\mathfrak{x}} : \mathbb{R}^n \to \mathbb{R}$ be given by $f_{\mathfrak{x}}(u) = \langle u, \mathfrak{x} \rangle$. Then $f_{\mathfrak{x}}$ is a continuous group homomorphism, and $\mathfrak{G}^* = \bigcap_{\mathfrak{x} \in \mathfrak{G}} f_{\mathfrak{x}}^{-1}(\mathbb{Z})$. Since $\mathbb{Z}$ is a closed subgroup, the same holds for $f_{\mathfrak{x}}^{-1}(\mathbb{Z})$, and therefore for $\mathfrak{G}^*$. Clearly $(\overline{\mathfrak{G}})^* \subseteq \mathfrak{G}^*$. Suppose conversely that $u \in \mathfrak{G}^*$ and $\mathfrak{x} \in \overline{\mathfrak{G}}$. Let $(\mathfrak{x}_n)_{n \geq 1}$ be a sequence in $\mathfrak{G}$ with $\mathfrak{x} = \lim_{n \to \infty} \mathfrak{x}_n$. Then for $n \in \mathbb{N}$ $\langle u, \mathfrak{x}_n \rangle \in \mathbb{Z}$, hence also $\langle u, \mathfrak{x} \rangle = \lim_{n \to \infty} \langle u, \mathfrak{x}_n \rangle \in \mathbb{Z}$, and $u \in (\overline{\mathfrak{G}})^*$.

7. We show first that, for $p < i \leq p+q, a_i^* \in \mathfrak{G}^*$ and that for $p+q < i \leq n$ $\mathbb{R}a_i^* \subseteq \mathfrak{G}^*$. Let $\mathfrak{x} = \sum_{k=1}^{p} t_k a_k + \sum_{k=p+1}^{p+1} n_k a_k \in \mathfrak{G}$. If $p < i \leq p+q, \langle \mathfrak{x}, a_i^* \rangle = n_i \in \mathbb{Z}$. If $p+q < i \leq n$, and $t \in \mathbb{R}$, then $\langle \mathfrak{x}, ta_i^* \rangle = t\langle \mathfrak{x}, a_i^* \rangle = 0 \in \mathbb{Z}$, so that $ta_i^* \in \mathfrak{G}^*$. Suppose conversely that $u = \sum_{i=1}^{n} \alpha_i a_i^* \in \mathfrak{G}^*$. For $1 \leq i \leq p$, it then follows that $t\alpha_i = \langle u, ta_i \rangle \in \mathbb{Z}$ for all $t \in \mathbb{R}$, which is only possible for $\alpha_i = 0$. If $p < i \leq p+q$ then $\alpha_i = \langle u, a_i \rangle \in \mathbb{Z}$. Therefore $\mathfrak{G}^* \subseteq \mathbb{Z}a_{p+1}^* + \ldots + \mathbb{Z}a_{p+q}^* + \mathbb{R}a_{p+q+1}^* + \ldots + \mathbb{R}a_n^*$.

8. Suppose first that $\mathfrak{G}$ is closed. Since $\{a_1, \ldots, a_n\}$ is the dual basis to $\{a_1^*, \ldots, a_n^*\}$, the assertion follows from (5) and (7). Since along with $\mathfrak{G}$ $\overline{\mathfrak{G}}$ is also a subgroup of $\mathbb{R}^n$, by (6) $\overline{\mathfrak{G}} = (\overline{\mathfrak{G}})^{**} = \mathfrak{G}^{**}$.

9. Let $\mathfrak{G} = L(\mathbb{Z}^n) + \mathbb{Z}^m$. Then $u \in \mathfrak{G}^* \Leftrightarrow$ for $\mathfrak{p} \in \mathbb{Z}^m$ $\langle u, \mathfrak{p} \rangle \in \mathbb{Z}$ and for $\mathfrak{q} \in \mathbb{Z}^n$ $\langle u, L(\mathfrak{q}) \rangle \in \mathbb{Z} \Leftrightarrow u \in \mathbb{Z}^m$ and for $1 \leq i \leq n$ $\langle u, L(e_i) \rangle \in \mathbb{Z}$. By (8) (i) holds $\Leftrightarrow$ $b \in \overline{\mathfrak{G}} = \mathfrak{G}^{**}$, therefore (i) holds $\Leftrightarrow$ for all $u \in \mathfrak{G}^*$ $\langle u, b \rangle \in \mathbb{Z}$.

10. With $\mathfrak{G} = L(\mathbb{Z}^n) + \mathbb{Z}^m$ we have (i) $\Leftrightarrow \overline{\mathfrak{G}} = \mathbb{R}^m = (\mathfrak{G}^*)^*$. Because $\mathbb{R}^{m*} = \{0\}, \mathfrak{G}^* = \{0\} \Leftrightarrow$ (i).

11. We have $\sum_{\mathfrak{p} \in \mathfrak{P} \cap \mathfrak{X}} \nu_{\mathfrak{p}}(\mathfrak{S}) = \sum_{\mathfrak{g} \in \mathfrak{G}} \sum_{\mathfrak{p} \in \mathfrak{P} \cap \mathfrak{X}} c_{\mathfrak{S}}(\mathfrak{p} + \mathfrak{g}) = \sum_{\mathfrak{g} \in \mathfrak{G}} \sum_{\mathfrak{p} \in \mathfrak{P} \cap \mathfrak{X}} c_{\mathfrak{S}-\mathfrak{g}}(\mathfrak{p}) =$
$\sum_{\mathfrak{g} \in \mathfrak{G}} |(\mathfrak{S}-\mathfrak{g}) \cap \mathfrak{P} \cap \mathfrak{X}| = \sum_{\mathfrak{g} \in \mathfrak{G}} |\mathfrak{S} \cap (\mathfrak{P}+\mathfrak{g}) \cap (\mathfrak{X}+\mathfrak{g})| = \sum_{\mathfrak{g} \in \mathfrak{G}} |\mathfrak{S} \cap \mathfrak{X} \cap (\mathfrak{P}+\mathfrak{g})| = |\mathfrak{S} \cap \mathfrak{X}|$,
because $\{\mathfrak{P} + \mathfrak{g} : \mathfrak{g} \in \mathfrak{G}\}$ is a partition of $\mathbb{R}^n$. Furthermore $\int_{\mathfrak{P}} \nu_{\mathfrak{p}}(\mathfrak{M} + \mathfrak{x}) d\mathfrak{x} =$
$\sum_{\mathfrak{g} \in \mathfrak{G}} \int_{\mathfrak{P}} c_{\mathfrak{M}}(\mathfrak{p} + \mathfrak{g} - \mathfrak{x}) d\mathfrak{x} = \sum_{\mathfrak{g} \in \mathfrak{G}} \int_{\mathfrak{P}-\mathfrak{g}} c_{\mathfrak{M}}(\mathfrak{p} - \mathfrak{x}) d\mathfrak{x} = \int_{\mathbb{R}^n} c_{\mathfrak{M}}(\mathfrak{p} - \mathfrak{x}) d\mathfrak{x} = \int_{\mathbb{R}^m} c_{\mathfrak{M}}(\mathfrak{x}) d\mathfrak{x} =$
$\lambda(\mathfrak{M})$, and therefore $\int_{\mathfrak{P}} |(\mathfrak{M} + \mathfrak{x}) \cap \mathfrak{X}| d\mathfrak{x} = |\mathfrak{P} \cap \mathfrak{X}| \lambda(\mathfrak{M}) < \infty$. If for all $\mathfrak{x}$ we had $|(\mathfrak{M} + \mathfrak{x}) \cap \mathfrak{X}| < |\mathfrak{P} \cap \mathfrak{X}| \lambda(\mathfrak{M})/\lambda(\mathfrak{P})$, then integration would lead to a contradiction.

12. If $\nu := \lambda(\mathfrak{M}) |\mathfrak{X} \cap \mathfrak{P}|/d(\mathfrak{G}) \notin \mathbb{Z}$, there is nothing to prove. Let $\lambda_k = 1 + \frac{1}{k}$ for $k \geq 1$. By (11) there exists some $\mathfrak{x}_k \in \mathfrak{P}$ with $|(\lambda_k \mathfrak{M} + \mathfrak{x}_k) \cap \mathfrak{X}| \geq \lambda_k^n \lambda(\mathfrak{M}) |\mathfrak{X} \cap \mathfrak{P}|/d(\mathfrak{G}) > \nu$. Let $(\mathfrak{x}_{k_j})_{j \geq 1}$ be a convergent subsequence, $\mathfrak{x}_0 = \lim_{k \to \infty} \mathfrak{x}_{k_j}$, and $\mathfrak{A}_j \subseteq (\lambda_{k_j} \mathfrak{M} + \mathfrak{x}_{k_j}) \cap \mathfrak{X}$ be so chosen, that $|\mathfrak{A}_j| = \nu + 1$. Then

$\bigcup\limits_{j=1}^{\infty} \mathfrak{A}_j \subseteq \mathfrak{X}$ is bounded, because $\lambda_{k_j} \leq 2$ and $(\mathfrak{x}_{k_j})_{j\geq 1}$ is bounded. Hence $\bigcup\limits_{j=1}^{\infty} \mathfrak{A}_j$ is finite. Therefore there exists a sequence $(j_l)_{l\geq 1}$ of natural numbers with $\mathfrak{A}_{j_l} = \mathfrak{A}_{j_1}$ and $\lim\limits_{l\to\infty} j_l = \infty$. We have $\mathfrak{A}_{j_1} \subseteq (\lambda_{k_{j_l}} \mathfrak{M} + x_{k_{j_l}}) \cap \mathfrak{X}$ for all $l \geq 1$. Now $\mathfrak{A}_{j_1} \subseteq (\mathfrak{M} + \mathfrak{x}_0) \cap \mathfrak{X}$, for if $\mathfrak{x} \in \mathfrak{A}_{j_1}$, for $l \geq 1$ there exists some $\mathfrak{m}_l \in \mathfrak{M}$ with $\mathfrak{x} = \lambda_{k_{j_l}} \mathfrak{m}_l + \mathfrak{x}_{k_{j_l}}$. Since $\mathfrak{M}$ is compact, $(\mathfrak{m}_l)_{l\geq 1}$ has a convergent subsequence $(\mathfrak{m}_{l_t})_{t\geq 1}$ and $\mathfrak{m}_0 = \lim\limits_{t\to\infty} \mathfrak{m}_{l_t} \in \mathfrak{M}$. It follows that $\mathfrak{x} = \mathfrak{m}_0 + \mathfrak{x}_0 \in \mathfrak{M} + \mathfrak{x}_0$. Therefore $|(\mathfrak{M} + \mathfrak{x}_0) \cap \mathfrak{X}| \geq \nu + 1$. If $\mathfrak{x}_0 \notin \mathfrak{P}$, choose $\mathfrak{x} \in \mathfrak{P}$ in such a way that $\mathfrak{x} - \mathfrak{x}_0 \in \mathfrak{G}$. It then follows that $\nu + 1 \leq |(\mathfrak{M} + \mathfrak{x}) \cap (\mathfrak{X} + (\mathfrak{x} - \mathfrak{x}_0))| = |(\mathfrak{M} + \mathfrak{x}) \cap \mathfrak{X}|$.

13. Let $\mathfrak{A} = \{\mathfrak{x} \in \mathbb{Z}^n : 0 \leq x_j < N^{m/n} \text{ for } 1 \leq j \leq n\}$. If $N^{m/n} \in \mathbb{Z}$, then $|\mathfrak{A}| = N^m$. If $N^{m/n} \notin \mathbb{Z}$, then $|\mathfrak{A}| = (1 + [N^{m/n}])^n > N^m$. For $\mathfrak{y} \in \mathbb{R}^m let \{\mathfrak{y}\} := (\{\mathfrak{y}_1\}, \ldots, \{\mathfrak{y}_m\})$. If there exists $\mathfrak{x}_1 \neq \mathfrak{x}_2 \in \mathfrak{A}$ with $\{L(\mathfrak{x}_1)\} = \{L(\mathfrak{x}_2)\}$, let $\mathfrak{x} = \mathfrak{x}_1 - \mathfrak{x}_2, \mathfrak{y} = L(\mathfrak{x}_1 - \mathfrak{x}_2)$. Then $0 < \|\mathfrak{x}\| < N^{m/n}$ and $L(\mathfrak{x}) = \mathfrak{y} \in \mathbb{Z}^m$. Otherwise we put $\mathfrak{X} = \mathbb{Z}^m + L(\mathfrak{A}) = \bigcup\limits_{\mathfrak{a}\in\mathfrak{A}} (\mathbb{Z}^m + L(\mathfrak{a}))$. Since $\mathfrak{A}$ is finite, $\mathfrak{X}$ is discrete. Let $\mathfrak{G} = \mathbb{Z}^m$, then $\mathfrak{X} + \mathfrak{g} = \mathfrak{X}$ for $\mathfrak{g} \in \mathfrak{G}$, and $d(\mathfrak{G}) = 1$. Let $\mathfrak{M} = [0, \frac{1}{N}]^m$. $\mathfrak{M}$ is compact and for $\mathfrak{a} \in \mathfrak{A}, \{L(\mathfrak{a})\} \in \mathfrak{P} \cap \mathfrak{X}$, so that $|\mathfrak{P} \cap \mathfrak{X}| \geq |\mathfrak{A}| \geq N^m$. By (12) there exists some $\mathfrak{x} \in \mathfrak{P}$ with $|(\mathfrak{M} + \mathfrak{x}) \cap \mathfrak{X}| > N^{-m} |\mathfrak{P} \cap \mathfrak{X}| \geq 1$. Let $\mathfrak{x}_1 \neq \mathfrak{x}_2 \in (\mathfrak{M} + \mathfrak{x}) \cap \mathfrak{X}$. Let $\mathfrak{a}_1, \mathfrak{a}_2 \in \mathfrak{A}$ be such that, for $i = 1, 2, \mathfrak{x}_i - L(\mathfrak{a}_i) \in \mathbb{Z}^m$, and $\mathfrak{m}_1, \mathfrak{m}_2 \in \mathfrak{M}$ be such that $\mathfrak{x}_i = \mathfrak{m}_i + \mathfrak{x}$. Then $\|\mathfrak{x}_1 - \mathfrak{x}_2\| = \|\mathfrak{m}_1 - \mathfrak{m}_2\| \leq \frac{1}{N} < 1$, from which $\mathfrak{a}_1 \neq \mathfrak{a}_2$ follows. If we put $\mathfrak{q} = \mathfrak{a}_1 - \mathfrak{a}_2$, then $0 < \|\mathfrak{q}\| < N^{m/n}$. And if $\mathfrak{p} = \mathfrak{x}_1 - L(\mathfrak{a}_1) - (\mathfrak{x}_2 - L(\mathfrak{a}_2)) \in \mathbb{Z}^m$, then $\|L(\mathfrak{q}) - \mathfrak{p}\| \leq \frac{1}{N}$.

14. First suppose that $\Delta(\mathfrak{K}) = \infty$. If $\mathfrak{G}$ is a lattice then (for $t > 0$) so is $\frac{1}{t}\mathfrak{G}$. If for each lattice $\mathfrak{G}, \mathfrak{G} \cap \mathfrak{K} \neq \{0\}$, then also $(\frac{1}{t}\mathfrak{G}) \cap \mathfrak{K} \neq \{0\}$, and $\mathfrak{G} \cap t\mathfrak{K} = t(\frac{1}{t}\mathfrak{G} \cap \mathfrak{K}) \neq \{0\}$. If $\Delta(\mathfrak{K}) < \infty$, then $\Delta(t\mathfrak{K}) = \inf\{d(\mathfrak{G}) : \mathfrak{G} \cap (t\mathfrak{K}) = \{0\}\} = \inf\{d(t\mathfrak{G}) : t\mathfrak{G} \cap t\mathfrak{K} = \{0\}\} = \inf\{t^n d(\mathfrak{G}) : \mathfrak{G} \cap \mathfrak{K} = \{0\}\} = t^n \Delta(\mathfrak{K})$.

15. Let $\varepsilon > 0$ and $\alpha = \varepsilon + \lambda(\mathfrak{K})$. By Lemma 3 there exists a lattice $\mathfrak{G}_0$ with $\alpha(\mathfrak{G}_0) = 1$, so that $\sum\limits_{\mathfrak{g}\in\mathfrak{G}_0, \mathfrak{g}\neq o} c_{\mathfrak{K}}(\alpha^{1/n}\mathfrak{g}) < \frac{\varepsilon}{\alpha} + \int\limits_{\mathbb{R}^n} c_{\mathfrak{K}}(\alpha^{1/n}\mathfrak{x})d\mathfrak{x} = \frac{\varepsilon}{\alpha} + \frac{\lambda(\mathfrak{K})}{\alpha} = 1$. Let $\mathfrak{G} = \alpha^{1/n}\mathfrak{G}_0$. Then $d(\mathfrak{G}) = \alpha$ and $\sum\limits_{\mathfrak{g}\in\mathfrak{G}, \mathfrak{g}\neq o} c_{\mathfrak{K}}(\mathfrak{g}) < 1$, i.e. $\sum\limits_{\mathfrak{g}\in\mathfrak{G}, \mathfrak{g}\neq o} c_{\mathfrak{K}}(\mathfrak{g}) = 0$. This implies $\mathfrak{K} \cap \mathfrak{G} = \{0\}$. It follows that, for each $\varepsilon > 0$, $\Delta(\mathfrak{K}) \leq \alpha = \varepsilon + \lambda(\mathfrak{K})$.

16. Clearly $I_n \subseteq I_{n-1} \subseteq \ldots \subseteq I_1$. We show first that $I_n \neq \emptyset$ (and hence $I_j \neq \emptyset$). Let $\{\mathfrak{e}_1, \ldots, \mathfrak{e}_n\}$ be the standard basis of $\mathbb{R}^n$. Since $o \in \mathfrak{K}^\circ$, there exists some $\lambda > 0$ with $\lambda\mathfrak{e}_i \in \mathfrak{K}^\circ$ for $1 \leq i \leq n$. It follows that, for $1 \leq i \leq n, \mathfrak{e}_i \in \frac{1}{\lambda}\mathfrak{K}^\circ \cap \mathbb{Z}^n \subseteq \frac{1}{\lambda}\mathfrak{K} \cap \mathbb{Z}^n$, so that $\frac{1}{\lambda} \in I_n$. If $\mu \geq \lambda$ and $\lambda \in I_j$, then $\mu\mathfrak{K} \supseteq \lambda\mathfrak{K}$, hence $\mu \in I_j$. Therefore I_j is an interval which is unbounded from above. Now let $(\mu_t)_{t\geq 1}$ be a decreasing sequence with $\mu_t > \lambda_j(\mathfrak{K})$ and $\lim\limits_{t\to\infty} \mu_t = \lambda_j(\mathfrak{K})$. There exists a linearly independent set $\mathfrak{B}_t \subseteq \mu_t\mathfrak{K} \cap \mathbb{Z}^n$, so that $|\mathfrak{B}_t| = j$. We have $\bigcup\limits_{t=1}^{\infty} \mathfrak{B}_t \subseteq \mu_1\mathfrak{K} \cap \mathbb{Z}^n$ as a bounded and discrete, hence finite subset. Therefore there exists a sequence $(t_k)_{k\geq 1}$ of natural numbers with $\lim\limits_{k\to\infty} t_k = \infty$ and $\mathfrak{B}_{t_k} = \mathfrak{B}_{t_1}$. It follows that $\mathfrak{B}_{t_1} \subseteq \mu_{t_k}\mathfrak{K} \cap \mathbb{Z}^n$. Since $\mathfrak{K}$ is compact, it follows that $\mathfrak{B}_{t_1} \subseteq \lambda_j(\mathfrak{K})\mathfrak{K} \cap \mathbb{Z}^n$, so that $\lambda_j(\mathfrak{K})\mathfrak{K} \cap \mathbb{Z}^n$ has rank $\geq j$. Therefore $\lambda_j(\mathfrak{K}) \in I_j$. The last assertion is trivial.

17. Let $\mu \geq a_{\sigma(j)}^{-1}$. Then $\mu\mathfrak{K} = \prod_{i=1}^{n}[-\mu a_i, \mu a_i]$. For $1 \leq i \leq j$, $\mu a_{\sigma(i)} \geq 1$, so that $\{e_{\sigma(1)}, \ldots, e_{\sigma(j)}\} \subseteq \mu\mathfrak{K} \cap \mathbb{Z}^n$. Therefore $\mu\mathfrak{K} \cap \mathbb{Z}^n$ has rank $\geq j$, so that $\lambda_j(\mathfrak{K}) \leq a_{\sigma(j)}^{-1}$. Let $\mu < a_{\sigma(j)}^{-1}$ and $\mathfrak{B} \subseteq \mu\mathfrak{K} \cap \mathbb{Z}^n$ be linearly independent. If $\mathfrak{x} \in \mathfrak{B}$, then for $i \geq j$ $|x_{\sigma(i)}| \leq \mu a_{\sigma(i)} \leq \mu a_{\sigma(j)} < 1$, so that $x_{\sigma(i)} = 0$. Therefore $\pi : \mathbb{R}^n \to \mathbb{R}^{j-1}$, $\pi(\mathfrak{x}) = (x_{\sigma(i)})_{1 \leq i < j}$ is linear and $\pi|\mathfrak{B}$ is injective, implying that $\mathfrak{B}$ and $\pi(\mathfrak{B}) \subseteq \mathbb{R}^{j-1}$ have the same rank. Therefore $\mu\mathfrak{K} + \mathbb{Z}^n$ has rank $< j$.

18. (i)   For $1 \leq j \leq n$ $\lambda_j(L(\mathfrak{K})) = \inf\{\lambda > 0 : \dim_{\mathbb{R}}\langle \lambda L(\mathfrak{K}) \cap \mathbb{Z}^n\rangle \geq j\}$
    $= \inf\{\lambda > 0 : \dim_{\mathbb{R}}\langle L(\lambda\mathfrak{K}) \cap \mathbb{Z}^n\rangle \geq j\} = \inf\{\lambda > 0 : \dim_{\mathbb{R}}\langle L(\lambda\mathfrak{K} \cap \mathbb{Z}^n)\rangle \geq j\}$
    $= \inf\{\lambda > 0 : \dim_{\mathbb{R}} L\langle \lambda\mathfrak{K} \cap \mathbb{Z}^n\rangle \geq j\} = \lambda_j(\mathfrak{K})$, since $\det L \neq 0$.

    (ii)   For $1 \leq j \leq n$ $\lambda_j(t\mathfrak{K}) = \inf\{\lambda > 0 : \dim_{\mathbb{R}}\langle \lambda t\mathfrak{K}, \mathbb{Z}^n\rangle \geq j\}$
    $= \inf\{\frac{\lambda}{t} : \lambda > 0, \dim_{\mathbb{R}}\langle \lambda\mathfrak{K} \cap \mathbb{Z}^n\rangle \geq j\} = \frac{1}{t}\lambda_j(\mathfrak{K})$.

19. By (16) $\lambda_j(\mathfrak{K}) \in I_j$. Let $\mathfrak{g}_1 \in \lambda_1(\mathfrak{K})\mathfrak{K} \cap \mathbb{Z}^n$, $\mathfrak{g}_1 \neq 0$, and $\mathfrak{g}_1, \ldots, \mathfrak{g}_{j-1}$ be already defined as a linearly independent set, so that, for $1 \leq i < j$, $\mathfrak{g}_i \in \lambda_i(\mathfrak{K})\mathfrak{K} \cap \mathbb{Z}^n$. Then $\{\mathfrak{g}_1, \ldots, \mathfrak{g}_{j-1}\} \subseteq \langle \lambda_j(\mathfrak{K})\mathfrak{K} \cap \mathbb{Z}^n\rangle$. Since this space has dimension $\geq j$, there exists some $\mathfrak{g}_j \in \lambda_j(\mathfrak{K})\mathfrak{K} \cap \mathbb{Z}^n$, so that $\{\mathfrak{g}_1, \ldots, \mathfrak{g}_j\}$ is linearly independent. Setting $j = n$ we obtain the assertion.

20. (i) $\Rightarrow$ (ii)   We have $\mathcal{O}_d = \mathbb{Z}(\sqrt{d})$. Let $d < -2$. We show that 2 is irreducible. Let $2 = \alpha\beta$, then $N(\alpha)N(\beta) = 4$. Let $\alpha = x + y\sqrt{d}$, $x, y \in \mathbb{Z}$. Since $x^2 - dy^2 = 2$ is not solvable, $N(\alpha) = 1$ or $N(\beta) = 1$. Therefore 2 is irreducible and hence prime. However $2|d(1 - d) = \sqrt{d^2}(1 - \sqrt{d})(1 + \sqrt{d})$, so that $2|\sqrt{d}$ or $2|1 \pm \sqrt{d}$. Let $\alpha \in \mathcal{O}_d$, $\alpha = x + y\sqrt{d}$, be such that $2\alpha = \sqrt{d}$ (resp. $2\alpha = 1 \pm \sqrt{d}$). Then $2y = \pm 1$, a contradiction.

    (ii) $\Rightarrow$ (i)   Let $\alpha, \beta \in \mathcal{O}_d$, $\beta \neq 0$. Since $\frac{\alpha}{\beta} \in \mathbb{Q}(\sqrt{d})$, there exist $r_1, r_2 \in \mathbb{Q}$ with $\frac{\alpha}{\beta} = r_1 + r_2\sqrt{d}$. Let $g_1, g_2 \in \mathbb{Z}$ be such that $|r_i - g_i| \leq 1/2$ for $i = 1, 2$, $q = g_1 + g_2\sqrt{d} \in \mathcal{O}_d$. Then $N(\alpha - \beta q) = N(\beta)|\frac{\alpha}{\beta} - q|^2 = N(\beta)((r_1 - g_1)^2 + |d|(r_2 - g_2)^2) \leq N(\beta)(\frac{1}{4} + \frac{|d|}{4}) \leq \frac{3}{4}N(\beta) < N(\beta)$, so that $\mathcal{O}_d$ is even Euclidean.

Chapter 4

1. It is clear that $(Z, +)$ is an abelian group. Let $f, g, h \in Z$. Then for $n \in \mathbb{N}$,
$$(f * g) * h(n) = \sum_{d|n}(f * g)(d)h(\tfrac{n}{d}) = \sum_{d|n}\sum_{t|d} f(t)g(\tfrac{d}{t})h(\tfrac{n}{d}) = \sum_{t|n}\sum_{u|n/t} f(t)g(u)h(\tfrac{n}{ut})$$
$$= \sum_{t|n} f(t)g * h(\tfrac{n}{t}) = f * (g * h)(n).$$

Clearly $f * (g + h) = f * g + f * h$ and $f * g = g * f$. Let $\underline{1}(n) = \begin{cases} 1 & n = 1 \\ 0 & n > 1 \end{cases}$.

Then $\underline{1} * f(n) = \sum_{d|n}\underline{1}(d)f(\tfrac{n}{d}) = f(n)$, so that $\underline{1} * f = f$. Finally suppose $f \neq 0$, $g \neq 0$. Let $n_0, m_0 \in \mathbb{N}$ be chosen minimal, so that $f(m_0)g(n_0) \neq 0$. Then

$$f * g(m_0 n_0) = \sum_{d|m_0 n_0} f(d)\left(\tfrac{m_0 n_0}{d}\right) = \sum_{\substack{d|m_0 n_0, d \geq m_0 \\ m_0 n_0/d \geq n_0}} f(d)g\left(\tfrac{m_0 n_0}{d}\right) = f(m_0)g(n_0) \neq 0.$$

Therefore $f * g \neq 0$.

2. (i) Let $f \in Z^*$. Then there exists some $g \in Z$ with $f * g = \underline{1}$. It follows that $1 = f(1)g(1)$, so that $f(1) \neq 0$. Suppose conversely that $f(1) \neq 0$. Define $g(n)$ by induction on n. Let $g(1) = \frac{1}{f(1)}$ and suppose that for $m < n$ $g(m)$ is already defined. Then write $g(n) = -\frac{1}{f(1)} \sum_{d|n, d<n} f(\tfrac{n}{d})g(d)$, so that

$$f * g(n) = \sum_{d|n} f(\tfrac{n}{d})g(d) = 0 \text{ for } n > 1, \ f * g(1) = f(1)g(1) = 1.$$

(ii) We show first that for $u \in Z$, $g_p|u \Leftrightarrow u(n) = 0$ for $p \nmid n$. Let $g_p|u$ and $f \in Z$ be such that $g_p * f = u$. Then for $p \nmid n$, $u(n) = \sum_{d|n} g_p(d)f(\tfrac{n}{d}) = 0$.

Suppose conversely that $u(n) = 0$ for $p \nmid n$. Then for $t \geq 1$ let $f(t) = u(pt)$. For $p|n$ we have $(g_p * f)(n) = \sum_{d|n} g_p(d)f(\tfrac{n}{d}) = f(\tfrac{n}{p}) = u(n)$ and for $p \nmid n$

$$(g_p * f)(n) = 0.$$

Now let $g_p|f * g$ and $g_p \nmid g$. Let $m_0 \in \mathbb{N}$ be chosen minimal, so that $p \nmid m_0, g(m_0) \neq 0$. If $g_p \nmid f$, then there would exist some minimal n_0 with $p \nmid n_0, f(n_0) \neq 0$. We have $p \nmid m_0 n_0$ and therefore $0 = (f * g)(m_0 n_0) = \sum_{\substack{d|n_0 m_0, d \geq n_0 \\ m_0 n_0/d \geq m_0}} f(d)g\left(\tfrac{m_0 n_0}{d}\right) = f(n_0)g(m_0) \neq 0$, a contradiction.

3. Let $\mathfrak{A} = \{\mathfrak{g} \in \mathbb{Z}^n : \Omega_\mathfrak{g} \subseteq \mathfrak{K}\}$ and $\mathfrak{A}' = \{\mathfrak{g} \in \mathbb{Z}^n : \Omega_\mathfrak{g} \cap \partial\mathfrak{K} \neq \emptyset\}$. Then $|\mathfrak{A}| = \sum_{\Omega_\mathfrak{g} \subseteq \mathfrak{K}} 1 = \sum_{\Omega_\mathfrak{g} \subseteq \mathfrak{K}} \lambda(\Omega_\mathfrak{g}) = \lambda(\bigcup_{\Omega_\mathfrak{g} \subseteq \mathfrak{K}} \Omega_\mathfrak{g}) \leq \lambda(\mathfrak{K})$. Since for $\mathfrak{g} \in \mathfrak{A}, \mathfrak{g} \in \Omega_\mathfrak{g} \subseteq \mathfrak{K}$, $\mathfrak{A} \subseteq \mathfrak{K} \cap \mathbb{Z}^n$, hence $|\mathfrak{A}| \leq |\mathfrak{K} \cap \mathbb{Z}^n|$. Now let $\mathfrak{g} \in \mathbb{Z}^n$, $\mathfrak{K} \cap \Omega_\mathfrak{g} \neq \emptyset$ and $\mathfrak{g} \notin \mathfrak{A}$. Then also $\Omega_\mathfrak{g} \setminus \mathfrak{K} \neq \emptyset$. Since $\Omega_\mathfrak{g}$ is connected, $\Omega_\mathfrak{g} \cap \partial\mathfrak{K} \neq \emptyset$, hence $\mathfrak{g} \in \mathfrak{A}'$. From this is follows on the one hand that $\mathfrak{K} \cap \mathbb{Z}^n \subseteq \mathfrak{A} \cup \mathfrak{A}'$, for with $\mathfrak{g} \in \mathfrak{K} \cap \mathbb{Z}^n$ $\mathfrak{K} \cap \Omega_\mathfrak{g} \neq \emptyset$. On the other hand $\mathfrak{K} = \bigcup_{\mathfrak{g} \in \mathbb{Z}^n} \mathfrak{K} \cap \Omega_\mathfrak{g} = \bigcup_{\mathfrak{g} \in \mathfrak{A} \cup \mathfrak{A}'} \mathfrak{K} \cap \Omega_\mathfrak{g}$, so that $\lambda(\mathfrak{K}) \leq |\mathfrak{A} \cup \mathfrak{A}'| \leq |\mathfrak{A}| + |\mathfrak{A}'|$. Therefore $|\mathfrak{A}| \leq \lambda(\mathfrak{K}) \leq |\mathfrak{A}| + |\mathfrak{A}'|$ and $|\mathfrak{A}| \leq |\mathfrak{K} \cap \mathbb{Z}^n| \leq |\mathfrak{A}| + |\mathfrak{A}'|$ from which the assertion follows.

4. Let $\mathfrak{K}_R = \{\mathfrak{x} \in \mathbb{R}^n : |\mathfrak{x}| \leq R\}$ and $\Omega_\mathfrak{g} \cap \partial\mathfrak{K}_R \neq \emptyset$. Let $\mathfrak{y} \in \Omega_\mathfrak{g} \cap \partial\mathfrak{K}_R$. We show that $\bigcup_{\Omega_\mathfrak{g} \cap \partial\mathfrak{K}_R \neq \emptyset} \Omega_\mathfrak{g} \subseteq \mathfrak{K}_{R+\sqrt{n}} \setminus \mathfrak{K}^\circ_{R-\sqrt{n}}$. Let $\mathfrak{x} \in \Omega_\mathfrak{g}$. Since $\Omega_\mathfrak{g}$ has diameter $\sqrt{n}$, $|\mathfrak{x} - \mathfrak{y}| \leq \sqrt{n}$. Because $|\mathfrak{y}| = R$, $R - \sqrt{n} \leq |\mathfrak{x}| \leq R + \sqrt{n}$. Therefore $\sum_{\Omega_\mathfrak{g} \cap \partial\mathfrak{K}_R \neq \emptyset} 1 \leq \lambda(\mathfrak{K}_{R+\sqrt{n}}) - \lambda(\mathfrak{K}_{R-\sqrt{n}}) = V_n((R+\sqrt{n})^n - (R-\sqrt{n})^n) = O(R^{n-1})$.

The assertion now follows from (3).

5. For $m \geq 1$ we have $\prod_{k=0}^{m-1}(X - e^{2\pi i k/m}) = X^m - 1$. Let $G = \mathbb{C}(X)^*$ be the multiplicative group of rational functions $\neq 0$ with complex coefficients. It is abelian. By Vinogradov's lemma, $\varphi_n(X) = \prod_{d|n} \prod_{\substack{d|k \\ 0 \leq k < n}} (X - e^{2\pi i k/n})^{\mu(d)} =$

$$\prod_{d|n} \prod_{k=0}^{n/d-1} (X - e^{2\pi i k d/n})^{\mu(d)} = \prod_{d|n} (X^{n/d} - 1)^{\mu(d)}. \text{ Now let } p(X) = \prod_{d|n,\mu(d)=1} (X^{n/d} - 1),$$

$$q(X) = \prod_{d|n,\mu(d)=-1} (X^{n/d} - 1). \text{ Then } \varphi_n(X) = \frac{p(X)}{q(X)}. \ q(X) \text{ is monic. Since the}$$

division algorithm holds in $\mathbb{Z}[X]$ for monic polynomials, $\varphi_n(X) \in \mathbb{Z}[X]$.

6. We have $\displaystyle\sum_{k=n}^{\infty} \frac{\varphi(k)}{k^3} = \sum_{k=n}^{\infty} \frac{1}{k^2} \sum_{d|k} \frac{\mu(d)}{d} = \sum_{d=1}^{\infty} \frac{\mu(d)}{d} \sum_{d|k\geq n} \frac{1}{k^2} = \sum_{d=1}^{\infty} \frac{\mu(d)}{d^3} \sum_{m\geq n/d} \frac{1}{m^2}$

$\displaystyle = \sum_{d=1}^{n} \frac{\mu(d)}{d^3} \sum_{m\geq n/d} \frac{1}{m^2} + \sum_{d=n+1}^{\infty} \frac{\mu(d)}{d^3} \frac{\pi^2}{6} = \sum_{d=1}^{n} \frac{\mu(d)}{d^3} \left(\frac{d}{n} + O(\frac{d^2}{n^2}) \right) + O\left(\sum_{d=n+1}^{\infty} d^{-3} \right)$

$\displaystyle = \frac{1}{n} \sum_{d=1}^{n} \mu(d) d^{-2} + O\left(\frac{1}{n^2} \sum_{d=1}^{n} d^{-1} \right) + O(n^{-2}) = \frac{1}{n} \left(\frac{6}{\pi^2} - \sum_{d=n+1}^{\infty} \mu(d) d^{-2} \right)$

$\displaystyle + O(n^{-2} \log n) = \frac{6}{\pi^2 n} + O(n^{-2}) + O(n^{-2} \log n) \text{ as } n \to \infty.$

7. By Vinogradov's Lemma $\displaystyle A_n = \sum_{d|n} \mu(d) \sum_{\substack{k=1 \\ d|k}}^{n} \frac{1}{k+n} = \sum_{d|n} \mu(d) \sum_{k\leq n/d} \frac{1}{kd+n} =$

$\displaystyle \sum_{d|n} \mu(\tfrac{n}{d}) \sum_{k=1}^{d} \frac{1}{k\frac{n}{d}+n} = \frac{1}{n} \sum_{d|n} \mu(\tfrac{n}{d}) \sum_{k=1}^{d} \frac{d}{k+d}.$

8. By (7) $\displaystyle A_n = \frac{1}{n} \sum_{d|n} d\mu(\tfrac{n}{d}) \sum_{k=d+1}^{2d} k^{-1} = \frac{1}{n} \sum_{d|n} d\mu(\tfrac{n}{d})(\log 2d - \log d + O(d^{-1})) =$

$\displaystyle \frac{1}{n} \sum_{d|n} d\mu(\tfrac{n}{d}) \log 2 + O(\tfrac{\tau(n)}{n}) = \sum_{d|n} d^{-1}\mu(d) \log 2 + O(n^{-1}\tau(n)) = n^{-1}\varphi(n) \log 2 +$

$O(n^{-1}\tau(n)).$

9. We have $\displaystyle S_n \leq n^{-1} \sum_{k=1}^{n} \sum_{l=1}^{n} k^{-1} l^{-1} = O(n^{-1} \log^2 n)$, so that $\displaystyle \lim_{n\to\infty} S_n = 0$. Fur-

thermore for $n > 1$, $\displaystyle S_n - S_{n-1} = 2 \sum_{\substack{k=1 \\ \text{g.c.d.}(k,n)=1}}^{n} (kn(k+n))^{-1}$

$\displaystyle + \sum_{\substack{k=1 \\ n<k+l, \text{g.c.d.}(k,l)=1}}^{n-1} \sum_{l=1}^{n-1} (kl(k+l))^{-1} - S_{n-1} = 2n^{-1} \sum_{\substack{k=1 \\ \text{g.c.d.}(k,n)=1}}^{n} (k(k+n))^{-1}$

$\displaystyle - \sum_{\substack{k=1 \\ k+l=n, \text{g.c.d.}(k,l)=1}}^{n-1} \sum_{l=1}^{n-1} (kln)^{-1} = 2n^{-1} \sum_{\substack{k=1 \\ \text{g.c.d.}(k,n)=1}}^{n} (k(k+n))^{-1}$

$\displaystyle - n^{-1} \sum_{\substack{k=1 \\ \text{g.c.d.}(k,n)=1}}^{n-1} (k(n-k))^{-1} = 2n^{-2} \sum_{\substack{k=1 \\ \text{g.c.d.}(k,n)=1}}^{n} k^{-1} - 2n^{-2} \sum_{\substack{k=1 \\ \text{g.c.d.}(k,n)=1}}^{n} (k+n)^{-1}$

$\displaystyle - n^{-2} \sum_{\substack{k=1 \\ \text{g.c.d.}(k,n)=1}}^{n-1} (n-k)^{-1} - n^{-2} \sum_{\substack{k=1 \\ \text{g.c.d.}(k,n)=1}}^{n-1} k^{-1} = -2n^{-2} A_n. \text{ From this follows}$

$\displaystyle S_N - S_M = \sum_{n=M+1}^{N} (S_n - S_{n-1}) = -2 \sum_{n=M+1}^{N} n^{-2} A_n. \text{ Allowing } N \to \infty \text{ we}$

obtain the assertion.

10. We have
$$\sum_{n>N} n^{-3}\tau(n) = \sum_{n>N} n^{-3}\sum_{d|n} 1 = \sum_{d=1}^{\infty}\sum_{d|n>N} n^{-3} = \sum_{d=1}^{\infty}\sum_{k>N/d}(kd)^{-3}$$
$$= \sum_{d=1}^{N} d^{-3}\sum_{k>N/d} k^{-3} + \sum_{d>N} d^{-3}\zeta(3) = O\Big(\sum_{d=1}^{N}(dN^2)^{-1}\Big) + O(N^{-2})$$
$$= O(N^{-2}\log N). \text{ Using (8), (9) und (6) it follows that } S_n = 2\sum_{k>n}(k^{-3}\varphi(k)\log 2$$
$$+ O(k^{-3}\tau(k))) = \pi^{-2}n^{-1}12\log 2 + O(n^{-2}\log n) \text{ as } n \to \infty.$$

11. We have $\sum_{n=1}^{N} n^{-s} = \frac{1}{2}(1 + N^{-s}) + \int_1^N x^{-s}dx + \int_1^N(\{x\} - \frac{1}{2})(-s)x^{-s-1}dx$. Because

$|(\{x\} - \frac{1}{2})x^{-s-1}| \le x^{-2}$ the integral $\int_1^{\infty}(\{x\} - \frac{1}{2})x^{-s-1}dx$ exists. We argue by

contradiction. We have $\lim_{N\to\infty}\int_1^N x^{-s}dx = \lim_{N\to\infty}(N^{-s+1} - 1)(1 - s)^{-1}$ (the case

$s = 1$ is trivial). Now let $t \in \mathbb{R}$, $t \ne 0$, be such that $s = 1 + it$. Therefore $\lim_{N\to\infty} N^{-it}$ exists and equals a. Let $q \in \mathbb{N}$. Then with $N = nq$ it follows that

$a = \lim_{n\to\infty}(nq)^{-it} = aq^{-it}$, so that $e^{-it\log q} = 1$. Therefore $\frac{1}{2\pi}t\log q \in \mathbb{Z}$ for all

$q \in \mathbb{N}$. This is not possible, since $\lim_{q\to\infty}\frac{t}{2\pi}(\log(q + 1) - \log q) = 0$.

12. For $\text{Re}(s) > 0$ $\sum_{n=1}^{\infty}(-1)^n n^{-s} = (2^{1-s} - 1)\zeta(s)$. If for $n \in \mathbb{Z}$, $n \ne 0$, one puts $s_n = 1 + \frac{2\pi in}{\log 2}$, then $2^{1-s_n} - 1 = 0$, from which the assertion follows.

13. (i) Let $\delta > 0$, $M < N$. Then for $|z| \le 1$, $|z - 1| \ge \delta$, $\Big|\sum_{n=M}^{N} n^{-1}z^n\Big|$

$$= \Big|\sum_{n=M}^{N-1}(\tfrac{1}{n} - \tfrac{1}{n+1})\sum_{j=M}^{n} z^j + \tfrac{1}{N}\sum_{j=M}^{N} z^j\Big| \le \sum_{n=M}^{N-1}(\tfrac{1}{n} - \tfrac{1}{n+1})\frac{|z^{n-M+1}-1|}{|z-1|} +$$

$\frac{|z^{N-M+1}-1|}{N|z-1|} \le \frac{2}{\delta}\Big(\sum_{n=M}^{N-1}(\tfrac{1}{n} - \tfrac{1}{n+1}) + \tfrac{1}{N}\Big) = \frac{2}{\delta M}$. This gives (i).

(ii) It suffices to prove the assertion for $0 < x < 1$. By (i) $-\log(1 - e^{2\pi ix}) = \sum_{n=1}^{\infty} n^{-1}e^{2\pi inx}$, and therefore $-\arg(1 - e^{2\pi ix}) = \sum_{n=1}^{\infty}\frac{\sin 2\pi nx}{n}$. We have $e^{i\arg(1-e^{2\pi ix})} = \frac{1-e^{2\pi ix}}{|1-e^{2\pi ix}|} = -ie^{i\pi x}\frac{\sin \pi x}{|\sin \pi x|} = e^{i\pi(x-1/2)}$, so that from $\text{Re}(1 - e^{2\pi ix}) \ge 0$ it follows that $\arg(1 - e^{2\pi ix}) = \pi(x - 1/2)$.

(iii) The inequality in (i) gives $\Big|\sum_{n=M}^{\infty}\frac{\sin 2\pi nx}{n}\Big| \le \frac{2}{M|1-e^{2\pi ix}|} = \frac{1}{M\sin \pi x}$ for

$0 < x \le 1/2$. Together with (ii) this implies that for $\frac{1}{M} \le x \le \frac{1}{2}$,
$$\Big|\sum_{n=1}^{M-1}\frac{\sin 2\pi nx}{n}\Big| \le \frac{\pi}{2} + \frac{1}{M\sin \pi x} \le \frac{\pi}{2} + \frac{1}{2Mx} \le \frac{\pi+1}{2}. \text{ If } 0 \le x < \frac{1}{M},$$
$$\Big|\sum_{n=1}^{M-1}\frac{\sin 2\pi nx}{n}\Big| \le \sum_{n=1}^{M-1}\frac{1}{n}|\sin 2\pi nx| \le 2\pi x(M - 1) < 2\pi.$$

14. (i) The point 0 is a simple zero of $z \mapsto e^z - 1$, so that $z \mapsto \frac{ze^{xz}}{e^z-1}$ is analytic in $|z| < 2\pi$.

(ii) We have $\sum_{n=0}^{\infty} \frac{B_n}{n!} z^n = \frac{z}{e^z-1}$ and thus $\sum_{n=0}^{\infty} \sum_{k=0}^{n} \binom{n}{k} B_k \frac{x^{n-k}}{n!} z^n$

$= \sum_{k=0}^{\infty} \sum_{n=k}^{\infty} \binom{n}{k} B_k \frac{1}{n!} x^{n-k} z^n = \sum_{k=0}^{\infty} \sum_{n=k}^{\infty} \frac{B_k}{k!} \frac{x^{n-k}}{(n-k)!} z^n = \sum_{k=0}^{\infty} \frac{B_k}{k!} e^{xz} z^k = \frac{z e^{xz}}{e^z-1}$,

so that $B_n(x) = \sum_{k=0}^{n} \binom{n}{k} B_k x^{n-k}$.

(iii) We have $\frac{z}{e^z-1} = \frac{z}{z+z^2/2+\dots} = \frac{1}{1+z/2+\dots} = 1-z/2\pm\dots$, so that $B_0 = 1, B_1 = -1/2$. Therefore $B_0(x) = 1, B_1(x) = x - 1/2$. The function $\sum_{n=2}^{\infty} \frac{B_n}{n!} z^n$

$= \frac{z}{e^z-1} - 1 + \frac{1}{2} z$ is even, since $\frac{-z}{e^{-z}-1} - 1 - \frac{1}{2} z - \frac{z}{e^z-1} + 1 - \frac{1}{2} z = \frac{z e^z}{e^z-1} - \frac{z}{e^z-1} - z = 0$. Therefore, for odd $n \geq 3, B_n = 0$.

(iv) By (iii) $B_0(x) = 1$, $B_1(x) = x - 1/2$. By (i) $\sum_{n=0}^{\infty} \frac{B_n(1)}{n!} z^n = \frac{-z}{e^{-z}-1}$

$= \sum_{n=0}^{\infty} \frac{B_n}{n!} (-1)^n z^n$, so that $B_n(1) = (-1)^n B_n$. Therefore for even n $B_n(1) = B_n$. If n is odd, $n > 1$, then from (iii) it follows that $B_n(1) = 0 = B_n$.

(v) By (ii) for $n > 1, B_n = B_n(1) = \sum_{k=0}^{n} \binom{n}{k} B_k$ therefore $\sum_{k=0}^{n-1} \binom{n}{k} B_k = 0$.

15. (i) This follows from (14v) and $B_0 = 1$ by induction on n.

(ii) We have $\sum_{n=0}^{\infty} \frac{B_n'(x) z^n}{n!} = \frac{z^2 e^{xz}}{e^z-1} = \sum_{n=0}^{\infty} \frac{B_n(x) z^{n+1}}{n!} = \sum_{n=1}^{\infty} \frac{B_{n-1}(x) z^n}{(n-1)!}$, so that for $n \geq 1$ $B_n'(x) = n B_{n-1}(x)$.

16. If the assertion is proved for the intervals $[a,b]$ and $[b,c]$, then by addition we obtain it for $[a,c]$. Therefore it is enough to prove it for $b-a < 1$. We desinguish between two cases.

(a) $(a,b] \cap \mathbb{Z} = \emptyset$. Then $x \mapsto \{x\}$ is differentiable in $[a,b]$ with derivative 1, so that by (15ii) $\frac{d}{dx} B_n(\{x\}) = n B_{n-1}(\{x\})$. The result follows by integration.

(b) $g \in (a,b] \cap \mathbb{Z}$. Then $(g,b] \cap \mathbb{Z} = \emptyset$, and we can apply case (a) for the interval $(g,b]$. Hence we can take $g = b$. Then we have $n \int_a^b B_{n-1}(\{x\}) dx$

$= \lim_{t \to b-0} n \int_a^t B_{n-1}(\{x\}) dx = n \lim_{t \to b-0} (B_n(\{t\}) - B_n(\{a\})) = n(B_n(1)$
$- B_n(\{a\})) = n(B_n(0) - B_n(\{a\})) = n(B_n(\{b\}) - B_n(\{a\}))$ for $n > 1$ because of (14iv).

We prove the last assertion by induction on m; $x \mapsto B_2(\{x\}) = B_2 + 2 \int_0^x B_1(\{t\}) dt$ is continuous. If the assertion has already been proved for m, then for $m + 1$ it follows from $B_{m+1}(\{x\}) = B_{m+1} + (m+1) \int_0^x B_m(\{t\}) dt$.

17. By induction on m. By (13ii) we have $-\frac{1}{2\pi i} \sum_{k \neq 0} \frac{1}{k} e^{2\pi i kx}$

$= -\frac{1}{2\pi i} \sum_{k=1}^{\infty} \frac{1}{k} \left(e^{2\pi i kx} - e^{-2\pi i kx} \right) = -\frac{1}{\pi} \sum_{k=1}^{\infty} \frac{1}{k} \sin 2\pi kx = \{x\} - \frac{1}{2} = B_1(\{x\})$.

Suppose that the assertion for m has already been proved. Because of (13iii)

$\sum_{0<|k|\leq N} k^{-1}e^{2\pi ikx}$ is uniformly bounded. For $m > 1$ $\sum_{0<|k|<\infty} k^{-m}e^{2\pi ikx}$ is uniformly convergent. From (16) it therefore follows that $B_{m+1}(\{x\})-B_{m+1}(\{0\}) =$

$$(m+1)\int_0^x B_m(\{t\})dt = -\frac{(m+1)!}{(2\pi i)^m}\sum_{k\neq 0}k^{-m}\frac{e^{2\pi ikx}-1}{2\pi ik} = -\frac{(m+1)!}{(2\pi i)^{m+1}}\sum_{k\neq 0}k^{-m-1}e^{2\pi ikx}$$

$+C_m$ for some constant C_m. Integrating once more it follows that $B_{m+2}(\{x\}) =$
$-\frac{(m+2)!}{(2\pi i)^{m+2}}\sum_{k\neq 0}k^{-m-2}e^{2\pi ikx} + (B_{m+1}(0) + C_m)x + B_{m+2}$. Since $B_{m+2}(\{x\})$ is bounded in x, it follows that $B_{m+1}(0) = -C_m$, therefore $B_{m+1}(\{x\}) =$
$-\frac{(m+1)!}{(2\pi i)^{m+1}}\cdot\sum_{k\neq 0}k^{-m-1}e^{2k\pi ix}$.

(ii) and (iii) follow from (i) by equating the real parts. (iv) follows from (i) with $x = 0$.

18. Induction on m. The Euler sum formula gives the case $m = 1$ directly. For the inductive step one has only to observe (because of (16)) that

$$\int_a^b B_m(\{x\})f^{(m)}(x)dx = \frac{1}{m+1}\left(B_{m+1}(\{b\})f^{(m)}(b) - B_{m+1}(\{a\})f^{(m)}(a)\right)$$

$$-\frac{1}{m+1}\int_a^b B_{m+1}(\{x\})f^{(m+1)}(x)dx .$$

19. This follows from (18) and the relation $B_n = B_n(0)$.

20. Let $m \in \mathbb{N}$, $m > 1$ and $\operatorname{Re} s > 1$. Let $f(x) = (x + \omega)^{-s}$. Then $f^{(k)}(x)$
$= k!\binom{-s}{k}(x + \omega)^{-s-k}$, hence from (19) for $N \in \mathbb{N}$ $\sum_{n=0}^N (n + \omega)^{-s} = \omega^{-s}$

$+\frac{1}{1-s}\left((N + \omega)^{-s+1} - \omega^{-s+1}\right)+\sum_{k=1}^m\frac{(-1)^k}{k}B_k\binom{-s}{k-1}\left((N + \omega)^{-s-k+1} - \omega^{-s-k+1}\right)$

$+ (-1)^{m+1}\binom{-s}{m}\int_0^N B_m(\{x\})\cdot(\omega + x)^{-s-m}dx$. As $N \to \infty$ then $\zeta(s,\omega) = \omega^{-s}$

$+ \frac{1}{s-1}\omega^{-s+1} + \sum_{k=1}^m\frac{(-1)^{k+1}}{k}B_k\binom{-s}{k-1}\omega^{-s-k+1} + (-1)^{m+1}\binom{-s}{m}\int_0^\infty B_m(\{x\})$

$(\omega + x)^{-s-m}dx$. The integral appearing on the right-hand side is analytic for $\operatorname{Re}(s) > 1 - m$, since firstly, for each $x \geq 0$, $s \mapsto B_m(\{x\})(x + \omega)^{-s-m}$ is analytic. Secondly $(x, s) \mapsto B_m(\{x\})(x + \omega)^{-s-m}$ is (because $m > 1$) continuous on $[0, \infty) \times \{s : \operatorname{Re} s > 1 - m\}$. Thirdly for each $\delta > 0$ $\lim_{k\to\infty}\int_k^\infty B_m(\{x\})(x + \omega)^{-s-m}dx = 0$ uniformly in s for $\operatorname{Re}(s) > 1 - m + \delta$. Namely there exists some $K_m > 0$ with $|B_m(\{x\})| \leq K_m$ for $x \in \mathbb{R}$. Therefore $\left|\int_k^\infty B_m(\{x\})(x + \omega)^{-s-m}dx\right| \leq K_m\int_k^\infty x^{-\operatorname{Re}s-m}dx \leq K_m\int_k^\infty x^{-1-\delta}dx = K_m\delta^{-1}k^{-\delta}$. Therefore $s \mapsto \zeta(s,\omega)$ is meromorphically continuable on $\operatorname{Re} s > 1 - m$ for each $m \in \mathbb{N}$, from which the first assertion follows. Moreover $\zeta(s,\omega)$ has only one pole at $s = 1$, and this is simple. The assertion now follows from $\lim_{s\to 1}\omega^{-s+1} = 1$ and $\zeta(s) = \zeta(s,1)$.

Chapter 5

1. For $n \geq 1$ $n = \pi(p_n)$, therefore by the prime number theorem $\lim_{n \to \infty} \frac{n \log p_n}{p_n} = 1$. It follows that $\lim_{n \to \infty} (\log n + \log \log p_n - \log p_n) = 0$, and hence that $\lim_{n \to \infty} \frac{\log n}{\log p_n} = \lim_{n \to \infty} \left(1 - \frac{\log \log p_n}{\log p_n}\right) = 1$. Multiplication gives $\lim_{n \to \infty} \frac{n \log n}{p_n} = 1$.

2. For $\operatorname{Re} s > 0$ $\zeta(s) = (1 - 2^{1-s}) \sum_{n=1}^{\infty} (-1)^{n+1} n^{-s}$. Since $(n^{-s})_{n \geq 1}$ is decreasing for $s > 0$, we have $\sum_{n=1}^{\infty} (-1)^{n+1} n^{-s} > 0$. Because $s < 1$, $2^{1-s} > 1$, and from this the assertion follows.

3. We argue by contradiction. We have $\sum_{p \leq x} \frac{\log p}{p} = \frac{\log x}{x} \pi(x) + \int_2^x \frac{\log t - 1}{t^2} \pi(t) dt = \frac{\log x}{x} \pi(x) + \int_2^x \frac{\log t - 1}{t^2} \cdot \frac{t}{\log t} dt + \int_2^x \frac{\log t - 1}{t^2} \left(\pi(t) - \frac{t}{\log t}\right) dt = O(1) + \int_2^x \frac{dt}{t} - \int_2^x \frac{1}{t \log t} dt + O\left(\int_2^x \frac{dt}{t \log^2 t}\right) = \log x - \log \log x + O(1)$, contradicting Mertens' theorem.

4. Let p_k be the kth prime number and $n = p_1 \ldots p_k$. By the prime number theorem $\log n_k = \sum_{p \leq p_k} \log p = \psi(p_k) + O\left(\sum_{2 \leq m \leq \frac{\log p_k}{\log 2}} \sum_{p^m \leq p_k} \log p\right) = \psi(p_k) + O\left(p_k^{1/2} \log^2 p_k\right) = p_k(1 + o(1))$. It follows that $\log \log n_k = \log p_k + o(1)$.

   (i) By Mertens' theorem $c = \lim_{k \to \infty} \log p_k \prod_{p \mid n_k} \left(1 - \frac{1}{p}\right) = \lim_{k \to \infty} \log p_k \frac{\varphi(n_k)}{n_k} = \lim_{k \to \infty} \log \log n_k \frac{\varphi(n_k)}{n_k}$.

   (ii) We have $\tau(n_k) = 2^{\pi(p_k)}$, hence $\log \tau(n_k) = \pi(p_k) \log 2$, so that by the prime number theorem $\overline{\lim}_{k \to \infty} \frac{\log \tau(n_k) \log \log n_k}{\log n_k} = \overline{\lim}_{k \to \infty} \frac{\pi(p_k) \log \log n_k}{\log n_k} \log 2 = \overline{\lim}_{k \to \infty} \frac{\pi(p_k) \log p_k}{p_k} \log 2 = \log 2$.

5. For $n \geq 1$ let $f(n) = \sum_{p \mid n, p > \log n} 1$. Then $n \geq \prod_{p \mid n} p \geq (\log n)^{f(n)}$, hence $f(n) \leq \frac{\log n}{\log \log n}$. For sufficiently large n we have $\log \left(1 - \frac{1}{\log n}\right) \geq -\frac{2}{\log n}$. From this it follows that $0 \geq \log P(n) = \sum_{p \mid n, p > \log n} \log \left(1 - \frac{1}{p}\right) \geq f(n) \log \left(1 - \frac{1}{\log n}\right) \geq -2 \frac{f(n)}{\log(n)} \geq -\frac{2}{\log \log n}$. From this it follows that $\lim_{n \to \infty} \log P(n) = 0$, or $\lim_{n \to \infty} P(n) = 1$.

6. For $n \geq 1$ let $P(n) = \prod_{p \mid n, p > \log n} \left(1 - \frac{1}{p}\right)$. Then $\frac{\varphi(n)}{n} = \prod_{p \mid n} \left(1 - \frac{1}{p}\right) = P(n) \prod_{\substack{p \mid n \\ p \leq \log n}} \left(1 - \frac{1}{p}\right) \geq P(n) \prod_{p \leq \log n} \left(1 - \frac{1}{p}\right)$. By Mertens' theorem $c = \lim_{n \to \infty} \log \log n \cdot \prod_{p \leq \log n} \left(1 - \frac{1}{p}\right)$, so that $\underline{\lim}_{n \to \infty} n^{-1} \varphi(n) \log \log n \geq \lim_{n \to \infty} P(n) \cdot c = c$ (by (5)). The assertion follows from this and (4i).

7. We have $\log f(n) \sum\limits_{f(n)\leq p} \alpha_p(n) \leq \sum\limits_{f(n)\leq p} \alpha_p(n)\log p \leq \sum\limits_{p}\log(p^{\alpha_p(n)}) = \log n$

and $\log P(n) = \sum\limits_{f(n)\leq p}\log(1+\alpha_p(n)) \leq \sum\limits_{f(n)\leq p}\log 2^{\alpha_p(n)} = \log 2 \cdot \sum\limits_{f(n)\leq p}\alpha_p(n)$

$\leq \log 2\frac{\log n}{\log f(n)}$.

8. We have $2^{\alpha_p(n)} \leq p^{\alpha_p(n)} \leq n$, therefore $\alpha_p(n) \leq \frac{\log n}{\log 2}$. From this it follows that

$\log Q(n) = \sum\limits_{p<f(n)}\log(1+\alpha_p(n)) = O\left(\log\log n \cdot \sum\limits_{p<f(n)} 1\right)$

$= O\left(\log\log n \cdot \pi(f(n))\right) = O\left(\log\log n \cdot \frac{f(n)}{\log f(n)}\right) = O(f(n)).$

9. For $n \geq 2$ write $f(n) = \frac{\log n}{\log^2\log n}$ and $n = \prod\limits_{p}p^{\alpha_p(n)}$, the prime factor de-

composition of n, $P(n) = \prod\limits_{f(n)\leq p}(1+\alpha_p(n))$, $Q(n) = \prod\limits_{p<f(n)}(1+\alpha_p(n))$. Then

$\tau(n) = P(n)Q(n)$ and therefore $\overline{\lim\limits_{n\to\infty}} \frac{\log\tau(n)\log\log n}{\log n} =$

$\overline{\lim\limits_{n\to\infty}} \frac{\log P(n)+\log Q(n)}{\log n} \cdot \log\log n \leq \log 2\,\overline{\lim\limits_{n\to\infty}} \frac{\log\log n}{\log f(n)} = \log 2$ by (7) and (8). The

assertion now follows from (4ii).

10. We show first that $\lim\limits_{|x|\to\infty} f(x) = 0$. It follows from $f(x) = f(0) + \int\limits_{0}^{x} f'(t)dt$

that $\lim\limits_{x\to\infty} f(x)$ exists. By the mean-value theorem from integral calculus there

exists some $\xi_x \in (x, x+1)$ with $f(\xi_x) = \int\limits_{x}^{x+1} f(t)dt$. From $\lim\limits_{x\to\infty} f(\xi_x) =$

$\lim\limits_{x\to\infty} \int\limits_{x}^{x+1} f(t)dt = 0$ it follows that $\lim\limits_{x\to\infty} f(x) = 0$. Analogously $\lim\limits_{x\to-\infty} f(x) = 0$.

By the Euler sum formula for $a,b \in \mathbb{Z}$, $a \leq b$, $\sum\limits_{n=a}^{b} f(n) = \frac{1}{2}(f(a)+f(b)) +$

$\int\limits_{a}^{b} f(x)dx + \int\limits_{a}^{b}(\{x\} - 1/2)f'(x)dx$. Then, as $b \to \infty$, $a \to -\infty$, we obtain

$\sum\limits_{n\in\mathbb{Z}} f(n) = \int\limits_{\mathbb{R}} f(x)dx + \int\limits_{\mathbb{R}}(\{x\} - 1/2)f'(x)dx$. By Exercise (13) in Chapter 4

$\sum\limits_{0<|k|\leq N} \frac{1}{k}e^{2\pi ikx}$ is uniformly bounded in N and x, from which it follows that

$\sum\limits_{n\in\mathbb{Z}} f(n) = \int\limits_{\mathbb{R}} f(x)dx - \frac{1}{2\pi i}\sum\limits_{k\neq 0}\frac{1}{k}\int\limits_{\mathbb{R}} e^{2\pi ikx}f'(x)dx$ – see Exercise (17i) for $m = 1$ in

Chapter 4. Finally for $a < b$ $\int\limits_{a}^{b} e^{2\pi ikx}f(x)dx = \frac{1}{2\pi ik}\left(f(b)e^{2\pi ikb} - f(a)e^{2\pi ika}\right) -$

$\frac{1}{2\pi ik}\int\limits_{a}^{b} e^{2\pi ikx}f'(x)dx$, from which, for $b \to \infty$, $a \to -\infty$, the existence of

$\int\limits_{\mathbb{R}} e^{2\pi ikx}f(x)dx = -\frac{1}{2\pi i}\int\limits_{\mathbb{R}} e^{2\pi ikx}f'(x)dx$ follows. Substitution of this above gives

the result.

11. For $z \in \mathbb{C}$, let $y(z) = \int\limits_{\mathbb{R}} e^{-\pi x^2+2\pi ixz}dx$. Since $\int\limits_{\mathbb{R}} 2\pi|x|e^{-\pi x^2+2\pi|xz|}dx < \infty$,

$y'(z) = \int\limits_{\mathbb{R}} 2\pi ixe^{-\pi x^2+2\pi ixz}dx = i\int\limits_{\mathbb{R}} 2\pi(x - zi)e^{-\pi x^2+2\pi ixz}dx$

$-\int_{\mathbb{R}} 2\pi z e^{-\pi x^2 + 2\pi i x z} dx = -i e^{-\pi x^2 + 2\pi i x z}\big|_{-\infty}^{\infty} - 2\pi z y(z) = -2\pi z y(z)$. The functions y and $z \mapsto e^{-\pi z^2}$ therefore satisfy the same linear differential equation, so that for some $C \in \mathbb{C}$ $y(z) = C e^{-\pi z^2}$. Now $y(0) = \int_{\mathbb{R}} e^{-\pi x^2} dx = \frac{2}{\sqrt{\pi}} \int_0^{\infty} e^{-u^2} du = \frac{1}{\sqrt{\pi}} \int_0^{\infty} e^{-v} v^{-1/2} dv = \frac{1}{\sqrt{\pi}} \Gamma(\frac{1}{2}) = 1$, hence $C = 1$.

12. We have $\int_{\mathbb{R}} 2|x| e^{-\pi x^2} dx < \infty$. It therefore follows from (10) and (11) that
$$\Theta(x) = \sum_{k \in \mathbb{Z}} \int_{\mathbb{R}} e^{-\pi x y^2 + 2\pi i k y} dy = \sum_{k \in \mathbb{Z}} x^{-1/2} \int_{\mathbb{R}} e^{-\pi u^2 + 2\pi i k u / \sqrt{x}} du$$
$$= \sum_{k \in \mathbb{Z}} x^{-1/2} e^{-\pi k^2 / x} = \frac{1}{\sqrt{x}} \Theta\left(\frac{1}{x}\right).$$

13. For all $x > 0$ we have $\Psi(x) \le \sum_{n=1}^{\infty} e^{-\pi n x} = \frac{1}{e^{\pi x} - 1}$. From $1 + 2\Psi(x)$
$= \frac{1}{\sqrt{x}}\left(1 + 2\Psi(\frac{1}{x})\right)$ it follows that $\Psi(x) \le (\frac{1}{\sqrt{x}} - 1)(\frac{1}{2}) + \frac{1}{\sqrt{x}} \cdot \frac{1}{e^{\pi/x} - 1}$
$\le \frac{1}{\sqrt{x}}\left(\frac{1}{2} + \frac{1}{e^{\pi/x} - 1}\right)$.

(i) Let $z \in \mathbb{C}$, then $\int_1^{\infty} |\Psi(x) x^z| \, dx \le \int_1^{\infty} \frac{x^{\mathrm{Re}\, z}}{e^{\pi x} - 1} dx < \infty$, from which (i) follows.

(ii) For $\mathrm{Re}\, s > 1$, $\int_0^1 \Psi(x) x^{-1 + \mathrm{Re}\, s/2} dx \le \left(\frac{1}{2} + \frac{1}{e^{\pi} - 1}\right) \int_0^1 x^{-1/2 - 1 + \mathrm{Re}\, s/2} dx < \infty$.

   Therefore we have the convergent integral $\int_0^{\infty} \Psi(x) x^{s/2} \frac{dx}{x}$
$$= \sum_{n=1}^{\infty} \int_0^{\infty} e^{-\pi n^2 x} x^{s/2} \frac{dx}{x} = \sum_{n=1}^{\infty} \frac{1}{\pi n^2} \int_0^{\infty} e^{-u} \left(\frac{u}{\pi n^2}\right)^{-1 + s/2} du$$
$$= \sum_{n=1}^{\infty} (\pi n^2)^{-s/2} \int_0^{\infty} e^{-u} u^{s/2} \cdot \frac{du}{u} = \Gamma\left(\frac{s}{2}\right) \pi^{-s/2} \zeta(s). \text{ (Here we use Lebesgue's}$$
   theorem on dominated convergence.)

14. By (12) for all $\mathrm{Re}\, s > 1$ we have $\int_0^{\infty} \Psi(x) x^{s/2} \frac{dx}{x} = \int_0^1 \Psi(x) x^{s/2} \frac{dx}{x} + \int_1^{\infty} \Psi(x) x^{s/2} \frac{dx}{x} =$
$\int_0^1 \frac{1}{2}\left(\frac{1}{\sqrt{x}} - 1\right) x^{s/2} \frac{dx}{x} + \int_0^1 \frac{1}{\sqrt{x}} \Psi(\frac{1}{x}) x^{s/2} \frac{dx}{x} + \int_1^{\infty} \Psi(x) x^{s/2} \frac{dx}{x} = \frac{1}{2} \int_0^1 x^{(s-3)/2} dx - $
$\frac{1}{2} \int_0^1 x^{(s-2)/2} dx + \int_0^1 \Psi\left(\frac{1}{x}\right) x^{(s-3)/2} dx + \int_1^{\infty} \Psi(x) x^{s/2} \frac{dx}{x} = \frac{1}{2} \frac{x^{(s-1)/2}}{(s-1)/2}\big|_0^1 - \frac{1}{2} x^{s/2} / \frac{s}{2}\big|_0^1 + $
$\int_1^{\infty} u^{-2} \Psi(u) u^{(3-s)/2} du + \int_1^{\infty} \Psi(x) x^{s/2} \frac{dx}{x} = \frac{1}{s-1} - \frac{1}{s} + \int_1^{\infty} \Psi(x) \left(x^{s/2} + x^{(1-s)/2}\right) \frac{dx}{x}$.

15. For $\mathrm{Re}\, s > 1$ it follows from (13) and (14) that $\Gamma(\frac{s}{2}) \pi^{-s/2} \zeta(s)$
$= \int_1^{\infty} \Psi(x) \left(x^{s/2} + x^{(1-s)/2}\right) \frac{dx}{x} + \frac{1}{s(s-1)}$. Because of (13) the right-hand side is meromorphic in $\mathbb{C}$ (hence also the left). Replacing s by $1 - s$ gives the result.

16. (i) By (15) for $n \in \mathbb{N}$ $\lim\limits_{s \to -2n} \pi^{-s/2} \Gamma\left(\frac{s}{2}\right) \zeta(s) = \pi^{-(1+2n)/2} \Gamma(n + \frac{1}{2})\zeta(2n + 1)$ $\neq 0$. Since Γ has a simple pole at $-n$, ζ has a simple zero at $-2n$.

(ii) By (15) again $\lim\limits_{s \to 0} \pi^{-s/2}\Gamma\left(\frac{s}{2}\right)\zeta(s)/\zeta(1 - s) = \pi^{-1/2}\Gamma\left(\frac{1}{2}\right) = 1$. Γ has a simple pole at 0 with residue 1, so that $\lim\limits_{s \to 0} \frac{s}{2}\Gamma\left(\frac{s}{2}\right) = 1$. Therefore $1 = \zeta(0)\lim\limits_{s \to 0} \frac{2}{s\zeta(1-s)} = 2\zeta(0)\lim\limits_{s \to 1}\frac{1}{(1-s)\zeta(s)} = -2\zeta(0)$. We prove the second assertion. For $\mathrm{Re}\, s \geq 1$ $\zeta(s) \neq 0$. Suppose now that $\mathrm{Re}\, s \leq 0$, $s \neq -2n$ for all $n \in \mathbb{Z}$, $n \geq 0$. If $\zeta(s)$ were to equal 0, then by (15) $0 = \pi^{-(1-s)/2}\Gamma\left(\frac{1-s}{2}\right) \cdot \zeta(1 - s)$, which is impossible, given that $\mathrm{Re}(1 - s) \geq 1$.

(iii) For $\mathrm{Re}\, s > 0$ we have $\zeta(s) = \left(1 - 2^{1-s}\right)^{-1} \sum\limits_{n=1}^{\infty}(-1)^{n+1}n^{-s}$. From this it follows that $\zeta(\bar{s}) = \overline{\zeta(s)}$. Therefore if $\zeta(s) = 0$, then also $\zeta(\bar{s}) = 0$. For $\mathrm{Re}\, s < 0$ the assertion follows from (ii).

17. Let $\varepsilon > 0$. Because $\lim\limits_{m \to \infty} \frac{m^{1-\varepsilon}}{\varphi(m)} = 0$ (Chapter 4) there exists some $c_\varepsilon > 0$, so that for all $m \in \mathbb{N}$, $m^{1-\varepsilon} \leq c_\varepsilon \varphi(m)$. Hence, if $\varphi(m) = n$, $m^{1-\varepsilon} \leq c_\varepsilon n$, from which the first assertion follows. For $s > 1$ we show that $\sum\limits_{m=1}^{\infty} \varphi(m)^{-s}$ converges. If we choose $\varepsilon = 1 - \frac{1}{\sqrt{s}}$, then because $\varepsilon > 0$, there exists some $K_s > 0$ with $m^{1/\sqrt{s}} \leq K_s\varphi(m)$, i.e. $m^{\sqrt{s}} \leq K_s^s\varphi(m)^s$ for all $m \in \mathbb{N}$. Therefore $\sum\limits_{m=1}^{\infty} \varphi(m)^{-s}$ converges. Suppose now that $N \in \mathbb{N}$ and $M_N = \{n \in \mathbb{N} : \varphi(n) \leq N\}$. Then $M_N \subseteq M_{N+1}$ and $\bigcup\limits_{N=1}^{\infty} M_N = \mathbb{N}$. For $\mathrm{Re}\, s > 1$ we have $\sum\limits_{m=1}^{N} \frac{r(m)}{m^s} = \sum\limits_{m=1}^{N} \frac{1}{m^s} \sum\limits_{\substack{n=1 \\ \varphi(n)=m}}^{\infty} 1 = \sum\limits_{n=1}^{\infty} \sum\limits_{\substack{m=1 \\ \varphi(n)=m}}^{N} \varphi(n)^{-s} = \sum\limits_{n \in M_N} \varphi(n)^{-s}$, so that $\sum\limits_{n=1}^{\infty} \frac{r(m)}{m^s} = \sum\limits_{n=1}^{\infty} \varphi(n)^{-s} = \prod\limits_{p} \sum\limits_{k=0}^{\infty} \varphi(p^k)^{-s} = \prod\limits_{p}\left(1 + \sum\limits_{k=1}^{\infty} p^{-ks}\left(1 - \frac{1}{p}\right)^{-s}\right) = \prod\limits_{p}\left(1 + \frac{1}{(p^s-1)(1-\frac{1}{p})^s}\right)$.

18. For $x, y > 0$ and $s = \sigma + it$, $\sigma \neq 0$, we have $|x^s - y^s| = \left|s \int\limits_{x}^{y} t^{s-1}dt\right| \leq$ $|s|\left|\int\limits_{x}^{y} t^{\sigma-1}dt\right| \leq \frac{|s|}{|\sigma|}|x^\sigma - y^\sigma|$. Hence for $\sigma > 0$, $|(p-1)^{-s} - p^{-s}| \leq$ $\frac{|s|}{\sigma}|(p-1)^{-\sigma} - p^{-\sigma}| = O(p^{-\sigma-1})$. Therefore for each $\delta > 0$ $\sum\limits_{p}\left(\frac{1}{(p-1)^s} - \frac{1}{p^s}\right)$ is absolutely and uniformly convergent in $\{s : \mathrm{Re}\, s \geq \delta\}$, so that $g(s) = \prod\limits_{p}\left(1 + \frac{1}{(p-1)^s} - \frac{1}{p^s}\right)$ is holomorphic in $\mathrm{Re}\, s > 0$. For $\mathrm{Re}\, s > 1$ we have $\frac{f(s)}{\zeta(s)} = \prod\limits_{p}\left(1 + \frac{1}{(p^s-1)(1-\frac{1}{p})^s}\right)\left(1 - \frac{1}{p^s}\right) = \prod\limits_{p}\left(1 + \frac{1}{(p-1)^s} - \frac{1}{p^s}\right) = g(s)$, i.e. $f(s) = g(s)\zeta(s)$. The first assertion follows from this. The second follows from $\lim\limits_{s \to 1}(s - 1)f(s) = g(1)\lim\limits_{s \to 1}(s - 1)\zeta(s) = g(1) = \prod\limits_{p}\left(1 + \frac{1}{p(p-1)}\right)$.

19. If $\operatorname{Re} s > 1$ let $f(s) = \sum\limits_{n=1}^{\infty} \frac{r(n)}{n^s}$ and $c := \prod\limits_{p}\left(1 + \frac{1}{p(p-1)}\right)$. By (18) f can be extended over $\{s \in \mathbb{C} : \operatorname{Re} s > 0\}$ and there $s \mapsto f(s) - \frac{c}{s-1}$ is holomorphic. The conclusion follows from the theorem of Wiener and Ikehara (see Addendum[20]).

20. First we show that as $N \to \infty$ $\int\limits_{2}^{N} \frac{1}{x} \left|\pi(x) - \frac{x}{\log x}\right| dx = o(\pi(N))$. Let $\varepsilon > 0$. By the prime number theorem there exists some $x_0(\varepsilon) \geq e$, so that for all $x \geq x_0(\varepsilon)$, $\left|\pi(x) - \frac{x}{\log x}\right| \leq \frac{\varepsilon x}{\log x}$. Now let $N \geq x_0(\varepsilon)^2$ and choose $K > 0$ such that, for all $x \geq 2$, $\frac{1}{x}\left|\pi(x) - \frac{x}{\log x}\right| \leq K$. It then follows that $\int\limits_{2}^{N} \frac{1}{x}\left|\pi(x) - \frac{x}{\log x}\right| dx \leq$

$$\int\limits_{2}^{x_0(\varepsilon)} \frac{1}{x}\left|\pi(x) - \frac{x}{\log x}\right| dx + \varepsilon \int\limits_{x_0(\varepsilon)}^{N} \frac{dx}{\log x} \leq K x_0(\varepsilon) + \varepsilon \int\limits_{x_0(\varepsilon)}^{\sqrt{N}} \frac{dx}{\log x} + \varepsilon \int\limits_{\sqrt{N}}^{N} \frac{dx}{\log x} \leq$$

$K\sqrt{N} + \varepsilon\sqrt{N} + \frac{2\varepsilon N}{\log N}$. Applying the prime number theorem it follows that

for each $\varepsilon > 0$, $\overline{\lim\limits_{N\to\infty}} \frac{1}{\pi(N)} \int\limits_{2}^{N} \frac{1}{x}\left|\pi(x) - \frac{x}{\log g}\right| dx \leq 2\varepsilon$.

Furthermore, $\int\limits_{2}^{N} \frac{e^{2\pi i \log x}}{\log x} dx = \int\limits_{2}^{N} x^{2\pi i} \frac{dx}{\log x} = \frac{x^{2\pi i+1}}{2\pi i+1} \cdot \frac{1}{\log x}\Big|_{2}^{N} + \int\limits_{2}^{N} \frac{x^{2\pi i}}{2\pi i+1} \cdot \frac{dx}{\log^2 x} =$

$\frac{N^{1+2\pi i}}{(2\pi i+1)\log N} + O(1) + O\left(\int\limits_{2}^{\sqrt{N}} \frac{dx}{\log^2 x} + \int\limits_{\sqrt{N}}^{N} \frac{dx}{\log^2 x}\right) = \frac{N^{1+2\pi i}}{(2\pi i+1)\log N} + O(\sqrt{N}) +$

$O\left(\frac{N}{\log^2 N}\right) = N^{2\pi i} \frac{\pi(N)}{2\pi i+1}(1 + o(1))$ for $N \in \mathbb{N}$ as $N \to \infty$.

From this it follows that $\sum\limits_{p\leq N} e^{2\pi i \log p} = e^{2\pi i \log N}\pi(N) - 2\pi i \int\limits_{2}^{N} \frac{e^{2\pi i \log x}}{x} \pi(x) dx$

$= e^{2\pi i \log N}\pi(N) - 2\pi i \int\limits_{2}^{N} \frac{e^{2\pi i \log x}}{\log x} dx + O\left(\int\limits_{2}^{N} \frac{1}{x}\left|\pi(x) - \frac{x}{\log x}\right| dx\right) = N^{2\pi i}\pi(N)$

$\times \left(1 - \frac{2\pi i}{1+2\pi i}\right) + o(\pi(N)) = \frac{1}{1+2\pi i}\pi(N)N^{2\pi i} + o(\pi(N))$. Thus it follows that $\lim\limits_{N\to\infty} \frac{1}{\pi(N)}\left|\sum\limits_{p\leq N} e^{2\pi i \log p}\right| = \frac{1}{\sqrt{1+4\pi^2}} \neq 0$.

Chapter 6

1. We show first that from $\chi \in \Gamma_m$, $\chi' \in \Gamma_n$ it follows that $\chi\chi' \in \Gamma_{mn}$. Clearly $\chi\chi'$ is strongly multiplicative and has period mn. If g.c.d.$(k, mn) > 1$, then either g.c.d.$(k, m) > 1$ or g.c.d.$(k, n) > 1$, hence $\chi(k) = 0$ or $\chi'(k) = 0$, and in either case $\chi\chi'(k) = 0$.
The map ω is clearly a homomorphism. If $\chi\chi' = \chi_0$, the principal character in Γ_{mn}, and $a \in \mathbb{Z}$, then we may choose some $x \in \mathbb{Z}$ with $x \equiv a(m)$, $x \equiv 1(n)$. It follows that $\chi(a) = \chi(x)\chi'(x) = \chi_0(x)$, so that χ is the principal character in Γ_m. Analogously χ' is the principal character in Γ_n. Hence ω is injective. Since $\Gamma_m \times \Gamma_n$ and Γ_{mn} have equal numbers of elements, ω is bijective.

2. (i) We show first that $m_{\chi\chi'} \geq m_\chi m_{\chi'}$. Because $m_{\chi\chi'}|mn$ and g.c.d.$(m, n) = 1$, there exist natural numbers d_1 and d_2 with $m_{\chi\chi'} = d_1 d_2$ and $d_1|m, d_2|n$.

Let $x \equiv 1 \pmod{d_1}$, g.c.d.$(x, m) = 1$. Choose an a with $x \equiv a \pmod{m}$, $a \equiv 1 \pmod{m}$. We have g.c.d.$(a, mn) = 1$, $a \equiv x \equiv 1(d_1)$, $a \equiv 1(d_2)$, hence $a \equiv 1(d_1 d_2)$. By Proposition 7 $\chi(a)\chi'(a) = 1$. From $\chi'(a) = 1$ it follows that $\chi(a) = 1$ and therefore $\chi(x) = 1$. By Proposition 8 d_1 is a defining modulus for χ. Analogously d_2 is one for χ', from which it follows that $m_\chi \le d_1$, $m_{\chi'} \le d_2$, hence also that $m_\chi m_{\chi'} \le d_1 d_2 = m_{\chi\chi'}$. Suppose now that conversely $x \equiv 1(m_\chi m_{\chi'})$ and g.c.d.$(x, mn) = 1$. Then g.c.d.$(x, m) = $ g.c.d.$(x, n) = 1$ and $x \equiv 1(m_\chi)$, $x \equiv 1(m_{\chi'})$. By Proposition 7 $\chi(x) = \chi'(x) = 1$, hence $\chi\chi'(x) = 1$. Also by Proposition 7 $m_\chi m_{\chi'}$ must be a defining modulus for $\chi\chi'$, so that $m_{\chi\chi'} \le m_\chi m_{\chi'}$.

(ii) The first statement follows immediately from (i). Now let $\chi\chi'$ be real, i.e. $\chi^2 \chi'^2$ be the principal character in Γ_{mn}. Then by (1) χ^2 and χ'^2 are the principal characters in Γ_m and Γ_n respectively, hence χ, χ' real. The converse is trivial.

3. For odd values of $n \in \mathbb{Z}$ let $\chi_1(n) = \left(\frac{-1}{n}\right)$ sgn n. Then χ_1 and χ_2 are multiplicative. Because $\chi_2(n) = (-1)^{(n^2-1)/8}$, χ_1 and χ_2 have period 8. Therefore $\chi_1, \chi_2 \in \Gamma_8$ and further $\chi_3 \in \Gamma_8$. χ_1 and χ_2 are not principal characters. We have $\chi_1(-1) = -1 \ne \chi_2(-1)$ and $\chi_3(-1) = -1$, so that the assertion follows from $|\Gamma_8| = 4$.

4. (i) With $\chi \ne \chi_0$, let g be a primitive root modulo p^k. If $\chi(g) = 1$ it would follow that $\chi(g^i) = 1$ for all $i \ge 0$ and thus $\chi = \chi_0$. Hence $\chi(g) = -1$, and there exists a unique real character $\chi \ne \chi_0 \bmod p^k$. Since $n \mapsto \left(\frac{n}{p}\right)$ is real, $\chi(n) = \left(\frac{n}{p}\right)$ for $n \in \mathbb{Z}$. From $n \equiv 1(p)$ it follows that $\chi(n) = 1$, so that p is a defining modulus for χ. Therefore $k = 1$. χ is primitive in Γ_p, because p is prime.

(ii) There exists no primitive character in Γ_2, so that 2 cannot be a defining modulus of a character $\ne \chi_0$. For odd n, $\chi(n) = (-1)^{(n-1)/2}$, and this gives the unique character $\ne \chi_0 \bmod 4$. This is necessarily primitive. Now let $\chi \in \Gamma_8$ be real and primitive. If $\chi(5) = 1$, then from $n \equiv 1(4)$ it would follow that $\chi(n) = 1$, so that 4 would be a defining modulus for χ. Hence $\chi(5) = -1$. By (3) for odd n, either $\chi(n) = \left(\frac{2}{n}\right)$ or $\chi(n) = (-1)^{(n-1)/2}\left(\frac{2}{n}\right)$. Because $\chi(5) = -1$, 4 fails to be a defining modulus for χ, so that χ is primitive.

Finally let $\chi \in \Gamma_{2^k}$ be primitive and real, $k > 3$ and $x \equiv 1(8)$. There exist $i, j \ge 0$ with $x \equiv (-1)^i 5^j (2^k)$ and therefore $1 = (-1)^i 5^j (8)$, from which $i \equiv j \equiv 0(2)$. Therefore $\chi(x) = \chi(-1)^i \chi(5)^j = 1$, so that 8 would be a defining modulus for χ. Hence $k \le 3$.

5. Let $m = p_1^{k_1} \ldots p_t^{k_t}$ be the prime factor decomposition of m. By (1) for $1 \le i \le t$ there exists $\chi_i \in \Gamma_{p_i^{k_i}}$ with $\chi = \chi_1 \ldots \chi_t$. By (2) each χ_i is primitive and real. By (4i) for $1 \le i \le t$, $k_i = 1$ and $\chi_i(n) = \left(\frac{n}{p_i}\right)$ for $n \in \mathbb{Z}$. Therefore m is square-free and $\chi(n) = \left(\frac{n}{p_1}\right) \ldots \left(\frac{n}{p_k}\right)$. Conversely it follows from (4) and (2) that χ must be primitive.

6. W.l.o.g. let n be odd. We distinguish between three cases: If m is odd, then for $k \in \mathbb{Z}$ (5) implies that $\chi(k) = \left(\frac{k}{m}\right)$, $\chi'(k) = \left(\frac{k}{n}\right)$. Therefore $\chi(n)\chi'(m) = -1 \Leftrightarrow \left(\frac{n}{m}\right)\left(\frac{m}{n}\right) = -1 \Leftrightarrow \frac{m-1}{2} \equiv \frac{n-1}{2} \equiv 1(2) \Leftrightarrow \chi(-1) = \chi'(-1) = -1$. Now let

$m = 2^t m'$ be even, $2 \nmid m'$. By (1) and (4) $t \in \{2,3\}$. Suppose first that $t = 2$. By (1), (2), (4) and (5) $\chi(k) = (-1)^{(k-1)/2}(\frac{k}{m'})$ for odd values of k. We have $\chi'(k) = (\frac{k}{n})$ for all $k \in \mathbb{Z}$. Therefore $\chi(n)\chi'(m) = -1 \Leftrightarrow (-1)^{(n-1)/2}(\frac{n}{m'})(\frac{m}{n}) = -1 \Leftrightarrow (-1)^{(n-1)/2}(\frac{n}{m'})(\frac{m'}{n}) = -1 \Leftrightarrow \frac{n-1}{2} \equiv \frac{m'+1}{2} \equiv 1(2) \Leftrightarrow -(\frac{-1}{m'}) = -1$ and $(\frac{-1}{n}) = -1 \Leftrightarrow \chi(-1) = \chi'(-1) = -1$.

Now let $t = 3$. By (1), (2), (4) and (5) for all odd k and for some $\alpha \in \{0,1\}$, $\chi(k) = (-1)^{\alpha(k-1)/2}(\frac{2}{k})(\frac{k}{m'})$ and we have $\chi'(k) = (\frac{k}{n})$ for all $k \in \mathbb{Z}$. Therefore $\chi(n)\chi'(m) = -1 \Leftrightarrow (\frac{m}{n})(\frac{n}{m'})(\frac{2}{n})(-1)^{\alpha(n-1)/2} = -1 \Leftrightarrow (\frac{m'}{n})(\frac{n}{m'})(-1)^{\alpha(n-1)/2} = -1 \Leftrightarrow -1 = (-1)^{\frac{m'-1}{2}\frac{n-1}{2}+\alpha(\frac{n-1}{2})} = (-1)^{(\frac{m'-1}{2}+\alpha)(\frac{n-1}{2})} \Leftrightarrow \frac{m'-1}{2} + \alpha \equiv \frac{n-1}{2} \equiv 1(2) \Leftrightarrow \chi(-1) = \chi'(-1) = -1$.

7. First let $p = 2$. Then by (4) $k \in \{2,3\}$. Let $k = 2$ and $\chi(n) = (-1)^{(n-1)/2}$ for all odd n. Then $\tau(\chi) = \chi(1)e^{2\pi i/4} + \chi(-1)e^{2\pi i \cdot 3/4} = e^{2\pi i/4} - e^{6\pi i/4} = i - (-i) = 2i = (-4)^{1/2}$. Let $k = 3$ and $\chi(n) = (\frac{2}{n})$ for all odd n. Then

$$\tau(\chi) = \sum_{j=0}^{7} \chi(j)e^{2\pi ij/8} = e^{2\pi i/8} + e^{-2\pi i/8} - e^{6\pi i/8} - e^{-6\pi i/8} = \frac{1+i}{\sqrt{2}} + \frac{1-i}{\sqrt{2}} - \frac{-1+i}{\sqrt{2}} - \frac{-1-i}{\sqrt{2}} = \frac{2}{\sqrt{2}} + \frac{2}{\sqrt{2}} = \sqrt{8} = (\chi(-1)8)^{1/2}.$$

Let $k = 3$, and $\chi(n) = (-1)^{(n-1)/2}(\frac{2}{n})$ for all odd n. Then $\tau(\chi) = e^{2\pi i/8} + e^{6\pi i/8} - e^{-6\pi i/8} - e^{-2\pi i/8} = \frac{1+i}{\sqrt{2}} + \frac{-1+i}{\sqrt{2}} - \frac{-1-i}{\sqrt{2}} - \frac{1-i}{\sqrt{2}} = \frac{2i}{\sqrt{2}} + \frac{2i}{\sqrt{2}} = i\sqrt{8} = (\chi(-1)8)^{1/2}$.

By (4) there exist no further real primitive characters mod 8. If $p > 2$, then by (4) $k = 1$, and the assertion follows from Theorem 4.

8. We note that $\{kn + lm : 0 \le k < m, 0 \le l < n\}$ is a complete system of residues mod mn. From this it follows that $\chi(n)\chi'(m)\tau(\chi)\tau(\chi') = \chi(n)\chi'(m) \cdot$

$$\sum_{k=0}^{m-1}\sum_{l=0}^{n-1} \chi(k)\chi'(l)e^{2\pi i(k/m+l/n)} = \sum_{k=0}^{m-1}\sum_{l=0}^{n-1} \chi(nk)\chi'(lm)e^{2\pi i(kn+lm)/mn}$$

$$= \sum_{k=0}^{m-1}\sum_{l=0}^{n-1} \chi(nk + lm)\chi'(nk + lm)e^{2\pi i(kn+lm)/nm} = \tau(\chi\chi').$$

9. Induction on the number of prime factors of m. If this equals 0, then $m = 1$, and the assertion is trivial. If this equals 1, then we apply (7). Suppose that the number of prime factors is greater than or equal to 2. Then there exist $m_1, m_2 \in \mathbb{N}$ with $m = m_1 m_2$, $m_1 > 1$, $m_2 > 1$ and g.c.d.$(m_1, m_2) = 1$. By (1) for $i = 1, 2$ there exists $\chi_i \in \Gamma_{m_i}$ with $\chi = \chi_1\chi_2$. For $i = 1, 2$ χ_i is primitive and real (2). By (8) and the inductive assumption, $\tau(\chi) = \chi_1(m_2)\chi_2(m_1)\tau(\chi_1)\tau(\chi_2) = \chi_1(m_2)\chi_2(m_1)(\chi_1(-1)m_1)^{1/2}(\chi_2(-1)m_2)^{1/2}$. If $\chi_1(m_2)\chi_2(m_1) = -1$, then by (6) $\chi_1(-1) = \chi_2(-1) = -1$, so that one obtains $\tau(\chi) = -i\sqrt{m_1} \cdot i\sqrt{m_2} = (\chi(-1)m)^{1/2}$. However, if $\chi_1(m_2)\chi_2(m_1) = 1$, then we can w.l.o.g. assume that $\chi_1(-1) = 1$, from which it follows that $\chi(-1) = \chi_2(-1)$. Therefore $\tau(\chi) = \sqrt{m_1}(\chi(-1)m_2)^{1/2} = (\chi(-1)m)^{1/2}$.

10. By Proposition 11, for $d \in \mathbb{N}$, $\sum_{k=0}^{4d-1} e^{2\pi ik/4d} = \frac{\sqrt{4d}}{1-i} = (1+i)\sqrt{d}$. Therefore $\sqrt{d} \in \mathbb{Q}(e^{2\pi i/4d}, i)$. But now $i = (e^{2\pi i/4d})^d$, so that $i \in \mathbb{Q}(e^{2\pi i/4d})$, from which both $\sqrt{d} \in \mathbb{Q}(e^{2\pi i/4d})$ and $\sqrt{-d} \in \mathbb{Q}(e^{2\pi i/4d})$ follow.

11. (i) We note that for $x \notin \mathbb{Z}$, $\sum\limits_{n=1}^{\infty} \frac{1}{n} e^{2\pi i n x} = -\log|1 - e^{2\pi i x}| + i \sum\limits_{n=1}^{\infty} \frac{\sin 2\pi x n}{n} = -\log 2(\sin \pi x) - \pi i(\{x\} - \frac{1}{2})$ (use Chapter 4, Exercise (13)). Therefore from Proposition 9 it follows that $L(\overline{\chi}, 1) = \sum\limits_{n=1}^{\infty} \frac{1}{n} \overline{\chi}(n) = \frac{1}{\tau(\chi)} \sum\limits_{n=1}^{\infty} \frac{1}{n} G(n, \chi)$

$$= \frac{1}{\tau(\chi)} \sum_{n=1}^{\infty} \frac{1}{n} \sum_{k=1}^{m-1} \chi(k) e^{2\pi i k n/m} = \frac{1}{\tau(\chi)} \sum_{k=1}^{m-1} \chi(k) \sum_{n=1}^{\infty} \frac{1}{n} e^{2\pi i k n/m}$$

$$= -\frac{1}{\tau(\chi)} \sum_{k=1}^{m-1} \chi(k) \left(\log 2 \sin \pi \tfrac{k}{m} + i\pi(\tfrac{k}{m} - \tfrac{1}{2}) \right)$$

$$= -\frac{1}{\tau(\chi)} \sum_{k=1}^{m-1} \chi(k) \left(\log \sin \pi \tfrac{k}{m} + i\pi \tfrac{k}{m} \right), \text{ because } \chi \neq \chi_0 \text{ implies that}$$

$$\sum_{k=1}^{m-1} \chi(k) = 0.$$

(ii) By (9) $\tau(\chi) = \sqrt{m}$. From $L(\chi, 1) \in \mathbb{R}$ it follows that $L(\chi, 1) = -\frac{1}{\sqrt{m}} \sum\limits_{k=1}^{m-1} \chi(k) \log \sin \frac{\pi k}{m}$ and $\sum\limits_{k=1}^{m-1} \chi(k) k = 0$.

(iii) By (9) $\tau(\chi) = i\sqrt{m}$. From $L(\chi, 1) \in \mathbb{R}$ it follows that $L(\chi, 1) = -\frac{\pi}{m^{3/2}} \sum\limits_{k=1}^{m-1} k\chi(k)$ and $\sum\limits_{k=1}^{m-1} \chi(k) \log \sin \frac{\pi k}{m} = 0$.

12. We have $L(\chi, s) = \prod\limits_{p} \left(1 - \frac{\chi(p)}{p^s}\right)^{-1} = \prod\limits_{p \nmid m} \left(1 - \frac{\chi(p)}{p^s}\right)^{-1} = \prod\limits_{p \nmid m} \left(1 - \frac{\chi'(p)}{p^s}\right)^{-1}$

$$= \prod_{p} \left(1 - \frac{\chi'(p)}{p^s}\right)^{-1} \prod_{p|m} \left(1 - \frac{\chi'(p)}{p^s}\right) = L(\chi', s) \prod_{p|m} \left(1 - \frac{\chi'(p)}{p^s}\right).$$

13. We have $m^{-s} \sum\limits_{k=1}^{m} \chi(k) \zeta(s, \frac{k}{m}) = m^{-s} \sum\limits_{k=1}^{m} \chi(k) \sum\limits_{n=0}^{\infty} (n + \frac{k}{m})^{-s}$

$$= \sum_{n=0}^{\infty} \sum_{k=1}^{m} \chi(k + nm)(k + nm)^{-s} = \sum_{p=1}^{\infty} \frac{\chi(p)}{p^s} = L(\chi, s). \text{ The second and third}$$

assertions follow from Chapter 4, (20), because for $\chi \neq \chi_0$, $L(\chi, s)$ has no pole at 1.

14. For $\mathrm{Re}\, s > 0$ $L'(\chi, s) = -\sum\limits_{n=1}^{\infty} \frac{\chi(n)}{n^s} \log n$, so that $\sum\limits_{n \leq x} \frac{\chi(n)}{n} \log n = O(1)$.

From this it follows that $O(1) = \sum\limits_{n \leq x} \frac{\chi(n)}{n} \log n = \sum\limits_{n \leq x} \frac{\chi(n)}{n} \sum\limits_{d|n} \Lambda(d)$

$$= \sum_{d \leq x} \Lambda(d) \sum_{d|n \leq x} \frac{\chi(n)}{n} = \sum_{d \leq x} \Lambda(d) \sum_{n \leq x/d} \frac{\chi(nd)}{nd} = \sum_{d \leq x} \tfrac{1}{d} \Lambda(d) \chi(d) \big(L(\chi, 1)$$

$$- \sum_{n > x/d} \tfrac{1}{n} \chi(n) \big) = \sum_{d \leq x} \tfrac{1}{d} \Lambda(d) \chi(d) L(\chi, 1) + O\big(\sum_{d \leq x} \tfrac{1}{d} \Lambda(d) \cdot \tfrac{d}{x} \big)$$

$$= \sum_{d \leq x} \tfrac{1}{d} \Lambda(d) \chi(d) L(\chi, 1) + O(\tfrac{1}{x} \Psi(x)), \text{ so that by Chebyshev's theorem } O(1) =$$

$$\sum_{d \leq x} \tfrac{1}{d} \Lambda(d) \chi(d) = \sum_{p \leq x} \tfrac{1}{p} \chi(p) \log p + O\big(\sum_{1 < m} \sum_{p^m \leq x} p^{-m} \log p \big) = \sum_{p \leq x} \tfrac{1}{p} \chi(p) \log p +$$

$$O\big(\sum_{k \leq x} \sum_{m=2}^{\infty} k^{-m} \log k \big) = \sum_{p \leq x} \tfrac{1}{p} \chi(p) \log p + O\big(\sum_{k \leq x} \tfrac{\log k}{k^2} \big) = \sum_{p \leq x} \tfrac{1}{p} \chi(p) \log p +$$

$$O(1).$$

15. If $\chi = \chi_0$, the proof will be found in Addendum[21].
Suppose now that $\chi \neq \chi_0$. Then $\sum\limits_{n \leq x} \mu(n)\chi(n)\frac{1}{n} \sum\limits_{k \leq x/n} \frac{\chi(k)}{k}$

$$= \sum_{n \leq x} \sum_{k \leq x/n} \frac{\mu(n)}{nk}\chi(nk) = \sum_{t \leq x} \sum_{d|t} \mu(d)\frac{\chi(t)}{t} = 1 \text{, from which it follows that}$$

$$1 = \sum_{n \leq x} \mu(n)\frac{\chi(n)}{n}\left(L(\chi,1) - \sum_{k > x/n} \frac{1}{k}\chi(k) \right) = \sum_{n \leq x} \frac{1}{n}\mu(n)\chi(n)L(\chi,1)$$

$$+ O\left(\sum_{n \leq x} \frac{1}{n} \cdot \frac{n}{x} \right). \text{ Because } L(\chi,1) \neq 0 \text{, we have } \sum_{n \leq x} \frac{1}{n}\mu(n)\chi(n) = O(1) \text{, so that}$$

$$\sum_{n \leq x} \sum_{d|n} \mu(d)\chi(d) = \sum_{d \leq x} \sum_{d|n \leq x} \mu(d)\chi(d) = \sum_{d \leq x} \left[\frac{x}{d} \right] \mu(d)\chi(d) = \sum_{d \leq x} \frac{x}{d}\mu(d)\chi(d) +$$

$$O(x) = O(x).$$

16. By (15) for $f(n) := \sum\limits_{k|n} \mu(k)\chi(k)$, as $x \to \infty$ we have $\sum\limits_{n \leq x} f(n) = O(x)$.

Addendum[21] implies that $f(n) \geq 0$. For $\mathrm{Re}\,s > 1$ $\sum\limits_{n=1}^{\infty} \frac{f(n)}{n^s} = \frac{\zeta(s)}{L(\chi,s)}$, and

$s \mapsto \frac{\zeta(s)}{L(\chi,s)} - \frac{1}{L(\chi,1)(s-1)}$ is holomorphic in $\{s : s \in \mathbb{C}, \mathrm{Re}\,s > 0\}$, if we put
$\frac{1}{\infty} = 0$. Now apply the Tauberian theorem of Ingham and Newman.

17. Consider Chapter 5, Cor. 3, and put $f(n) = 1$, $g(n) = \mu(n)\chi(n)$ for $n \in \mathbb{N}$.
Then $F(s) = \zeta(s)$ and $G(s) = \frac{1}{L(\chi,s)}$. The functions $s \mapsto F(s) - \frac{1}{s-1}$ and
$s \mapsto \frac{1}{L(\chi,s)}$ are holomorphic in some region containing $\{s \in \mathbb{C} : \mathrm{Re}\,s \geq 1\}$. The
assertion now follows.

18. The function $\chi\lambda$ is strongly multiplicative. It follows that for $\mathrm{Re}\,s > 1$ we have
$$\sum_{n=1}^{\infty} \chi(n)\lambda(n)n^{-s} = \prod_{p} \sum_{j=0}^{\infty} \chi(p)^j \lambda(p)^j p^{-js} = \prod_{p}(1 - \chi(p)\lambda(p)p^{-s})^{-1} = \prod_{p}(1 + $$
$$\chi(p)p^{-s})^{-1} = \prod_{p}(1 - \chi(p)p^{-s}) \prod_{p}(1 - \chi^2(p)p^{-2s})^{-1} = L(\chi^2, 2s)/L(\chi,s). \text{ Once}$$
more consider Chapter 5, Cor. 3, and put $f(n) = 1$, $g(n) = \chi(n)\lambda(n)$ for $n \in \mathbb{N}$.
Then $F(s) = \zeta(s)$ and $G(s) = \frac{L(\chi^2, 2s)}{L(\chi,s)}$. G and $F(s) - \frac{1}{s-1}$ are holomorphic in
some region containing $\{s \in \mathbb{C} : \mathrm{Re}\,s \geq 1\}$. From this the assertion follows.

19. (i) Suppose first that g.c.d.$(a,m) = 1$. Then $\sum\limits_{n \leq x, n \equiv a(m)} \mu(n)$

$$= \frac{1}{\varphi(m)} \sum_{\chi \in \Gamma_m} \overline{\chi}(a) \sum_{n \leq x} \mu(n)\chi(n) = \frac{1}{\varphi(m)} \sum_{\chi \in \Gamma_m} \overline{\chi}(a)o(x) = o(x) \text{ (by (17))}.$$

Now let a be an arbitrary element of $\mathbb{Z}$, g.c.d.$(a,m) = d$, g.c.d.$(\frac{m}{d},d) = g$
and l.c.m.$(\frac{m}{d},d) = v$. If $1 \leq k \leq d$, g.c.d.$(k,d) = 1$ and $k \equiv \frac{a}{d}(\mathrm{mod}\,g)$, then
because $g = $ g.c.d.$(\frac{m}{d},d)$ divides $\frac{a}{d} - k$, there exists some $u_k \in \mathbb{Z}$ with $u_k \equiv$
$\frac{a}{d}(\mathrm{mod}\,\frac{m}{d})$ and $u_k \equiv k(\mathrm{mod}\,d)$. Then g.c.d.$(u_k, v) = 1$. From this we have

$$\sum_{\substack{n \leq x, n \equiv a(m)}} \mu(n) = \sum_{\substack{n \leq x/d \\ n \equiv a/d(m/d)}} \mu(dn) = \sum_{\substack{k=1 \\ \text{g.c.d.}(k,d)=1}}^{d} \mu(d) \sum_{\substack{n \leq x/d, n \equiv k(d) \\ n \equiv a/d(m/d)}} \mu(n) =$$

$$\sum_{\substack{k=1 \\ \text{g.c.d.}(k,d)=1 \\ k \equiv a/d(g)}}^{d} \mu(d) \sum_{\substack{n \leq x/d \\ n \equiv u_k(v)}} \mu(n) = o(x) \text{ by the case above.}$$

(ii) Let $d = \text{g.c.d.}(a, m)$. Then we have
$$\sum_{\substack{n \equiv a(m) \\ n \leq x}} \lambda(n) = \sum_{\substack{n \equiv a/d(m/d) \\ n \leq x/d}} \lambda(d)\lambda(n) =$$
$$\lambda(d)\frac{1}{\varphi(m/d)} \sum_{\chi \in \Gamma_{m/d}} \overline{\chi}(\tfrac{a}{d}) \sum_{n \leq x/d} \lambda(n)\chi(n) = o(x) \text{ (by (18))}.$$

20. We note that $\{jn + lm : 0 \leq j < m, 0 \leq l < n, \text{g.c.d.}(j, m) = \text{g.c.d.}(l, n) = 1\}$ is a complete system of residues prime modulo mn. From this it follows that

$$c_m(k)c_n(k) = \sum_{\substack{j=0 \\ \text{g.c.d.}(j,m)=\text{g.c.d.}(l,n)=1}}^{m-1} \sum_{l=0}^{n-1} e^{2\pi i k(j/m + l/n)}$$

$$= \sum_{\substack{j=0 \\ \text{g.c.d.}(j,m)=\text{g.c.d.}(l,n)=1}}^{m-1} \sum_{l=0}^{n-1} e^{2\pi i k(jn + lm)/mn} = \sum_{\substack{t=0 \\ \text{g.c.d.}(t,mn)=1}}^{mn-1} e^{2\pi i k t/mn} = c_{mn}(k).$$

Index of Names

K. Ireland, University of New
Brunswick, Fredericton, N.B.,
M. Rosen, Brown University,
Providence, RI

A Classical Introduction to Modern Number Theory

2nd ed. 1990. XIV, 394 pp. 1 fig.
(Graduate Texts in Mathematics, Vol. 84) Hardcover
DM 98,– ISBN 3-540-97329-X

Contents: Unique Factorization. – Applications of Unique
Factorization. – Congruence. –
The Structure of $U(Z/nZ)$. –
Quadratic Reciprocity. – Quadratic Gauss Sums. – Finite
Fields. – Gauss and Jacobi
Sums. – Cubic and Biquadratic
Reciprocity. – Equations Over
Finite Fields. – The Zeta Function. – Algebraic Number
Theory. – Quadratic and Cyclotomic Fields. – The Stickelberger Relation and the Eisenstein Reciprocity Law. –
Bernoulli Numbers. – Dirichlet
L-Functions. – Diophantine
Equations. – Elliptic Curves. –
The Mordell-Weil Theorem. –
New Progress in Arithmetic
Geometry. – Selected Hints for
the Exercises. – Bibliography. –
Index.

N. Koblitz, University of
Washington, Seattle, WA

A Course in Number Theory and Cryptography

1987. VIII, 208 pp. 5 figs.
(Graduate Texts in Mathematics, Vol. 114)
Hardcover DM 74,–
ISBN 3-540-96576-9

Contents: Some Topics in
Elementary Number Theory. –
Finite Fields and Quadratic
Residues. – Cryptography. –
Public Key. – Primality and
Factoring. – Elliptic Curves. –
Answers to Exercises. –
Index.

GPSR Compliance
The European Union's (EU) General Product Safety Regulation (GPSR) is a set
of rules that requires consumer products to be safe and our obligations to
ensure this.

If you have any concerns about our products, you can contact us on

ProductSafety@springernature.com

In case Publisher is established outside the EU, the EU authorized
representative is:

Springer Nature Customer Service Center GmbH
Europaplatz 3
69115 Heidelberg, Germany

www.ingramcontent.com/pod-product-compliance
Ingram Content Group UK Ltd.
Pitfield, Milton Keynes, MK11 3LW, UK
UKHW052345070726
473059UK00009B/2546